5th International Workshop on Nuclear Data Covariances (CW2022)

Held online

EPJ Web of Conferences Volume 281 (2023)

Tokyo, Japan
26 - 30 September 2022

Editors:

Go Chiba
Osamu Iwamoto

ISBN: 978-1-7138-7250-4

Printed from e-media with permission by:

Curran Associates, Inc.
57 Morehouse Lane
Red Hook, NY 12571

Some format issues inherent in the e-media version may also appear in this print version.

No changes have been made to the content of these proceedings. There may be changes to pagination and minor adjustments for aesthetics.

Printed with permission by Curran Associates, Inc. (2023)

For additional information, please contact EDP Sciences – Web of Conferences
at the address below.

EDP Sciences – Web of Conferences
17, Avenue du Hoggar
Parc d'Activité de Courtabœuf
BP 112
F-91944 Les Ulis Cedex A
France

Phone: +33 (0) 1 69 18 75 75
Fax: +33 (0) 1 69 28 84 91

contact-edps@webofconferences.org

Additional copies of this publication are available from:

Curran Associates, Inc.
57 Morehouse Lane
Red Hook, NY 12571 USA
Phone: 845-758-0400
Fax: 845-758-2633
Email: curran@proceedings.com
Web: www.proceedings.com

TABLE OF CONTENTS

edp sciences

Web of Conferences

Statement of Peer review

In submitting conference proceedings to *Web of Conferences*, the editors of the proceedings certify to the Publisher that

1. They adhere to its **Policy on Publishing Integrity** in order to safeguard good scientific practice in publishing.
2. All articles have been subjected to peer review administered by the proceedings editors.
3. Reviews have been conducted by expert referees, who have been requested to provide unbiased and constructive comments aimed, whenever possible, at improving the work.
4. Proceedings editors have taken all reasonable steps to ensure the quality of the materials they publish and their decision to accept or reject a paper for publication has been based only on the merits of the work and the relevance to the journal.

Title, date and place of the conference

5th International Workshop on Nuclear Data Covariances - CW2022 -
September 26 to 30, 2022, Online by Tokyo Institute of Technology

Proceedings editor(s):

Dr. Go Chiba, Hokkaido University
Dr. Osamu Iwamoto, Japan Atomic Energy Agency

Date and editor's signature

On behalf of the editors, Osamu Iwamoto
March 22, 2023

Importance of cross reaction covariance data for user applications

Patrick Griffin[1*]

[1]Sandia National Laboratories, Advanced Science and Technology Division, Albuquerque, NM, USA

Abstract. The characterization of the uncertainty in radiation damage metrics presents many challenges. This paper examines the current approaches to characterizing radiation damage metrics such as hydrogen and helium gas production, material heating, trapped charge in microelectronics, and lattice displacement damage. Critical uncertainty aspects go beyond just the material cross sections and involve the consideration of energy-dependent cross reaction correlations, the recoil ion energy spectrum, and models used for the partitioning of the recoil ion energy into various forms of energy deposition. This paper starts with a review of terminology and then examines the current approaches in the characterization of uncertainty in radiation damage metrics for several applications. The major deficiencies in the uncertainty of the damage metric characterization are also identified.

1 Introduction

The nuclear data community has made great strides in improving the characterization of the uncertainty in the recommended nuclear data and in the modelling of various radiation metrics. The baseline expectation in an uncertainty characterization for an energy-dependent quantity, like a neutron cross section, includes presentation of the energy-dependent correlation matrix as well as the traditional presentation of the energy-dependent standard deviations. The challenge our community now faces is in extending our high-fidelity uncertainty characterization to derived quantities/metrics where there can be correlations between the various nuclear data inputs and models when representing the desired quantity.

This paper examines two classes of metrics where cross correlation between nuclear data quantities that describe the metric may be important in the uncertainty characterization of the metric. The first class consists of dosimetry metrics. Examples of dosimetry metrics where correlations between input terms are important include: a) neutron-induced activities in dosimeters where elemental, rather than isotopically pure, materials are used; b) gas production in materials, e.g., hydrogen and helium accumulation, bubble formation, and outgassing. The second class consists of radiation damage metrics. Example of radiation damage metrics include: a) ionizing dose and trapped charge in dielectrics found in microelectronics; b) Frenkel pair generation and evolved defects that affect carrier lifetime in semiconductors; c) material embrittlement metrics in materials, e.g., welds in iron pressure vessels used in light water reactors. A mathematical formulation for calculating a wide variety of damage metrics is introduced and examples of the importance in correlations between input terms are addressed.

2 Dosimetry metrics

Metrics used for dosimetry purposes most commonly correspond to reactions where a residual activity in an isotope is used as the relevant metric. The residual activity can be measured in different manners; emitted gammas are typically measured though the use of high purity germanium detectors (HPGe); emitted betas are typically measured using organic scintillators; hydrogen and helium generation is typically measured using atomic mass spectroscopic analysis. The following sections provide a more detailed discussion of how the covariance considerations are treated in the metrics used to represent some of these dosimetry application areas.

2.1 Activation sensors

Legacy dosimetry files, e.g., DOSDAM77 [1] and IRDF-90 [2], addressed isotopic-specific reactions of interest for dosimetry monitoring applications. However, most applications use elemental, and not isotopically enriched, activation sensors. Thus, new dosimetry libraries, e.g., IRDFF-II [3], report activation cross sections for elemental sensors. In addition, nuclear standards organizations, in standards for light water reactor dosimetry, are now referring to activities in elemental sensors, with their natural isotopic abundances, e.g., ASTM E526-22, Standard Test Method for Measuring Fast-Neutron Reaction by Radioactivation of Titanium [4].

Figure 1 shows the dosimetry cross section for the above E526 example of an elemental activation, i.e.,

* Corresponding author: pjgriff@sandia.gov

^{nat}Ti(n,X)^{46}Sc. For elemental sensors the adopted convention is to report the cross section normalized per atom of the target element. Since this ASTM standard only addresses applications for fission-based reactors, the data in this standard only goes up to 20 MeV – and only two reaction channels are relevant. However, the IRDFF-II dosimetry library uses the ENDF-6 format [5] MF10/MT5 file to report this cross section and it includes **all** relevant reaction channels up to the maximum energy in the library, 60 MeV. In this example, the IRDFF-II library entry includes cross section contributions from the following 11 channels [notated with relevant target isotopic abundance and ENDF-6 reaction identifier]:

- ^{46}Ti(n,p) {8.25%; MAT 2225, MF3/MT103}
- ^{47}Ti(n,d+pn+np) {7.44%; MAT 2228, MF10/MT5}
- ^{48}Ti(n,nd) {73.72%; MAT 2231, MF3/MT32}
- ^{48}Ti(n,2np) {73.72%; MAT 2231, MF3/MT41}
- ^{48}Ti(n,t) {73.72%; MAT 2231, MF3/MT105}
- ^{49}Ti(n,2nd) {5.41%; MAT 2234, MF3/MT11}
- ^{49}Ti(n,nt) {5.41%; MAT 2234, MF3/MT33}
- ^{49}Ti(n,3np) {5.41%; MAT 2234, MF3/MT42}
- ^{50}Ti(n,2nt) {5.18%; MAT 2237, MF3/MT154}
- ^{50}Ti(n,4np) {5.18%; MAT 2237, MF3/MT156}

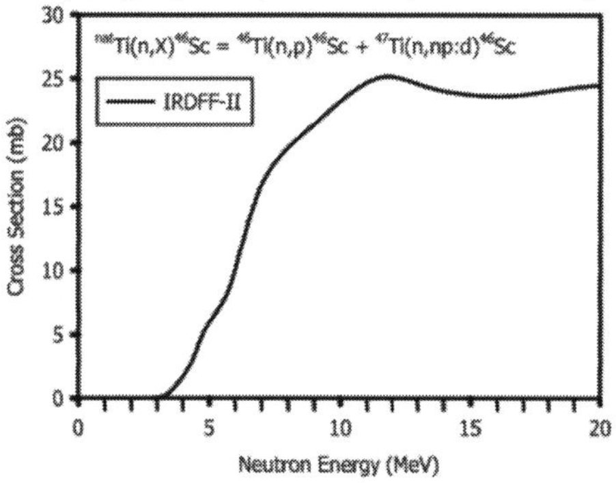

Fig. 1. ^{nat}Ti(n,X)^{46}Sc cross section (normalized per elemental Ti atom) [4]

The IRDFF-II library also includes the covariance matrix for this composite elemental activation reaction. The IRDFF-II covariance file is obtained by combining, with proper weighting, the covariance matrices for each of the individual reaction channels. The IRDFF-II covariance combination makes the assumption that the reactions are uncorrelated. However, there exists strong correlations between the separate reaction channels – and this correlation, based within the experimental data and in the model calculations, is not currently treated in generating the associated covariance matrix for the ^{nat}Ti(n,X)^{46}Sc reaction.

Many of the high threshold reactions (> 20 MeV) in the IRDFF-II library are taken from the TENDL library. The TENDL nuclear data evaluations depend heavily upon the TALYS-calculated cross sections. So, one approach to address the cross section correlation for these high threshold reaction channels, as represented in the TENDL random libraries, is to generate a covariance matrix based on a Total Monte Carlo (TMC) methodology. This approach generates a covariance matrix based upon a sampling of the underlying nuclear model data, e.g., the optical model parameters, DWBA deformations, and level density parameters, to capture the cross-reaction correlations. However, most of the IRDFF-II low energy (<20 MeV) reaction data comes from isotope-specific and reaction-specific evaluations that are produced explicitly for dosimetry applications – and which do not have associated model-based random libraries that can be sampled. One challenge in representing the uncertainty comes in how to combine the low threshold calculated covariance files, which are dominated by a consideration of the available experimental data, with the high threshold covariance files, which are determined from the parameter variation in the nuclear reaction models. A second challenge in representing the uncertainty comes from how to combine the separate isotopic-specific reaction evaluations when done for low threshold reactions, e.g., in the above example, to combine the ^{46}Ti(n,p) covariance with the ^{47}Ti(n,np) and ^{47}Ti(n,d) covariance matrices.

These three data-driven dosimetry reactions were separately derived/evaluated in the IRDFF-II evaluations, but strong correlations are expected to exist between experimental datasets used in the individual evaluations, e.g., most experiments used natural titanium samples and, as seen in Figure 2, some datapoints were taken by the same experimenters.

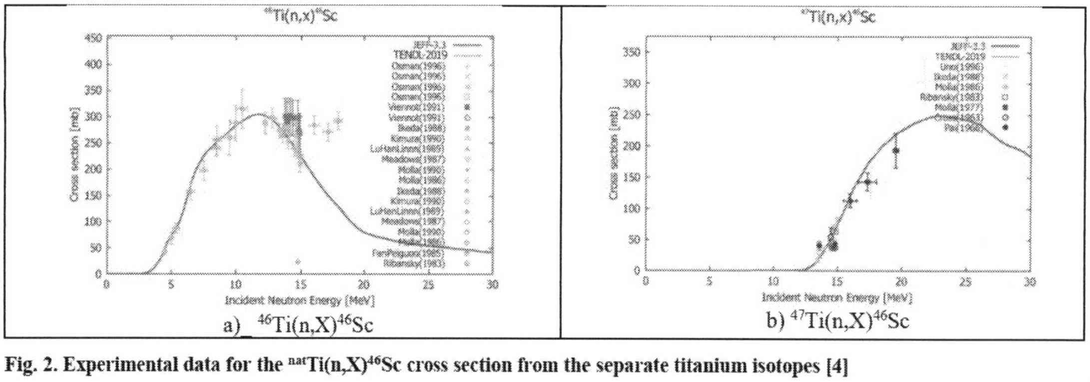

a) ^{46}Ti(n,X)^{46}Sc b) ^{47}Ti(n,X)^{46}Sc

Fig. 2. Experimental data for the ^{nat}Ti(n,X)^{46}Sc cross section from the separate titanium isotopes [4]

Cross reactions correlations can be important in addressing activation sensors, and there is room for improvement in how we address this.

2.2 Gas production

The area of gas production metrics can be divided into three categories. The first category is when the metric is used as a simple reaction identifier. An example of this category is the use of a helium accumulation fluence monitor (HAFM) as a sensor. A sample application here is found in ASTM E910 Standard Test Method for Application and Analysis of Helium Accumulation Fluence Monitors for Reactor Vessel Surveillance [6]. The covariance considerations here are nearly identical to those addressed for the activation sensors in Section 2.1. Figure 3 shows the threshold energy and cross section energy-dependence for some of the many reactions that need to be considered for helium production in elemental boron. Helium can be produced by both the ^{10}B and ^{11}B naturally occurring isotopes. Figure 4, work by A. Trkov and documented in the IRDFF-II library [3], shows that complex pathways must be considered in order to account for all the helium production mechanisms, e.g., consideration of the production of ^{8}Be and its breakup into two alpha particles. Thus, correlations in the energy-dependence of a HAFM sensor can depend upon correlations between: a) cross sections for the ^{10}B and ^{11}B naturally occurring isotopes; b) different reactions channels within a given isotope; and c) decay paths for metastable products that also produce helium, e.g., ^{10}B(n,^{3}He) → ^{3}He + ^{8}Li* → ^{3}He + 2 ^{4}He, where subsequent decay paths produce additional helium that would be totally correlated with the direct helium production from the diect reaction channel.

The second category of gas production metrics takes the various gas production reaction channels and addresses the diffusion and evolution of the generated gas as it forms voids, more complicated defects, and over-pressurized bubbles that may be subject to sudden release mechanisms, i.e., out-gassing. The normal radiation-induced lattice defects, e.g., vacancies and interstitials, can interact with released helium and affect the microstructure of evolved defects in materials such as metals or carbides. Helium-vacancy defects have been seen in helium implanted samples – and this interaction of helium with the annealing of the lattice defects results in a very complex dynamics that requires

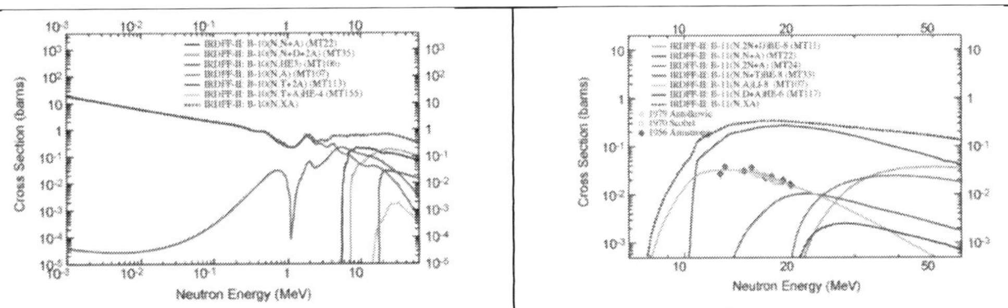

a) ^{10}B Helium Production   b) ^{11}B Helium Production

Fig. 3. Helium production in the boron isotopes [3]

MT	LR	Gas	Reaction	Products	Source evaluations of IRDFF-II	
105		^{3}H	^{10}B(n,t)	→ **^{3}H** + (^{8}Be* → 2 ^{4}He)	EAF-2010(r) (E > 20)	
700			^{10}B(n,t)	→ **^{3}H** + (^{8}Be* → 2 ^{4}He)		ENDF/B-VIII.0 (
113			^{10}B(n,t2α)	→ **^{3}H** + 2 ^{4}He	EAF-2010(r) MT105	ENDF/B-VIII.0 M
155			^{10}B(n,tα)	→ **^{3}H** + 2 ^{4}He	EAF-2010	
106		^{3}He	^{10}B(n,^{3}He)	→ **^{3}He** + (^{8}Li → ^{8}Be* → 2 ^{4}He)	EAF-2010	
22		^{4}He	^{10}B(n,nα)	→ n + **^{4}He** + ^{6}Li	EAF-2010(r) (E > 20)	ENDF/B-VIII.0 M
32			^{10}B(n,nd)	→ n + ^{2}H + (^{8}Be* → **2 ^{4}He**)	EAF-2010(r) (E > 20)	
35			^{10}B(n,nd2α)	→ n + ^{2}H + **2 ^{4}He**	EAF-2010(r) MT32	ENDF/B-VIII.0 M
55,56,...	22		^{10}B(n,n)	→ n + **^{4}He** + ^{6}Li		ENDF/B-VIII.0 (
55,56,...	35		^{10}B(n,n)	→ n + ^{2}H + **2 ^{4}He**		ENDF/B-VIII.0 (
105			^{10}B(n,t)	→ ^{3}H + (^{8}Be* → **2 ^{4}He**)	EAF-2010(r) (E > 20)	
700			^{10}B(n,t)	→ ^{3}H + (^{8}Be → **2 ^{4}He**)		ENDF/B-VIII.0 (
106			^{10}B(n,^{3}He)	→ ^{3}He + (^{8}Li* →^{8}Be* → **2 ^{4}He**)	EAF-2010	
107			^{10}B(n,α)	→ **^{4}He** + ^{7}Li	EAF-2010(r) (E > 20)	ENDF/B-VIII.0 M
800			^{10}B(n,α)	→ **^{4}He** + ^{7}Li		ENDF/B-VIII.0 (
801			^{10}B(n,α)	→ **^{4}He** + ^{7}Li*		ENDF/B-VIII.0 (
113			^{10}B(n,t2α)	→ ^{3}H + **2 ^{4}He**	EAF-2010(r) MT105	ENDF/B-VIII.0 M
155			^{10}B(n,tα)	→ ^{3}H + **2 ^{4}He**	EAF-2010	

Fig. 4. Complex pathway for helium production in ^{10}B [3]

multi-scale physics models to capture the detailed behavior. In addition to this metric involving modelling of the defect evolution, the correlation between the helium release, its rate of release, and the temperature (both in the cascade where the helium is generated and in the subsequent annealing stages) are all important considerations in representing this damage metric.

The third category of gas production metrics builds upon the category two microscale description of the evolved defects and addresses the implications of the evolved gas on macroscale applications, e.g., changes in the material properties that may have implications within the application area. The realm of important correlations encompasses the correlations with the gas release rate, the application-induced temperature of the sample (as opposed to the temperature spikes due to the damage cascade), and synergies between different gasses, e.g., synergistic damage has been observed to be associated with the production of both hydrogen and helium. This synergistic effect means that empirical models for changes in material mechanical properties, due to either hydrogen or helium generation, when combined, fail to represent the observed damage when both hydrogen and helium are simultaneously introduced (this has been simulated using implantation methods).

3 Damage metrics

A discussion of the importance of correlations between different nuclear data input parameters requires us to first establish a common terminology for characterizing how these damage metrics are modelled. Section 3.1 establishes a basic framework and terminology that can represent most of the simpler damage metrics that are currently modelled and correlated with observed damage modes. The following sections look more in-depth at the correlations that can be important to a consideration of specific damage modes – and within the context of specific applications.

3.1 Terminology for calculated metrics

Common damage metrics are often modelled using easily calculated quantities such as the total kerma, displacement kerma, ionizing kerma, and displacement-per-atom (dpa). Most high-level damage metrics can be expressed as a scalar function of these quantities and can be represented as in Equation 1. This scalar metric can be modelled as a normalization factor multiplied by the convolution of the radiation field [characterized by a fluence, Φ, and a unit normalized energy-dependent spectrum, $\phi(E)$] and a response function, $\Re(E)$.

$$^{facility}D_{type} = {}^{type}\mathbb{C} \bullet {}^{facility}\Phi \bullet \int_0^\infty \phi^{facility}(E) \bullet \Re_{type}(E) \bullet dE \tag{1}$$

Neutron damage is typically correlated with the details for the energy deposition in a material – and this energy deposition is related to details of the specific neutron interactions, e.g., the energy deposition, that comes from the outgoing recoil ions resulting from the neutron interaction. Because of this, most scalar response functions can be expressed, as in Equation 2, as a summation over the various reaction channels and an integration over the response due to the energy and angle of the resulting recoil particles.

$$\Re_{type}(E) = \sum_{i,j_i} \sigma_{i,j_i}(E) \int_0^\infty dT_{R,j_i} \int_{-1}^{1} d\mu \cdot f\left(E,\mu,T_{R,j_i}\right) \cdot {}^{type-A}\Lambda\left(E_d, {}^{ion}_{type-B,C} T_{R,j_i}\right) \bullet {}^{type-B}\zeta\left(E_d, T_{R,j_i}, {}^{type-D}T_{dam}\right) \bullet {}^{type-C}\xi\left(T_{R,j_i}\right) \tag{2}$$

As described in Reference 7, in this expression: f describes the energy- and angle-dependent reaction-specific recoil spectrum; Λ, the damage partition function, describes the partitioning of energy into the relevant deposition process, e.g., ionization or displacement for some damage metrics; ζ, the threshold treatment modifier, describes the probability of Frenkel pair recombination for displacement energies near the displacement threshold energy – an angle-dependent value; and ξ, a recoil ion damage efficiency factor, describes the efficiency function relevant for a specific type of defect and typically comes from molecular dynamics code calculations. The "type-A", "type-B", "type-C", and "type-D" qualifiers that appear in the superscripts for the various terms of Equation 2 denote various conventions for those terms that have been used in the published literature. Reference 7 provides details for some examples for these designators.

Some damage metrics can be more complex than what can be captured in scalar expressions such as that shown in equations 1 and 2. One such damage metric is the linear transfer energy (LET), $\mathscr{L}(E)$. The LET can be represented as in Equation 3 [8]. This type of damage metric represents a probability distribution function for charge released by a recoil ion based on its energy at the time of its creation. Rather than being characterized as a scalar quantity, it must be described as a function of a scaled response from the angle-integrated recoil spectrum, Θ, as shown in Equation 4.

$$\mathscr{L}(E) = \sum_{i,j_i} {}^{GaAs}S\left[T_{R,j_i} \Rightarrow LET\left({}^{A^{j_i}}_{Z^{j_i}}P\right)\right] \bullet {}^{O_{i,j_i}}\Theta(T_{R,j_i}) \tag{3}$$

$$^{O_{i,x}}\Theta(T_{R,j,x}, E) = \sum_{i,j_i} \delta\left[j_i, O_{i,x}\right] \cdot \phi(E) \cdot \sigma_{i,j_i}(E) \cdot \int_{-1}^{1} d\mu \cdot f\left(E,\mu,T_{R,j_i}\right) \tag{4}$$

Note, Equation 3 represents the LET damage metric, a <u>calculated</u> metric which is an approximation for the actual damage in processes such as neutron-induced upset, where it is the actual collected charge that affects the damage in a semiconductor. This calculated damage metric neglects the change in the recoil ion LET as it down-scatters within the target lattice material and the effective electric field in a semiconductor that defines the sensitive volume over which the generated charge is collected.

This section only addresses the generic terminology for expressing the damage metrics. The details of which input quantities, as represented as in Equations 2 and 3, are correlated, and the importance of that correlation in the use of the damage metric, are addressed in later sections. This correlation between the different nuclear data input quantities, seen in Equations 2 and 3, has a significant effect on the uncertainty for the resulting damage metric. As seen in the Section 2 discussion of activation sensors, there can be a strong correlation between reaction-channel-specific neutron cross sections within a given isotope – and between different isotopes in a naturally occurring element. What is new in the discussion of damage metrics is the correlation between additional nuclear data quantities that are addressed by the various damage-specific functions that are part of Equation 2, e.g., the recoil spectra, ion stopping power, and the damage partition function.

An additional consideration in the uncertainty characterization, based on use of the above expressions, is that the uncertainty propagation process often makes the assumption that the representation of the component-level uncertainty can be used in the form of a covariance matrix – and this often makes the assumption that the uncertainty can be represented by a normal/Gaussian distribution. This can be a significant assumption – and it is often not supported by a careful consideration of the physics behind the definition of the parameter or the sampled data. For example, when large uncertainties are present, the physical constraints of non-negativity in some quantities mean that the uncertainty often should not be described as a normal/Gaussian distribution, e.g., in the cases where a log-normal distribution or a Poisson distribution are indicated by the physical basis. If one knows the probability distribution function (pdf) associated with a parameter that influences the damage metric, it can be properly propagated through a nonlinear process into a damage metric – even if the approach is through the use of a Total Monte Carlo (TMC) sampling. Complications arise when insufficient data, or a physics-based argument, are not available to support use of the assumption that a parameter is normally distributed. Some observations on the expected probability distribution, when made at a fundamental physics-based level, can be inferred. However, when a representation of the probability distribution is made at a higher level of representation, the distribution for quantities, e.g., the Frenkel pair distribution, the ion stopping power, or the damage partition function, can be much more complicated than a normal distribution.

3.2 Nuclear data

The starting point for an investigation into the uncertainty in most radiation damage metrics, in either equation 2 or 3, is a consideration of the neutron cross section. For simplicity in this depiction/analysis, the set of reaction channels is divided into elastic, inelastic, disappearance channels (where there is no outgoing neutron in the residual channel), and "other" (reaction channels where there is an outgoing neutron in the residual channel as well as other recoil particles). This partitioning into "classes" of reactions was used because it represents the "classes" of output damage metrics automatically provided in the NJOY analysis code [9].

Using ^{28}Si as a representative case for a material of interest to the user community, in this case the semiconductor radiation damage community, Figure 5a shows the neutron energy-dependent fractional contribution from these various "classes" of reaction channels to the total kerma. Figure 5b shows the overall energy-dependent uncertainty (a 1-sigma standard deviation) in the damage energy for these various classes of reactions. In Figure 5b, in addition to depicting the various reaction "classes", the red curve shows the uncertainty for the total damage energy – as captured using a Total Monte Carlo calculation based upon sampling from the set of TENDL-2012 random evaluations. The TENDL-2012 random evaluations were used in this analysis because they were provided by the TENDL developers at the time of the initial analysis. The following discussion reflects the consistent use of these TENDL-2012 random nuclear data evaluations. These results are expected to be comparable to the results if the more recent TENDL-2015 or TENDL-2017 random files were to be used in a repeat of this analysis. The most recent TENDL library, TENDL-2021, does not provide random files for the silicon isotopes, but, a sophisticated user could use the T6 system to generate their own random libraries based on the documented TENDL-2021 input files. Figure 5 clearly shows that the total uncertainty, at high incident neutron energies (>5 MeV), is <u>much less</u> than the weighted sum of the uncertainty in the damage energy for the various "classes" of reactions.

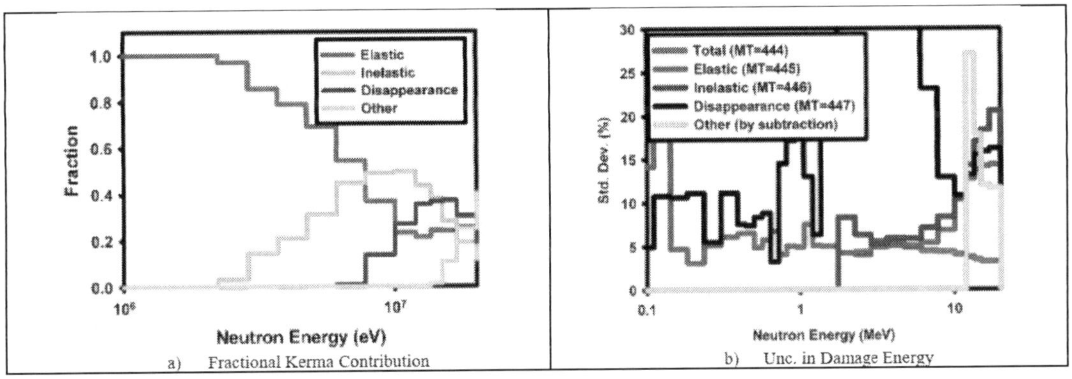

a) Fractional Kerma Contribution b) Unc. in Damage Energy

Fig. 5. Energy-dependent components of the ^{28}Si kerma (left) and the associated energy-dependent uncertainty in the damage energy (right)

 https://doi.org/10.1051/epjconf/202328100001

This is clear evidence of the importance of including the cross-reaction covariance of different reaction channels in evaluating damage metrics.

Most current high fidelity cross section evaluations provide a good characterization of the cross-section uncertainty in the form of an energy-dependent covariance matrix in the ENDF-6 MF33/MF40 file. The covariance matrix is best considered by decomposing it into an energy-dependent standard deviation and a correlation matrix – a matrix whose entries are normalized to be between -1 and +1 and, hence, more meaningfully interpreted. Figure 6 shows how this cross-section correlation matrix for the displacement kerma can be decomposed into energy-dependent correlations for the various classes of reaction channels. Figure 6 shows that there is a very large difference in the energy-dependent correlation in these classes of reaction channels. This difference in the energy-dependent correlation underlines why a careful consideration of the covariance in the various reaction channels must be included in the derivation of the uncertainty of radiation damage metrics.

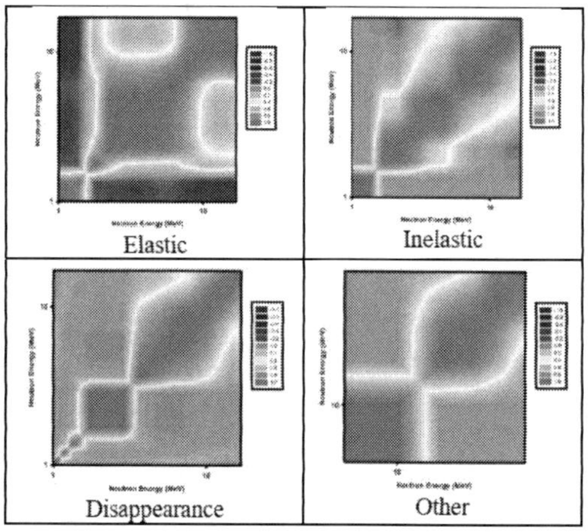

Fig. 6. Correlation matrices in the ^{28}Si displacement kerma for various classes of reactions

In order to provide a context for the importance of addressing the energy-dependent and cross-reaction correlation, one must look at the damage metric relative to a given neutron spectrum. As a representative case, we looked at the uncertainty in the ^{28}Si total kerma, ionizing kerma and displacement kerma when a material is exposed to the ^{252}Cf spontaneous fission neutron spectrum, notated as ^{252}Cf(s.f.). Table 1 shows the resulting uncertainty in the damage metrics due only to the nuclear data, i.e., the cross sections, when treated in a properly correlated manner using the Total Monte Carlo approach, versus when the various "classes" of reaction channels are treated as fully correlated (in energy) and summed, i.e., the case where the proper energy-dependent correlation is not taken into account and a proper correlation matrix is replaced by a correlation matrix where all entries are unity (this case represents consideration of the energy-dependent standard deviations, but not the energy-dependent correlations). A factor of two difference is seen in the resulting spectrum-averaged uncertainty for both damage metrics in this neutron field. A similar analysis, shows that cross-reaction correlations are also important for the damage metrics [14].

Table 1. Uncertainty contributions from the nuclear data for ^{28}Si damage metrics in the ^{252}Cf(s.f.) neutron field.

Metric	Fully Energy Correlated	Properly Correlated
Total Kerma	6.283%	3.817%
Ionizing Kerma	7.299%	4.946%
Displacement Kerma	4.815%	2.653%

3.3 Recoil spectra

After the cross section, the next most important quantity in assessing the uncertainty in damage metrics in either equation 2 or 3, is a consideration of the neutron-induced residual atom recoil spectra. The need here goes beyond a desire to evaluate the importance of cross reaction recoil spectra. The current ENDF-6 format [5] does not even support capturing the uncertainty in the recoil atom spectra.

In the absence of a nuclear data format that addresses the uncertainty in the recoil spectra, the only available approach to shed light on this uncertainty is to compare the recoil spectra from different evaluations. Figure 7a compares the elastic recoil spectra for 10 MeV incident neutrons on ^{28}Si for four modern nuclear data evaluations. For elastic scattering, the kinematics totally control the recoil spectrum and the observed variation is due to differences in the angular distribution. Figure 7b shows the variation in the ENDF/B-VII angular distribution as a function of neutron energy.

For transmutation reactions, there can be a very large variation in the recoil atom spectrum – especially near the reaction threshold energy. Figure 8 shows the large variation seen in the shape of the primary recoil atom spectra. As is seen in Fig. 8b for the (n,2n) reaction at 17.8 MeV incident neutron energy, this difference in recoil spectrum results is a factor of two difference in the resulting ion recoil energy that goes into the creation of Frenkel pairs. The ENDF/B-VII recoil spectrum came from GNASH code calculations, while the TENDL-2012 recoil spectra come from TALYS code calculations. Future analysis needs to look at the independence between the modelling of the recoil atom spectra for transmutation reaction channels in various Hauser-Feshbach codes such as GNASH, TALYS, EMPIRE, CCONE, and CoH3. Is there a strong correlation between computational models in the different codes? When Total Monte Carlo analysis is done by varying the nuclear data parameters, e.g., as guided by the uncertainty in the RIPL-3 database, is the variation in the model parameters sufficient to capture the degrees of freedom in the underlying physics that controls the reaction processes? Significant work remains to be done in understanding the influence of uncertainty and correlations in the recoil spectrum on the damage metrics used by the various applications.

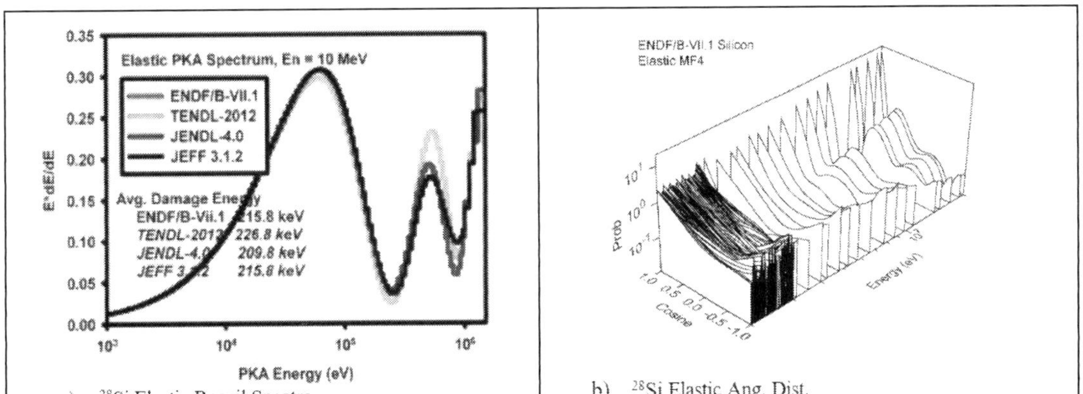

a) ^{28}Si Elastic Recoil Spectra
b) ^{28}Si Elastic Ang. Dist.

Fig. 7. ^{28}Si elastic scattering

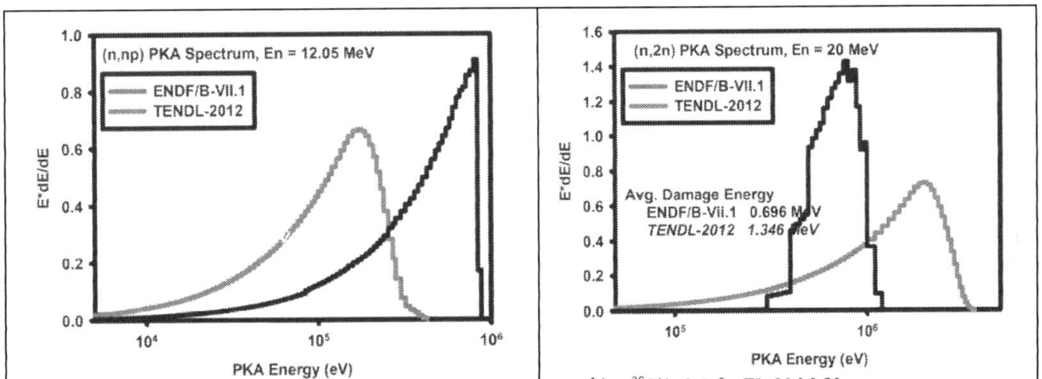

a) ^{28}Si(n,np) for E_n=12 MeV
b) ^{28}Si(n,2n) for E^n=20 MeV

Fig. 8. ^{28}Si primary recoil atom spectra from transmutation reaction channels

3.4 Stopping power

After the uncertainty in the recoil spectrum has been addressed, attention then goes to how the energy from the recoil atoms is deposited in a material. This energy deposition is given by the stopping power for an ion in a material – and the stopping power can be divided into an electronic (ionizing) and a nuclear (lattice displacement, including lattice phonon generation) component.

The stopping power in silicon is depicted in Figure 9 [10]. The symbols represent experimental datapoints and the curve represent model-based calculations. Interatomic potentials are usually calibrated to measurements, including the stopping power data. Figure 10a shows the variation that can result in the damage partition function, the division of the stopping power into ionization and displacement, when integrated along the ion slowing down trajectory, into the total fraction of the energy that goes into displacements rather than into electron production. This figure shows the difference that results from use of the Akkerman damage partition function, which uses the Ziegler, Biersack, and Littmark (ZBL) potential [11], versus the Robinson partition function [12], that uses the Lindhard, Scharff, and Schiott (LSS) approach with the Thomas-Fermi screening function over the Coulomb potential and the Firsov non-local free uniform gas model for the inelastic energy loss.

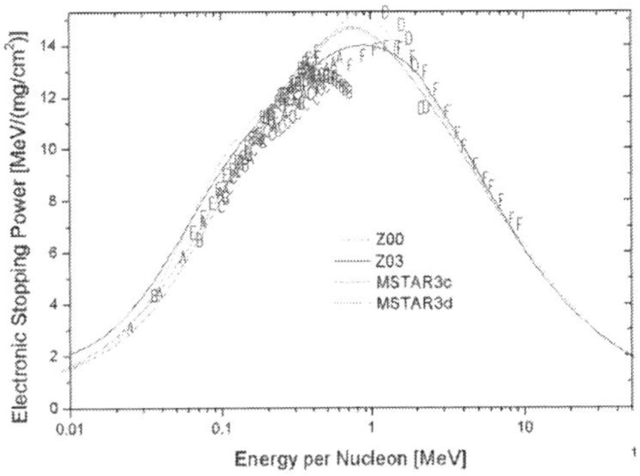

Fig. 9. Silicon stopping power [10]

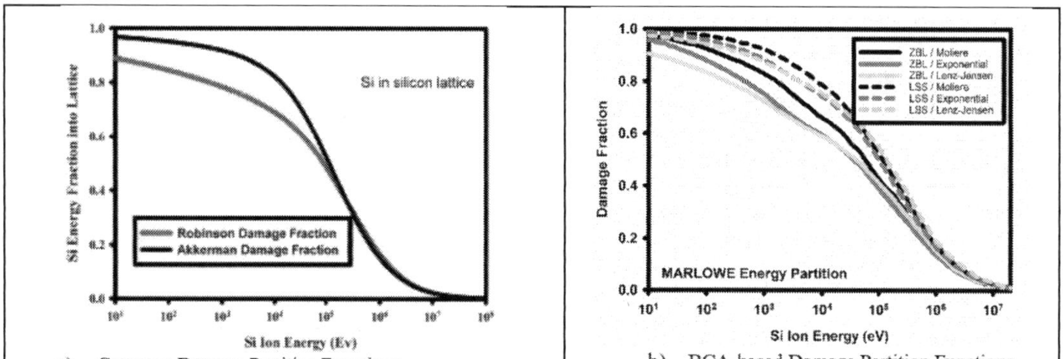

a) Common Damage Partition Functions b) BCA-based Damage Partition Functions

Fig. 10. Variation in the silicon partition function [13]

Figure 10b shows the variation in the damage partition function that can result when binary collision approximation (BCA) theory codes are used, with a range of interatomic potentials, to capture this division of deposited energy.

4 Application to the behavior of semiconductors

Section 3 addressed the various uncertainty contributors that affect damage metrics that are sensitive to the details of the energy deposited by a neutron interaction. The Section 4 focus is on the application area of electronics damage to semiconductors and examines three generic categories of damage: a) permanent damage; b) single event effects; and c) stochastic effects in small feature size microelectronics.

4.1 Permanent neutron damage metrics

To place this in context, we consider a representative application – permanent neutron radiation damage to silicon semiconductors. If we look at the semiconductor minority carrier recombination lifetime, the damage metric that represents the change in gain of a bipolar transistor, the calculated metric is the damage energy, i.e., the energy that goes into the creation of primary Frenkel pair defects. Figure 11 shows the correlation matrices for the various uncertainty contributors to the silicon damage energy, i.e., nuclear cross section data, interaction potentials for the recoil ion energy deposition, and close Frenkel pair recombination aspects for threshold displacement energies near the lattice binding energy. Each of these uncertainty contributors has a different energy-dependent weighting and a different correlation matrix. These three different contributors have a different physics basis and, within the modelling approach, can be considered to be uncorrelated. The bottom-right plot in Figure 11 shows the resulting total correlation matrix for the composite effect. The strong energy-dependent correlation clearly shows that this correlation in the energy-dependent response must be taken into account when using modelling to predict radiation damage that occurs in real-world applications.

The strong energy-dependent correlation seen in Figure 11, especially for the uncertainty due to the interaction potential, shows why application-specific metrics often use a normalized damage metric to compare neutron irradiation in different environments, e.g., the use of the 1-MeV equivalent fluence for radiation damage to silicon or GaAs electronics. When a normalization constraint is applied, there can be a very large decrease in the resulting uncertainty when comparing damage between different radiation environments. For other applications, such as embrittlement of the pressure vessel in pressurized water reactors (PWR), an effective normalization takes place by constraining the application of the metric to a narrow range of irradiation conditions and applying a damage metric that addresses correlations due to material impurities.

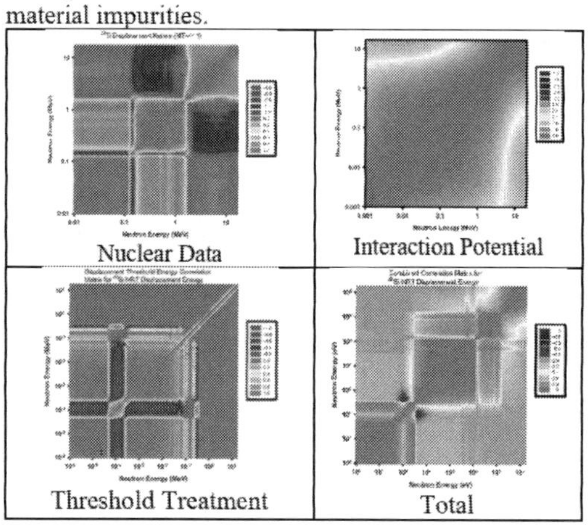

Fig. 11. Correlation matrices various uncertainty contributors in the ^{28}Si damage energy [7]

4.2 nSEEs

Neutron single event effects (nSEE) is a category of neutron damage where the physics behind the damage is related to the charge generated within a "sensitive volume" due to a single neutron interaction and collected within the clock cycle for the device operation. Since the device-specific charge collection volume can have a complex geometry, a geometry that is affected by changes in the local electric field resulting from a high

 https://doi.org/10.1051/epjconf/202328100001

energy ion moving through the volume, the more accurate charge collection metric of collected charge is often replaced through the use of the threshold value for the LET of the recoil ion. Since the recoil ion can have a range of energies, based on the stochastic consideration of the neutron interaction (the type of reaction and the angular distribution for reaction products), the commonly used damage metric is the pdf for the LET, as seen in Equation 3.

Figure 12a shows the LET pdf for a 14 MeV neutron and for the Annular Core Research Reactor (ACRR), a pool-type fission reactor. Figure 12b shows the cumulative probability distribution (cdf) for these same neutron spectra. Since the damage state is related to the probability of the LET being above a given threshold value, the cdf is the useful damage metric.

Since the damage is related to the probability of a neutron interaction resulting in a recoil ion with an LET above a threshold value, and since the same recoil ion LET can be generated from different incident neutron energies, the interpretation of the uncertainty analysis dependents upon both the neutron spectrum and the response function. The uncertainty in the LET response relates back to the uncertainty in the both the recoil atom energies and in the ion stopping powers. Figure 13a shows this correlation in the LET distribution when a device is exposed to the ACRR free field neutron environment. Figure 13b compares the energy-dependent standard deviations in the pdf and cdf for both the ACRR and the ^{252}Cf(s.f.) neutron environments. Both the pdf and cdf are seen to have a similar standard deviation in these neutron spectra. The difference in the magnitude of the standard deviations between the two spectra is due to the uncertainty in the underlying neutron spectrum characterization, i.e., the ^{252}Cf(s.f.) neutron spectrum is much better characterized.

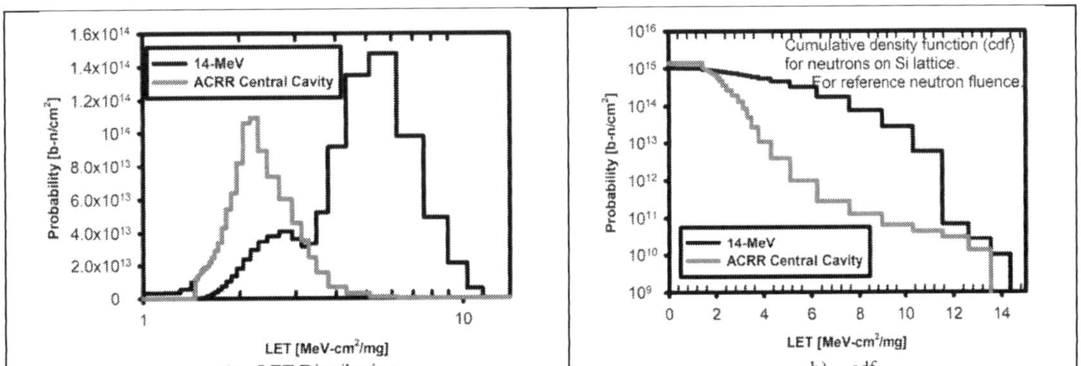

a) LET Distribution (linear y-axis; log x-axis)
b) cdf (log y-axis; linear x-axis)

Fig. 12. Silicon LET distribution for two neutron spectra [8]

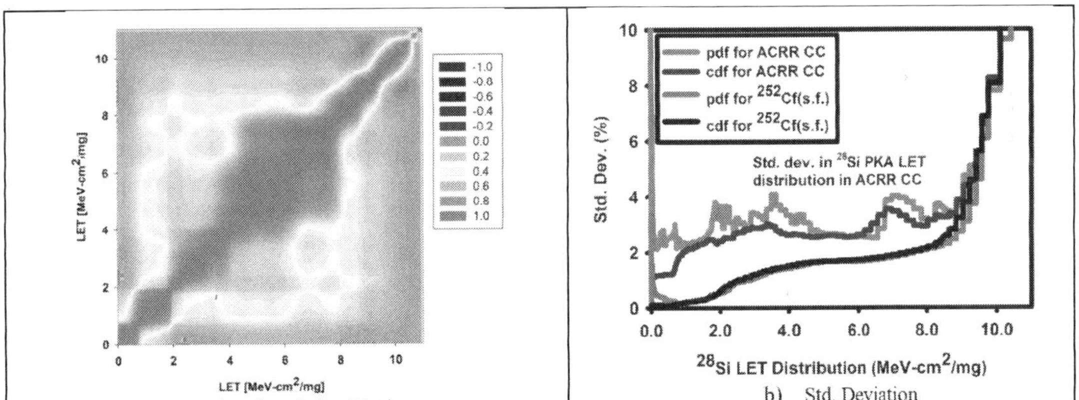

a) Correlation Matrix
b) Std. Deviation

Fig. 13. Uncertainty in the cdf for silicon LET in a pool-type reactor spectrum and in ^{252}Cf(s.f.) spectrum [8]

4.3 Stochastic considerations

As the state-of-the-art in silicon electronics goes towards the use of smaller feature sizes (<12 nm), when the damage mechanism depends upon energy deposition within a sensitive device volume, the traditional use of calculated damage metrics is no longer sufficient. When the number of neutron-induced interactions within the sensitive volume is a few rather than hundreds, the use of average damage metrics, such as an average number of Frenkel pairs, an average damage energy or an average kerma becomes less applicable. Now the stochastic variation in the damage must be considered.

Figure 14a shows the pdf for the number of Frenkel pairs generated due to an incident 100-keV silicon ion on a silicon lattice. This figure plots the results of simulating 897 cascades. The total number of Frenkel pairs generated in the cascade is depicted on the x-axis and the frequency of this occurrence is plotted on the y-axis. This 100-keV silicon ion in a silicon lattice results in an average of 680 Frenkel pairs with a pdf having a full-width-at-half-maximum (FWHM) of 129. While this pdf can be considered to approximate a Gaussian distribution, it does have a low tail in the distribution. Figure 14b shows the results for a 1 MeV neutron, which samples over a range of neutron reaction channels and the resulting recoil atom energies. This case results in an average of 244 Frenkel pairs with a FWHM of 216. This is a very non-Gaussian distribution. The source of the shape in this distribution is traced directly back to the distribution of recoil ion energies from a 1 MeV neutron interaction, i.e., a 1 MeV neutron has an average recoil ion energy of 38-keV but with a FWHM of 31 keV for the recoil energy.

Modern electronic circuits often employ a balancing of transistors in the front-end electronics in order to extend the applicable operation range of the circuit. When these devices have a small feature size, the use of calculated damage metrics based on average interactions may no longer be justified. For example, consider an operational amplifier design where the front-end transistors typically incorporate this type of design and the applicable performance metrics are offset voltage, input bias current, and slew rate. These circuit-level performance metrics are calculated from the device-level displacement damage and charge release metrics. In order to properly predict the reliable operation of these devices in radiation, the stochastic level variation in the device-level damage metrics must be sampled – and this sampling must take into account the non-Gaussian nature of the damage distributions. Use of damage metrics based on an "average" neutron interaction is no longer sufficient. While the initial neutron interactions in the different frontend electronic devices can be considered to be uncorrelated, the correlations in the underlying damage mechanisms must be applied to the pdfs such as that shown in Figure 14b.

5 Conclusion

Th importance of correlations in modelling the uncertainty in radiation damage metrics is shown to be very dependent upon the application. This paper outlined a mathematical formulation that can be applied to most currently used device-level radiation damage metrics. Some uncertainty contributions are strongly correlated, e.g., different reaction channels, and should always be treated with high fidelity. High fidelity modelling can be a challenge – and one must often use a TMC approach in order to capture the critical correlations. Other uncertainty contributions can, based on physics considerations, be considered to be uncorrelated and are more easily treated, e.g., for permanent neutron damage modes, the uncertainty in the neutron spectrum is generally uncorrelated with the uncertainty in the response metric. An exception is addressed here for some damage metrics, such as nSEE, where the relevant physics depends upon the distribution of recoils and the damage is parametrized in terms of a distribution and a threshold vulnerability value.

Results of a careful treatment of correlations between uncertainty contributors were summarized for a range of damage modes in modern electronics. Some deficiencies in current treatment were identified, e.g., a calculated damage metric is based upon a Frenkel pair generation whereas observed damage is based upon evolved defects. For emerging small feature-size electronics, the importance of the stochastic variation in the neutron radiation damage was highlighted – exposing a deficiency in our use of average damage metrics.

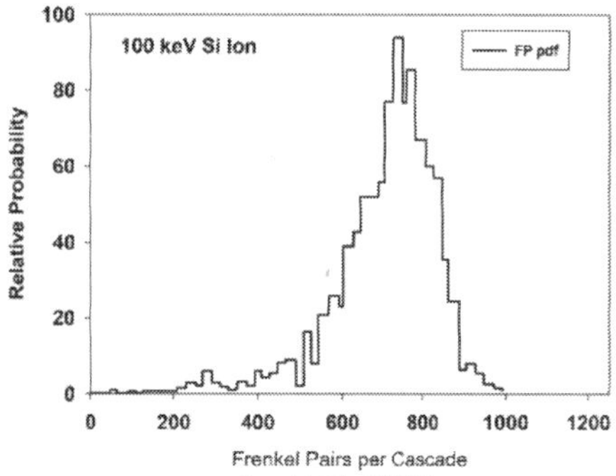

a) 100-keV Si Ions

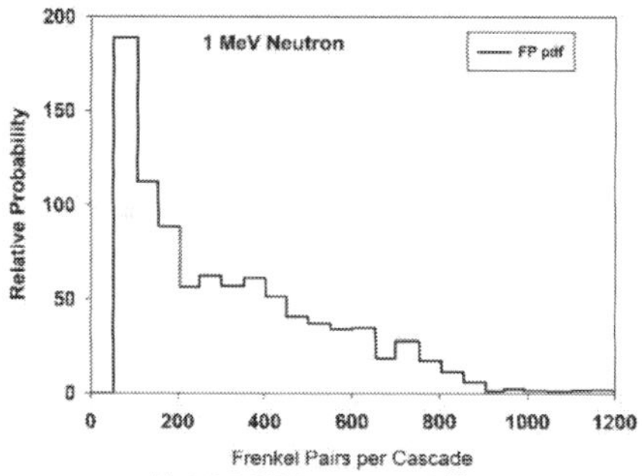

b) 1-MeV Neutron on Silicon

Fig. 14. Probability distributions for the number of Frenkel pairs created

Acknowledgement

The author wished to acknowledge the IAEA for their leadership in organizing the CRP that focused the international contributions that led to the development of the IRDFF-II library and for the separate technical contributions their staff made to the library contents. Many examples of that effort are cited in this paper as examples of the importance of addressing correlations in metrics.

This work was supported in part by the U.S. Department of Energy under contract DE-NA0003525. Sandia National Laboratories is a multi-mission laboratory managed and operated by National Technology and Engineering Solutions of Sandia, LLC, a wholly owned subsidiary of Honeywell International, Inc., for the U.S. Department of Energy's

National Nuclear Security Administration under contract DE-NA0003525.

The views expressed in the paper do not necessarily represent the views of the U.S. Department of Energy or the United States Government.

References

1. W. L. Zijp, K. H. Appelman, H. J. Nolthenius, and H. CH. Rieffe, *Damage Cross Section Library (DAMSIG77)*, report **ECN-36** (Feb. 1978).
2. N.P. Kocherov, P.K. McLaughlin, *The International Reactor Dosimetry File (IRDF-90, Version 2)*, Report **IAEA-NDS-141**, Rev. 2, International Atomic Energy Agency, (Oct. 1992).
3. A. Trkov, P.J. Griffin, et al., "IRDFF-II: A New Neutron Metrology Library", Nuclear Data Sheets, **163**, 1-108 (2020).
4. ASTM E526-22, Standard Test Method for Measuring Fast-Neutron Reaction by Radioactivation of Titanium, ASTM International, West Conshohocken, PA (2022)
5. A. Trkov, M. Herman and D. A. Brown (Eds.), *ENDF-6 Formats Manual*, CSEWG Document **ENDF-102**, Report **BNL-203218-2018-INRE**, Brookhaven National Laboratory, Upton, NY (2018) https://www.osti.gov/biblio/1425114/
6. ASTM E910, Standard Test Method for Application and Analysis of Helium Accumulation Fluence Monitors for Reactor Vessel Surveillance, ASTM International, West Conshohocken, PA (2022)
7. P.J. Griffin "Uncertainty characterization of silicon damage metrics", IEEE Trans. Nucl. Sci., **66**, pp. 327-336 (2019)
8. P.J. Griffin, A.J. Konig, D. Rochman, "Impact of Nuclear Data Uncertainty in the Modelling of Neutron-Induced Recoil Atom Energy Distributions in Silicon", IEEE Trans. Nucl. Sci., **66**, pp. 1719-1729 (2019)
9. R.E. MacFarlane, D.W. Muir, R.M. Boicourt, A.C. Kahler, J.L. Conlin, W. Haeck, *The NJOY nuclear data processing system, version 2016*, Los Alamos National Laboratory, report **LA-UR-17-20093** (2016) https://github.com/njoy/
10. H. Paul, *Stopping Power of Matter for Ions: Graphs Data Comments and Programs*, (Sep. 2018) Available: https://www-nds.iaea.org/stopping/stopping_heav.html.
11. J. F. Ziegler, J. P. Biersack and U. Littmark, *The Stopping and Range of Ions in Solids*, Pergamon Press, Inc., New York (1985) https://link.springer.com/chapter/10.1007/978-3-642-68779-2_5
12. M. T. Robinson, *Nuclear Fusion Reactors, Nuclear Fusion Reactors*, Proceedings of a conference held at UKAEA Culham Laboratory, Abingdon, 17-19 Sept. 1969, British Nuclear Energy Soc., pp. 364-378 (1970) https://www.icevirtuallibrary.com/doi/abs/10.1680/nfr.44661.0025.
13. P.J. Griffin, P.J. Cooper "Influence of the Damage Partition Function on the Uncertainty of the Silicon Displacement Damage Metric", IEEE Trans. Nucl. Sci., **64**, pp. 574-581 (2016)
14. Nuclear Reaction Data and Uncertainties for Radiation Damage, IAEA report INDC(NDS)-0719, August, 2016. https://www-nds.iaea.org/publications/indc/indc-nds-0719.pdf

Application of the extended bias factor method for highly reliable benchmark suite

Mikio Tokashiki[1*], *Tadashi* Ikehara[2**], *Kenichi* Tada[2], *Tohru* Egawa[1], *Kenji* Yokoyama[3] and *Osamu* Iwamoto[2]

[1]Nuclear Fuel Industries, Ltd., Shinsugita-cho, Isogo, Yokohama 235-8523, Japan
[2]Japan Atomic Energy Agency, Tokai-mura, Ibaraki 319-1195, Japan
[3]Japan Atomic Energy Agency, Oarai-machi, Ibaraki 311-1393, Japan

Abstract. It was tried to get useful information feedback on cross section and covariance by application of extended bias factor method with experimental benchmarks' C/E, sensitivity coefficient and covariance data. As a result, it was succeeded to distinguish "the case unrecommended by specialists" as outlier case. And it was succeeded to get the hints that there is room for improvement in cross section or covariance of some nuclides by investigating outlier cases judged as recommended case by specialists.

1 Introduction

Differences in the keff values between continuous energy Monte Carlo calculations (C) and critical experiments (E) potentially contain hidden information for better making adjustment of cross sections and their covariance data. In this study, a possible approach, which is formed by mediating the bias factor corrected E (for short E') and combining the specialist judgement with this, is explained.

2 Used benchmarks and covariance data

2.1 Overview of used benchmarks

238 cases (UO_2:173, MOX:65) in ICSBEP[1] /IRPhEP[2] that were investigated in the JAEA report[3], edited by Reactor Integral Test Working Group (WG), JENDL Committee, were used. In the report, benchmarks similar to LWR were selected from ICSBEP/IRPhEP, and those keff were calculated by MVP[4] (continuous energy Monte Carlo code developed by JAEA). On the basis of the calculation results, recommended case and unrecommended case for JENDL development were judged by specialists in the WG. Table 1 is the list of the benchmarks.

Sensitivity coefficients of 212 cases (UO_2:147, MOX:65) calculated by MCNP[5], that were prepared by the WG (See 2.2), were available. The sensitivity coefficients were calculated with available MCNP input files in ICSBEP, IRPhEP and MCNP Whisper[6].

In this study, 185 cases (UO_2:131, MOX:54) whose differences of keff between MVP and MCNP are less than about 0.1%dk were used. It is expected that the input files of 185 cases do not contain significant input error, because keff of MVP and MCNP were consistent though each of input file was made independently. Therefore, the input files of these cases are expected to be reliable. These cases are shown in Table 1 with "✓".

In the following, JENDL-4.0u were used consistently for cross section and covariance data.

2.2 Used multi-group covariance data, sensitivity coefficients and tools

Multi-group covariance data, sensitivity coefficients and tools were made by aforementioned WG, as an activity of the WG.

Multi-group covariance data were processed by FRENDY[7,8] Ver.2.00.012 (for MODER, RECONR, BROADR, THERMR and GROUPR) and NJOY2016[9] (for ERRORR), and those file formats are COVERX[10] of SCALE[11]. Available group structures are SCALE's 44, 56, 252 and so on (In this study, 56 group structure was used).

Sensitivity coefficients of 212 cases (UO_2:147, MOX:65) are available that are part of benchmarks investigated in the JAEA report[3]. Additionally, sensitivity coefficients of undisclosed 28 experiments (UO_2:23, MOX:5), that were submitted by the WG members (These 28 cases were not used in this study), are available. Those file formats are Sensitivity Data File (SDF) of SCALE, and available group structures are aforementioned.

The tools, file format changer (NJOY2016 output to COVERX and MCNP output to SDF) and analysis tool of extended bias factor (PE method type. In the following, "extended bias factor" means "PE method type")[12], were available and used in this study.

* Corresponding author: tokasiki@nfi.co.jp
** Present affiliation: Toshiba Energy Systems & Solutions Corporation

The above mentioned database and tools will be opened as JAEA report in FY 2022.

Table 2 shows available covariance data in JENDL-4.0u for used benchmarks in this study.

Table 1 (1/3). Used benchmarks.

Benchmark Name	No.	Measured Value	Measured Uncertainty	Calculated (MVP) Value	Calculated (MVP) Uncertainty	Calculated (MCNP) Value	Calculated (MCNP) Uncertainty	MCNP -MVP %dk	|MCNP-MVP| < ~0.1%dk (Used benchmark)	Reference for Used MCNP Input
KRITZ2-01	1	1.0025	0.0020	0.99837	0.00015	0.99750	0.00005	-0.1%	✓	IRPhEP
	2	1.0024	0.0028	0.99667	0.00016	0.99891	0.00005	0.2%		IRPhEP
KRITZ2-13	1	1.0013	0.0022	1.00076	0.00015	0.99993	0.00005	-0.1%	✓	IRPhEP
	2	1.0019	0.0021	0.99927	0.00016	0.99877	0.00005	-0.1%	✓	IRPhEP
LCT-001	1	0.9998	0.0030	0.99984	0.00017	0.99931	0.00005	-0.1%	✓	MCNP6.2 Whisper
	2	0.9998	0.0030	0.99870	0.00017	0.99856	0.00005	0.0%	✓	MCNP6.2 Whisper
	3	0.9998	0.0030	0.99839	0.00017	0.99779	0.00005	-0.1%	✓	MCNP6.2 Whisper
	4	0.9998	0.0030	0.99902	0.00017	0.99860	0.00005	0.0%	✓	MCNP6.2 Whisper
	5	0.9998	0.0030	0.99654	0.00017	0.99625	0.00005	0.0%	✓	MCNP6.2 Whisper
	6	0.9998	0.0030	0.99879	0.00017	0.99838	0.00005	0.0%	✓	MCNP6.2 Whisper
	7	0.9998	0.0031	0.99784	0.00016	0.99763	0.00004	0.0%	✓	MCNP6.2 Whisper
	8	0.9998	0.0030	0.99713	0.00017	0.99666	0.00005	0.0%	✓	MCNP6.2 Whisper
LCT-002	1	0.9997	0.0020	0.99741	0.00019	0.99784	0.00005	0.0%	✓	MCNP6.2 Whisper
	2	0.9997	0.0020	0.99910	0.00017	0.99921	0.00005	0.0%	✓	MCNP6.2 Whisper
	3	0.9997	0.0020	0.99885	0.00016	0.99851	0.00005	0.0%	✓	MCNP6.2 Whisper
	4	0.9997	0.0018	0.99775	0.00016	0.99792	0.00005	0.0%	✓	MCNP6.2 Whisper
	5	0.9997	0.0019	0.99621	0.00016	0.99665	0.00005	0.0%	✓	MCNP6.2 Whisper
LCT-005	1	1.0000	0.0023	1.00237	0.00019	1.00203	0.00005	0.0%	✓	MCNP6.2 Whisper
	2	1.0000	0.0021	1.00025	0.00020	1.00010	0.00006	0.0%	✓	MCNP6.2 Whisper
	3	1.0000	0.0029	1.00120	0.00017	1.00184	0.00005	0.1%	✓	MCNP6.2 Whisper
	4	1.0000	0.0025	1.00002	0.00016	1.00068	0.00005	0.1%	✓	MCNP6.2 Whisper
	5	1.0000	0.0047	1.00407	0.00019	1.00344	0.00005	-0.1%	✓	MCNP6.2 Whisper
	6	1.0000	0.0042	1.00551	0.00019	1.00485	0.00006	-0.1%	✓	MCNP6.2 Whisper
	7	1.0000	0.0043	1.00261	0.00018	1.00217	0.00006	0.0%	✓	MCNP6.2 Whisper
	8	1.0000	0.0021	1.00406	0.00016	1.00396	0.00006	0.0%	✓	MCNP6.2 Whisper
	9	1.0000	0.0040	1.00557	0.00016	1.00516	0.00005	0.0%	✓	MCNP6.2 Whisper
	10	1.0000	0.0028	1.00498	0.00016	1.00487	0.00005	0.0%	✓	MCNP6.2 Whisper
	11	1.0000	0.0043	1.00550	0.00016	1.00543	0.00005	0.0%	✓	MCNP6.2 Whisper
	12	1.0000	0.0066	1.00606	0.00018	1.00507	0.00005	-0.1%	✓	MCNP6.2 Whisper
	13	1.0000	0.0064	1.01256	0.00017	1.01134	0.00005	-0.1%	✓	MCNP6.2 Whisper
	14	1.0000	0.0020	0.99815	0.00018					(None)
	15	1.0000	0.0020	1.01818	0.00017					(None)
	16	1.0000	0.0032	1.01253	0.00018					(None)
LCT-006	1	1.0000	0.0020	0.99984	0.00018	0.99892	0.00005	-0.1%	✓	MCNP6.2 Whisper
	2	1.0000	0.0020	1.00027	0.00018	0.99939	0.00005	-0.1%	✓	MCNP6.2 Whisper
	3	1.0000	0.0020	1.00032	0.00018	0.99924	0.00005	-0.1%	✓	MCNP6.2 Whisper
	4	1.0000	0.0020	0.99948	0.00018	0.99906	0.00005	0.0%	✓	MCNP6.2 Whisper
	5	1.0000	0.0020	0.99971	0.00019	0.99896	0.00005	-0.1%	✓	MCNP6.2 Whisper
	6	1.0000	0.0020	1.00026	0.00018	0.99947	0.00005	-0.1%	✓	MCNP6.2 Whisper
	7	1.0000	0.0020	0.99998	0.00018	0.99942	0.00005	-0.1%	✓	MCNP6.2 Whisper
	8	1.0000	0.0020	1.00001	0.00018	0.99940	0.00005	-0.1%	✓	MCNP6.2 Whisper
	9	1.0000	0.0020	0.99976	0.00018	0.99944	0.00005	0.0%	✓	MCNP6.2 Whisper
	10	1.0000	0.0020	0.99968	0.00018	0.99937	0.00005	0.0%	✓	MCNP6.2 Whisper
	11	1.0000	0.0020	1.00003	0.00018	0.99945	0.00005	-0.1%	✓	MCNP6.2 Whisper
	12	1.0000	0.0020	0.99943	0.00019	0.99929	0.00005	0.0%	✓	MCNP6.2 Whisper
	13	1.0000	0.0020	0.99936	0.00018	0.99914	0.00005	0.0%	✓	MCNP6.2 Whisper
	14	1.0000	0.0020	0.99935	0.00018	0.99943	0.00005	0.0%	✓	MCNP6.2 Whisper
	15	1.0000	0.0020	0.99971	0.00018	0.99945	0.00005	0.0%	✓	MCNP6.2 Whisper
	16	1.0000	0.0020	0.99930	0.00018	0.99942	0.00005	0.0%	✓	MCNP6.2 Whisper
	17	1.0000	0.0020	0.99958	0.00018	0.99927	0.00005	0.0%	✓	MCNP6.2 Whisper
	18	1.0000	0.0020	0.99948	0.00018	0.99926	0.00005	0.0%	✓	MCNP6.2 Whisper
LCT-007	1	1.0000	0.0014	0.99687	0.00020	0.99621	0.00006	-0.1%	✓	MCNP6.2 Whisper
	2	1.0000	0.0008	0.99865	0.00020	0.99848	0.00005	0.0%	✓	MCNP6.2 Whisper
	3	1.0000	0.0007	0.99742	0.00019	0.99773	0.00005	0.0%	✓	MCNP6.2 Whisper
	4	1.0000	0.0008	0.99765	0.00017	0.99800	0.00004	0.0%	✓	MCNP6.2 Whisper
	5	1.0000	0.0014	0.99653	0.00020	0.99581	0.00006	-0.1%	✓	MCNP6.2 Whisper
	6	1.0000	0.0008	0.99889	0.00020	0.99841	0.00005	0.0%	✓	MCNP6.2 Whisper
	7	1.0000	0.0007	0.99898	0.00018	0.99874	0.00005	0.0%	✓	MCNP6.2 Whisper
	8	1.0000	0.0014	0.99819	0.00020	0.99707	0.00006	-0.1%	✓	MCNP6.2 Whisper
	9	1.0000	0.0008	0.99831	0.00020	0.99803	0.00006	0.0%	✓	MCNP6.2 Whisper
	10	1.0000	0.0007	0.99849	0.00018	0.99879	0.00005	0.0%	✓	MCNP6.2 Whisper
LCT-008	1	1.0007	0.0006	0.99892	0.00012	1.00052	0.00005	0.2%		MCNP6.2 Whisper
	2	1.0007	0.0006	0.99924	0.00012	1.00057	0.00005	0.1%	✓	MCNP6.2 Whisper
	3	1.0007	0.0006	0.99966	0.00012	1.00115	0.00005	0.1%	✓	ICSBEP DICE
	4	1.0007	0.0006	0.99932	0.00012	1.00056	0.00005	0.1%	✓	ICSBEP DICE
	5	1.0007	0.0006	0.99903	0.00012	1.00023	0.00005	0.1%	✓	MCNP6.2 Whisper
	6	1.0007	0.0006	0.99922	0.00013	1.00083	0.00005	0.2%		ICSBEP DICE
	7	1.0007	0.0006	0.99894	0.00013	1.00013	0.00005	0.1%	✓	MCNP6.2 Whisper
	8	1.0007	0.0006	0.99865	0.00014	0.99980	0.00005	0.1%	✓	MCNP6.2 Whisper
	9	1.0007	0.0006	0.99917	0.00014	1.00012	0.00005	0.1%	✓	ICSBEP DICE
	10	1.0007	0.0006	0.99931	0.00012	1.00044	0.00005	0.1%	✓	ICSBEP DICE
	11	1.0007	0.0006	0.99935	0.00012	1.00120	0.00005	0.2%		MCNP6.2 Whisper
	12	1.0007	0.0006	0.99914	0.00012	1.00076	0.00005	0.2%		ICSBEP DICE
	13	1.0007	0.0006	0.99911	0.00012	1.00072	0.00005	0.2%		ICSBEP DICE
	14	1.0007	0.0006	0.99879	0.00012	1.00059	0.00005	0.2%		ICSBEP DICE
	15	1.0007	0.0006	0.99887	0.00012	1.00031	0.00005	0.1%	✓	ICSBEP DICE
	16	1.0007	0.0006	0.99866	0.00012	0.99999	0.00005	0.1%	✓	ICSBEP DICE
	17	1.0007	0.0006	0.99691	0.00013	0.99881	0.00005	0.2%		ICSBEP DICE

Table 1 (2/3). Used benchmarks. (Continued)

Benchmark Name	No.	Measured Value	Measured Uncertainty	Calculated (MVP) Value	Calculated (MVP) Uncertainty	Calculated (MCNP) Value	Calculated (MCNP) Uncertainty	MCNP -MVP %dk	|MCNP-MVP| < ~0.1%dk (Used benchmark)	Reference for Used MCNP Input
LCT-010	1	1.0000	0.0021	1.00844	0.00020	1.00805	0.00005	0.0%	✓	MCNP6.2 Whisper
	2	1.0000	0.0021	1.00774	0.00019	1.00787	0.00005	0.0%	✓	MCNP6.2 Whisper
	3	1.0000	0.0021	1.00595	0.00019	1.00610	0.00005	0.0%	✓	MCNP6.2 Whisper
	4	1.0000	0.0021	0.99600	0.00019	0.99615	0.00005	0.0%	✓	MCNP6.2 Whisper
	5	1.0000	0.0021	0.99888	0.00018	0.99869	0.00005	0.0%	✓	MCNP6.2 Whisper
	6	1.0000	0.0021	0.99934	0.00018	0.99953	0.00005	0.0%	✓	MCNP6.2 Whisper
	7	1.0000	0.0021	1.00018	0.00018	1.00049	0.00005	0.0%	✓	MCNP6.2 Whisper
	8	1.0000	0.0021	0.99699	0.00018	0.99706	0.00005	0.0%	✓	MCNP6.2 Whisper
	9	1.0000	0.0021	0.99936	0.00019	0.99972	0.00005	0.0%	✓	MCNP6.2 Whisper
	10	1.0000	0.0021	1.00020	0.00019	1.00011	0.00005	0.0%	✓	MCNP6.2 Whisper
	11	1.0000	0.0021	1.00062	0.00020	1.00036	0.00005	0.0%	✓	MCNP6.2 Whisper
	12	1.0000	0.0021	0.99940	0.00020	0.99936	0.00005	0.0%	✓	MCNP6.2 Whisper
	13	1.0000	0.0021	0.99656	0.00019	0.99657	0.00005	0.0%	✓	MCNP6.2 Whisper
	14	1.0000	0.0028	1.00088	0.00020	1.00036	0.00006	-0.1%	✓	ICSBEP DICE
	15	1.0000	0.0028	1.00177	0.00019	1.00113	0.00006	-0.1%	✓	ICSBEP DICE
	16	1.0000	0.0028	1.00246	0.00019	1.00161	0.00006	-0.1%	✓	ICSBEP DICE
	17	1.0000	0.0028	1.00177	0.00019	1.00112	0.00006	-0.1%	✓	ICSBEP DICE
	18	1.0000	0.0028	1.00133	0.00019	1.00075	0.00005	-0.1%	✓	ICSBEP DICE
	19	1.0000	0.0028	1.00092	0.00019	1.00032	0.00005	-0.1%	✓	ICSBEP DICE
	20	1.0000	0.0028	1.00545	0.00019	1.00479	0.00005	-0.1%	✓	ICSBEP DICE
	21	1.0000	0.0028	1.00486	0.00020	1.00478	0.00005	0.0%	✓	ICSBEP DICE
	22	1.0000	0.0028	1.00399	0.00020	1.00336	0.00005	-0.1%	✓	ICSBEP DICE
	23	1.0000	0.0028	1.00101	0.00019	1.00042	0.00005	-0.1%	✓	ICSBEP DICE
	24	1.0000	0.0028	0.99939	0.00018	0.99837	0.00005	-0.1%	✓	ICSBEP DICE
	25	1.0000	0.0028	1.00044	0.00019	0.99990	0.00005	-0.1%	✓	ICSBEP DICE
	26	1.0000	0.0028	1.00098	0.00018	1.00052	0.00005	0.0%	✓	ICSBEP DICE
	27	1.0000	0.0028	1.00178	0.00018	1.00083	0.00005	-0.1%	✓	ICSBEP DICE
	28	1.0000	0.0028	1.00152	0.00018	1.00094	0.00005	-0.1%	✓	ICSBEP DICE
	29	1.0000	0.0028	1.00153	0.00019	1.00096	0.00005	-0.1%	✓	ICSBEP DICE
	30	1.0000	0.0028	0.99951	0.00018	0.99887	0.00005	-0.1%	✓	ICSBEP DICE
LCT-017	1	1.0000	0.0031	1.00359	0.00017	1.00319	0.00005	0.0%	✓	MCNP6.2 Whisper
	2	1.0000	0.0031	1.00268	0.00017	1.00250	0.00005	0.0%	✓	MCNP6.2 Whisper
	3	1.0000	0.0031	1.00022	0.00016	1.00014	0.00005	0.0%	✓	MCNP6.2 Whisper
	4	1.0000	0.0031	0.99863	0.00016	0.99768	0.00004	-0.1%	✓	MCNP6.2 Whisper
	5	1.0000	0.0031	0.99988	0.00016	0.99954	0.00004	0.0%	✓	MCNP6.2 Whisper
	6	1.0000	0.0031	0.99995	0.00016	0.99976	0.00004	0.0%	✓	MCNP6.2 Whisper
	7	1.0000	0.0031	1.00005	0.00016	0.99952	0.00004	-0.1%	✓	MCNP6.2 Whisper
	8	1.0000	0.0031	0.99785	0.00016	0.99758	0.00004	0.0%	✓	MCNP6.2 Whisper
	9	1.0000	0.0031	0.99710	0.00016	0.99691	0.00004	0.0%	✓	MCNP6.2 Whisper
	10	1.0000	0.0031	0.99891	0.00017	0.99804	0.00005	-0.1%	✓	MCNP6.2 Whisper
	11	1.0000	0.0031	0.99873	0.00017	0.99834	0.00005	0.0%	✓	MCNP6.2 Whisper
	12	1.0000	0.0031	0.99872	0.00017	0.99839	0.00005	0.0%	✓	MCNP6.2 Whisper
	13	1.0000	0.0031	0.99905	0.00017	0.99856	0.00005	0.0%	✓	MCNP6.2 Whisper
	14	1.0000	0.0031	0.99856	0.00017	0.99869	0.00005	0.0%	✓	MCNP6.2 Whisper
	15	1.0000	0.0028	0.99736	0.00017					(None)
	16	1.0000	0.0028	0.99885	0.00017					(None)
	17	1.0000	0.0028	0.99987	0.00017	0.99908	0.00005	-0.1%	✓	ICSBEP Handbook
	18	1.0000	0.0028	0.99894	0.00018					(None)
	19	1.0000	0.0028	0.99921	0.00018					(None)
	20	1.0000	0.0028	0.99776	0.00018					(None)
	21	1.0000	0.0028	0.99775	0.00017					(None)
	22	1.0000	0.0028	0.99681	0.00017					(None)
	23	1.0000	0.0028	1.00182	0.00017					(None)
	24	1.0000	0.0028	1.00229	0.00017					(None)
	25	1.0000	0.0028	0.99884	0.00017					(None)
	26	1.0000	0.0028	0.99584	0.00016					(None)
	27	1.0000	0.0028	0.99798	0.00016					(None)
	28	1.0000	0.0028	0.99861	0.00016					(None)
	29	1.0000	0.0028	0.99836	0.00016					(None)
LCT-018	1	1.0000	0.0020	1.00263	0.00021					(None)
LCT-019	1	1.0000	0.0063	1.01493	0.00018					(None)
	2	1.0000	0.0058	1.00932	0.00019					(None)
	3	1.0000	0.0061	1.00602	0.00016					(None)
LCT-022	1	1.0000	0.0046	1.00208	0.00019	1.00120	0.00006	-0.1%	✓	MCNP6.2 Whisper
	2	1.0000	0.0046	1.00658	0.00020	1.00554	0.00006	-0.1%	✓	MCNP6.2 Whisper
	3	1.0000	0.0036	1.00701	0.00021	1.00631	0.00006	-0.1%	✓	MCNP6.2 Whisper
	4	1.0000	0.0037	1.00779	0.00020	1.00756	0.00005	0.0%	✓	MCNP6.2 Whisper
	5	1.0000	0.0038	1.00324	0.00019	1.00313	0.00005	0.0%	✓	MCNP6.2 Whisper
	6	1.0000	0.0046	1.00072	0.00018	1.00122	0.00005	0.0%	✓	MCNP6.2 Whisper
	7	1.0000	0.0046	1.00354	0.00017	1.00375	0.00005	0.0%	✓	MCNP6.2 Whisper
LCT-024	1	1.0000	0.0054	1.00058	0.00018	0.99953	0.00006	-0.1%	✓	MCNP6.2 Whisper
	2	1.0000	0.0040	1.00776	0.00019	1.00696	0.00006	-0.1%	✓	MCNP6.2 Whisper
LCT-025	1	1.0000	0.0041	0.98710	0.00019	0.98622	0.00005	-0.1%	✓	MCNP6.2 Whisper
	2	1.0000	0.0044	0.99471	0.00019	0.99403	0.00005	-0.1%	✓	MCNP6.2 Whisper
	3	1.0000	0.0047	0.99947	0.00019	0.99916	0.00006	0.0%	✓	MCNP6.2 Whisper
	4	1.0000	0.0052	1.00193	0.00019	1.00194	0.00005	0.0%	✓	MCNP6.2 Whisper
LCT-026	1	1.0000	0.0034	1.00221	0.00021	1.00233	0.00006	0.0%	✓	ICSBEP Handbook
	2	0.9996	0.0034	1.00185	0.00020	1.00222	0.00006	0.0%	✓	ICSBEP Handbook
	3	1.0018	0.0062	1.00503	0.00018	1.00515	0.00005	0.0%	✓	ICSBEP Handbook
	4	0.9978	0.0062	1.00062	0.00018	1.00122	0.00005	0.1%	✓	ICSBEP Handbook
LCT-048	1	1.0000	0.0025	1.00145	0.00017					(None)
	2	1.0000	0.0025	1.00120	0.00017					(None)
	3	1.0000	0.0025	1.00121	0.00017					(None)
	4	1.0000	0.0025	1.00136	0.00017					(None)
	5	1.0000	0.0025	1.00166	0.00017					(None)

Table 1 (3/3). Used benchmarks. (Continued)

Benchmark Name	No.	Measured Value	Measured Uncertainty	Calculated (MVP) Value	Calculated (MVP) Uncertainty	Calculated (MCNP) Value	Calculated (MCNP) Uncertainty	MCNP -MVP %dk	|MCNP-MVP| < ~0.1%dk (Used benchmark)	Reference for Used MCNP Input
LCT-079	1	0.9999	0.0016	0.99892	0.00019	0.99713	0.00006	-0.2%		MCNP6.2 Whisper
	2	1.0002	0.0016	0.99965	0.00020	0.99772	0.00006	-0.2%		MCNP6.2 Whisper
	3	1.0005	0.0016	0.99983	0.00020	0.99823	0.00006	-0.2%		MCNP6.2 Whisper
	4	1.0004	0.0016	1.00025	0.00019	0.99880	0.00006	-0.1%	✓	MCNP6.2 Whisper
	5	1.0004	0.0016	1.00073	0.00020	0.99893	0.00006	-0.2%		MCNP6.2 Whisper
	6	0.9994	0.0008	0.99878	0.00020	0.99825	0.00005	-0.1%	✓	MCNP6.2 Whisper
	7	1.0003	0.0008	0.99950	0.00020	0.99752	0.00005	-0.2%		MCNP6.2 Whisper
	8	1.0008	0.0008	1.00089	0.00020	0.99894	0.00005	-0.2%		MCNP6.2 Whisper
	9	1.0003	0.0008	1.00086	0.00019	0.99873	0.00005	-0.2%		MCNP6.2 Whisper
	10	1.0009	0.0008	1.00208	0.00019	1.00020	0.00005	-0.2%		MCNP6.2 Whisper
KRITZ2-19	1	1.0063	0.0030	1.00384	0.00017	1.00348	0.00005	0.0%	✓	IRPhEP
	2	1.0045	0.0027	1.00364	0.00017	1.00110	0.00005	-0.3%		IRPhEP
MCT-001	1	0.9999	0.0025	1.00073	0.00020	1.00046	0.00006	0.0%	✓	MCNP6.2 Whisper
	2	1.0002	0.0026	1.00018	0.00020	0.99958	0.00006	-0.1%	✓	MCNP6.2 Whisper
	3	1.0003	0.0032	0.99881	0.00020	0.99971	0.00006	0.1%	✓	MCNP6.2 Whisper
	4	1.0005	0.0039	1.00055	0.00020	1.00172	0.00006	0.1%	✓	MCNP6.2 Whisper
MCT-002	1	1.0005	0.0059	1.00150	0.00018	1.00068	0.00005	-0.1%	✓	MCNP6.2 Whisper
	2	1.0007	0.0045	1.00327	0.00017	1.00226	0.00005	-0.1%	✓	MCNP6.2 Whisper
	3	1.0015	0.0029	1.00321	0.00018	1.00293	0.00005	0.0%	✓	MCNP6.2 Whisper
	4	1.0030	0.0021	1.00813	0.00016	1.00809	0.00005	0.0%	✓	MCNP6.2 Whisper
	5	1.0030	0.0022	1.00476	0.00019	1.00467	0.00005	0.0%	✓	MCNP6.2 Whisper
	6	1.0033	0.0024	1.00716	0.00015	1.00751	0.00005	0.0%	✓	ICSBEP Handbook (pdf)
MCT-003	1	1.0004	0.0017	1.00181	0.00018	0.99973	0.00006	-0.2%		MCNP6.2 Whisper
	2	1.0006	0.0016	1.00222	0.00020	1.00047	0.00006	-0.2%		MCNP6.2 Whisper
	3	1.0001	0.0015	1.00223	0.00019	1.00372	0.00006	0.1%	✓	MCNP6.2 Whisper
	4	1.0008	0.0013	1.00256	0.00020	1.00051	0.00006	-0.2%		MCNP6.2 Whisper
	5	1.0010	0.0012	1.00257	0.00020	1.00063	0.00006	-0.2%		MCNP6.2 Whisper
	6	1.0018	0.0014	1.00270	0.00020	1.00185	0.00005	-0.1%	✓	MCNP6.2 Whisper
MCT-004	1	1.0005	0.0046	0.99724	0.00018	0.99692	0.00005	0.0%	✓	*1
	2	1.0005	0.0046	0.99815	0.00018	0.99776	0.00005	0.0%	✓	*1
	3	1.0005	0.0046	0.99884	0.00018	0.99778	0.00005	-0.1%	✓	*1
	4	1.0008	0.0039	0.99804	0.00018	0.99773	0.00005	0.0%	✓	*1
	5	1.0008	0.0039	0.99896	0.00017	0.99865	0.00005	0.0%	✓	*1
	6	1.0008	0.0039	0.99906	0.00018	0.99861	0.00005	0.0%	✓	*1
	7	1.0010	0.0040	0.99929	0.00017	0.99887	0.00005	0.0%	✓	*1
	8	1.0010	0.0040	0.99942	0.00017	0.99930	0.00005	0.0%	✓	*1
	9	1.0010	0.0040	0.99975	0.00017	0.99981	0.00005	0.0%	✓	*1
	10	1.0018	0.0051	0.99986	0.00016	0.99970	0.00005	0.0%	✓	*1
	11	1.0018	0.0051	0.99963	0.00016	0.99953	0.00005	0.0%	✓	*1
MCT-005	1	1.0003	0.0022	1.00276	0.00018	1.00200	0.00005	-0.1%	✓	ICSBEP Handbook (pdf)
	2	1.0004	0.0026	1.00027	0.00018	0.99975	0.00005	-0.1%	✓	ICSBEP Handbook (pdf)
	3	1.0009	0.0029	1.00737	0.00019	1.00693	0.00005	0.0%	✓	ICSBEP Handbook (pdf)
	4	1.0015	0.0028	1.00362	0.00019	1.00337	0.00005	0.0%	✓	ICSBEP Handbook (pdf)
	5	1.0018	0.0036	1.00548	0.00018	1.00546	0.00005	0.0%	✓	ICSBEP Handbook (pdf)
	6	1.0020	0.0042	1.00398	0.00017	1.00401	0.00005	0.0%	✓	ICSBEP Handbook (pdf)
	7	1.0022	0.0042	1.00495	0.00016	1.00485	0.00004	0.0%	✓	ICSBEP Handbook (pdf)
MCT-006	1	1.0009	0.0051	0.99835	0.00018	0.99784	0.00005	-0.1%	✓	ICSBEP Handbook (pdf)
	2	1.0012	0.0036	1.00235	0.00019	1.00179	0.00005	-0.1%	✓	ICSBEP Handbook (pdf)
	3	1.0018	0.0036	0.99854	0.00019	0.99812	0.00005	0.0%	✓	ICSBEP Handbook (pdf)
	4	1.0042	0.0044	1.00450	0.00018	1.00433	0.00005	0.0%	✓	ICSBEP Handbook (pdf)
	5	1.0024	0.0054	1.00387	0.00017	1.00430	0.00005	0.0%	✓	ICSBEP Handbook (pdf)
	6	1.0041	0.0051	1.00168	0.00017	1.00181	0.00004	0.0%	✓	ICSBEP Handbook (pdf)
MCT-007	1	1.0012	0.0035	1.00498	0.00018	1.00505	0.00005	0.0%	✓	ICSBEP Handbook
	2	1.0012	0.0039	1.00038	0.00018	1.00052	0.00005	0.0%	✓	ICSBEP Handbook
	3	1.0024	0.0046	1.00217	0.00017	1.00247	0.00005	0.0%	✓	ICSBEP Handbook
	4	1.0025	0.0057	1.00166	0.00016	1.00157	0.00004	0.0%	✓	ICSBEP Handbook
	5	1.0029	0.0061	0.99917	0.00015	0.99971	0.00004	0.1%	✓	ICSBEP Handbook
MCT-008	1	0.9993	0.0032	0.99933	0.00017	0.99852	0.00005	-0.1%	✓	ICSBEP Handbook
	2	1.0002	0.0030	1.00038	0.00018	0.99961	0.00005	-0.1%	✓	ICSBEP Handbook
	3	1.0013	0.0038	1.00002	0.00017	0.99956	0.00005	0.0%	✓	ICSBEP Handbook
	4	1.0007	0.0047	1.00224	0.00017	1.00239	0.00005	0.0%	✓	ICSBEP Handbook
	5	1.0015	0.0056	1.00227	0.00015	1.00241	0.00004	0.0%	✓	ICSBEP Handbook
	6	1.0020	0.0065	1.00174	0.00015	1.00176	0.00004	0.0%	✓	ICSBEP Handbook
MCT-009	1	0.9998	0.0054	1.00348	0.00017	1.00224	0.00005	-0.1%	✓	ICSBEP Handbook (pdf)
	2	1.0014	0.0049	0.99950	0.00017	0.99863	0.00005	-0.1%	✓	ICSBEP Handbook (pdf)
	3	1.0024	0.0050	0.99985	0.00017	0.99955	0.00005	0.0%	✓	ICSBEP Handbook (pdf)
	4	1.0034	0.0062	0.99915	0.00017	0.99844	0.00005	-0.1%	✓	ICSBEP Handbook (pdf)
	5	1.0044	0.0074	1.00078	0.00016	1.00037	0.00004	0.0%	✓	ICSBEP Handbook (pdf)
	6	1.0051	0.0080	1.00255	0.00015	1.00227	0.00004	0.0%	✓	ICSBEP Handbook (pdf)
MCT-011	1	0.9998	0.0024	1.00107	0.00021	1.00454	0.00006	0.3%		ICSBEP Handbook
	2	0.9998	0.0024	1.00147	0.00021	1.00456	0.00006	0.3%		ICSBEP Handbook
	3	0.9998	0.0025	1.00242	0.00021	1.00537	0.00006	0.3%		ICSBEP Handbook
	4	0.9999	0.0018	1.00076	0.00021	1.00457	0.00006	0.4%		ICSBEP Handbook
	5	0.9999	0.0016	0.99909	0.00021	1.00243	0.00006	0.3%		ICSBEP Handbook
	6	0.9999	0.0016	0.99932	0.00021	1.00244	0.00006	0.3%		ICSBEP Handbook

*1: MCNP input file courtesy of Dr. YAMAMOTO Toshihiro.

Table 2. Available covariance data in JENDL-4.0u for used benchmarks in this study.

Nuclides	Existence of Covariance
H-1	
B-10	√
B-11	√
C-0	
N-14	√
N-15	√
O-16	√
Na-23	√
Mg-24	
Mg-25	
Mg-26	
Al-27	
Si-28	
Si-29	
Si-30	
P-31	
S-32	
S-33	
S-34	
S-36	
K-39	
K-40	
K-41	
Ca-40	
Ca-42	
Ca-43	
Ca-44	
Ca-46	
Ca-48	

Nuclides	Existence of Covariance
Ti-46	
Ti-47	
Ti-48	√
Ti-49	
Ti-50	
Cr-50	
Cr-52	√
Cr-53	√
Cr-54	
Mn-55	√
Fe-54	
Fe-56	√
Fe-57	
Fe-58	
Ni-58	√
Ni-60	√
Ni-61	
Ni-62	
Ni-64	
Cu-63	
Cu-65	
Zn-64	
Zn-66	
Zn-67	
Zn-68	
Zn-70	
Zr-90	√
Zr-91	
Zr-92	
Zr-94	
Zr-96	

Nuclides	Existence of Covariance
Nb-93	
Mo-92	
Mo-94	
Mo-95	
Mo-96	
Mo-97	
Mo-98	
Mo-100	
Rh-103	
Sn-112	
Sn-114	
Sn-115	
Sn-116	
Sn-117	
Sn-118	
Sn-119	
Sn-120	
Sn-122	
Sn-124	
Gd-152	
Gd-154	
Gd-155	
Gd-156	
Gd-157	
Gd-158	
Gd-160	

Nuclides	Existence of Covariance
Hf-174	
Hf-176	
Hf-177	
Hf-178	
Hf-179	
Hf-180	
Pb-204	√
Pb-206	√
Pb-207	√
Pb-208	√
U-234	√
U-235	√
U-236	√
U-238	√
Pu-238	√
Pu-239	√
Pu-240	√
Pu-241	√
Pu-242	√
Am-241	√
H_2O S(α,β)	

3 Evaluation of the benchmarks

3.1 C/E-1 of keff

Figure 1 shows C/E-1 (C: calculated value by MVP, E: experiment value) of keff for aforementioned 185 cases (UO_2:131, MOX:54). Uncertainty of C includes uncertainty evaluated from covariance data of cross section and statistical uncertainty of Monte Carlo calculation of MVP. Uncertainty of E includes uncertainty of experiment.

In Fig. 1, useful information feedback to cross section is not clear due to large uncertainty from C and E. Therefore, extended bias factor method was applied to reduce uncertainty of cross section.

3.2 Extended bias factor method

Bias factor method[12] evaluates prediction value E_k' as C_k (calculated value of system k) corrected by multiplying with f_i, where f_i is bias factor $f_i = E_i/C_i$ of experiment i. As a premise, each sensitivity coefficient of k and i is required to be similar.

In this evaluation process, uncertainty of cross section can be reduced as following equations[12].

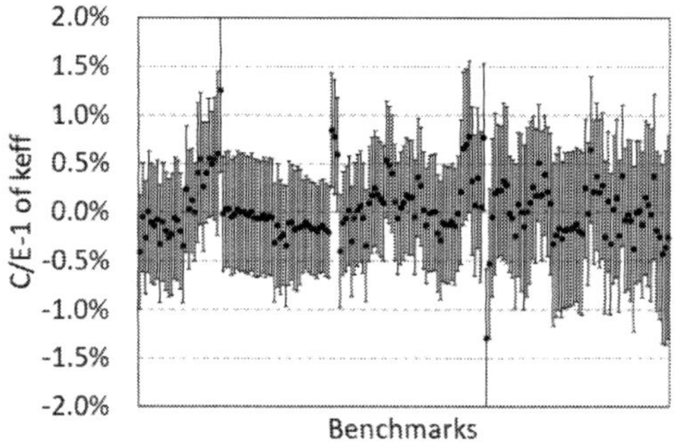

Fig. 1. C/E-1 of keff.

$$\begin{aligned} E_k' &= C_k \cdot f_i = C_k \cdot (E_i/C_i) \\ &= T_k \cdot (1 + \boldsymbol{S}_k\,\Delta\boldsymbol{\sigma} + \Delta M_k) \cdot \frac{T_i \cdot (1 + \Delta E_i)}{T_i \cdot (1 + \boldsymbol{S}_i\,\Delta\boldsymbol{\sigma} + \Delta M_i)} \\ &\cong T_k \cdot [1 + (\boldsymbol{S}_k - \boldsymbol{S}_i)\,\Delta\boldsymbol{\sigma} + (\Delta M_k - \Delta M_i) + \Delta E_i] \end{aligned} \quad (1)$$

where

T : True value

$\boldsymbol{S}$: Sensitivity coefficient

$\Delta\boldsymbol{\sigma}$: Error of cross section

ΔM : Error of calculation method †
ΔE : Error of experiment

Variance of E'_k/T_k can be written as following[12].

$$V(E'_k/T_k) = (\boldsymbol{S}_k - \boldsymbol{S}_i)\, \boldsymbol{V}_\sigma\, (\boldsymbol{S}_k - \boldsymbol{S}_i)^t + V(\Delta M_k - \Delta M_i) + V(\Delta E_i) \quad (2)$$

where $\boldsymbol{V}_\sigma$ is covariance data of cross section.

If each sensitivity coefficient of k and i is similar, uncertainty of cross section $(\boldsymbol{S}_k - \boldsymbol{S}_i)\, \boldsymbol{V}_\sigma\, (\boldsymbol{S}_k - \boldsymbol{S}_i)^t$ becomes small.

In the extended bias factor method, uncertainty of cross section can be reduced further by using various experiments[12].

$$E'_k = C_k \cdot f_k, \qquad f_k = \left(\prod_{i=1}^{N} E_i^{w_i}\right) / \left(\prod_{i=1}^{N} C_i^{w_i}\right) \quad (3)$$

where w_i are obtained by solving the following simultaneous linear equation determined to minimize variance of E'_k/T_k.

$$\begin{bmatrix} \boldsymbol{S}_1 \boldsymbol{V}_\sigma \boldsymbol{S}_1^t + \mathrm{cov}(\Delta M_1, \Delta M_1) + \mathrm{cov}(\Delta E_1, \Delta E_1) & \cdots & \boldsymbol{S}_1 \boldsymbol{V}_\sigma \boldsymbol{S}_N^t + \mathrm{cov}(\Delta M_1, \Delta M_N) + \mathrm{cov}(\Delta E_1, \Delta E_N) \\ \vdots & \ddots & \vdots \\ & \cdots & \boldsymbol{S}_N \boldsymbol{V}_\sigma \boldsymbol{S}_N^t + \mathrm{cov}(\Delta M_N, \Delta M_N) + \mathrm{cov}(\Delta E_N, \Delta E_N) \end{bmatrix} \begin{bmatrix} w_1 \\ \vdots \\ w_N \end{bmatrix} = \begin{bmatrix} \boldsymbol{S}_k \boldsymbol{V}_\sigma \boldsymbol{S}_1^t + \mathrm{cov}(\Delta M_k, \Delta M_1) \\ \vdots \\ \boldsymbol{S}_k \boldsymbol{V}_\sigma \boldsymbol{S}_N^t + \mathrm{cov}(\Delta M_k, \Delta M_N) \end{bmatrix} \quad (4)$$

where "cov" means covariance of each term. Variance of E'_k/T_k can be written as following for extended bias factor method[12].

$$V(E'_k/T_k) = \left(\boldsymbol{S}_k - \sum_{i=1}^{N} w_i \cdot \boldsymbol{S}_i\right) \boldsymbol{V}_\sigma \left(\boldsymbol{S}_k - \sum_{i=1}^{N} w_i \cdot \boldsymbol{S}_i\right)^t + V\left(\Delta M_k - \sum_{i=1}^{N} w_i \cdot \Delta M_i\right) + V\left(\sum_{i=1}^{N} w_i \cdot \Delta E_i\right) \quad (5)$$

3.3 E'/E-1 of keff

Figure 2 shows E'/E-1 of keff for aforementioned 185 cases (UO_2:131, MOX:54), where E' means prediction value by extended bias factor method with MVP's calculation results of 184 cases. For the prediction, a target case is excluded from aforementioned 185 cases. And E means experiment value. Uncertainty of E'/E-1 was calculated by the following equation (In the equation, $V(E'_k/T_k)$ was used in equation (5), and covariance between $\Delta E'_k$ and ΔE_k was not considered for convinience).

$$V(E'_k/E_k - 1) = \left[V(E'_k/T_k) \cdot {T_k}^2\right]/{E_k}^2 + {E'_k}^2 \cdot V(\Delta E_k)/{E_k}^4 \cong \left[V(E'_k/T_k) \cdot {E'_k}^2\right]/{E_k}^2 + {E'_k}^2 \cdot V(\Delta E_k)/{E_k}^4 \quad (6)$$

In Fig. 2, useful information feedback to cross section has been clearer than Fig.1 due to reduction of uncertainty of cross section. Figure 3 shows histogram of "normalized E'/E-1 by uncertainty of E'/E-1" to judge outlier case. The cases whose distance from median > 3σ (σ: standard deviation evaluated by quartile deviation) were judged as outlier.

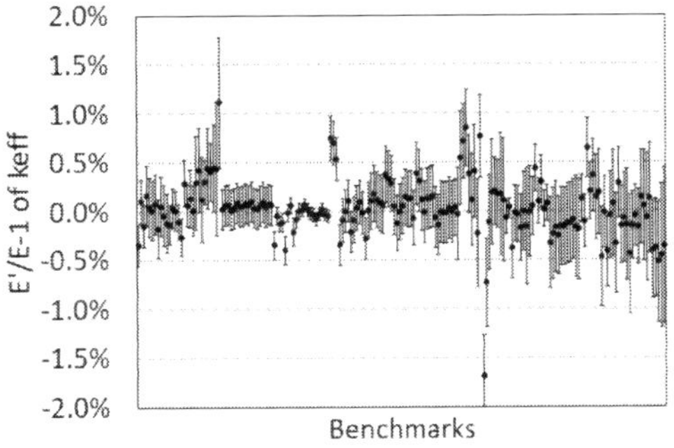

Fig. 2. E'/E-1 of keff.

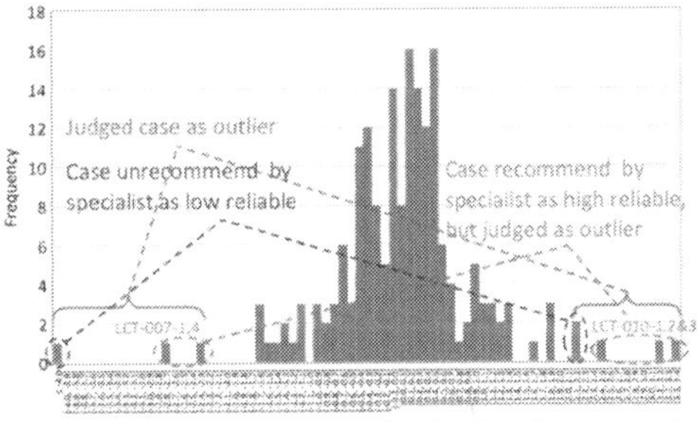

Fig. 3. Histogram of "normalized E'/E-1 by uncertainty of E'/E-1" to judge outlier case.

8 cases (LCT-007-1,5, LCT-010-1,2,3, LCT-022-4, LCT-025-1 and MCT-005-3) were judged as outlier, in which 3 cases (LCT-022-4, LCT-025-1 and MCT-005-3) were unrecommend cases judged by the specialists to be unreliable in the JAEA report[3]. Remaining 5 cases (LCT-007-1, 5, LCT-010-1, 2 and 3) were judged outlier even though the cases were recommended cases judged by the specialists to be highly reliable in the JAEA report[3]. Therefore, these 5 cases are investigated for the possibilities of containing useful information feedback to cross section in the following.

† In this study, the errors were assumed zero except for statistical uncertainty of Monte Carlo calculation, because MVP and MCNP calculation models are sufficiently precise.

4 Discussion

4.1 Outlier cause of LCT-007

Benchmark LCT-007 (LEU-COMP-THERM-007) in ICSBEP is the experiment of water-reflected 4.738-wt.%-enriched uranium dioxide fuel-rod arrays[1]. Table 3 and Fig. 4 show array geometry of each case in the experiment.

Figure 5(b) shows E'/E-1 of keff with X-axis as pin pitch. Figure 5(b) implies dependency on pin pitch (= dependency on H/U). It was considered that the dependency implies that there is room for improvement in H-1 and S(α,β) of H_2O. Incidentally, Fig.5(a), that shows C/E - 1 before reduction of cross section's uncertainty, does not show clear dependency on pin pitch.

Table 3. Array geometry of LCT-007.

Case Number	Experiment Number	Pitch Type	Pitch (cm)
1	1662	Square	1.26 (a)
2	1659	Square	1.60 (b)
3	1655	Square	2.10 (c)
4	1825	Square	2.52 (d)
5	1682	Triangular	1.35 (a)
6	1690	Triangular	1.72 (b)
7	1694	Triangular	2.26 (c)
8	1689	Triangular	1.35 (a)
9	1693	Triangular	1.72 (b)
10	1697	Triangular	2.26 (c)

(a) Moderator-to-fuel ratio (Vm/Vu) ≈ 1.82
(b) Moderator-to-fuel ratio (Vm/Vu) ≈ 3.81
(c) Moderator-to-fuel ratio (Vm/Vu) ≈ 7.58
(d) Moderator-to-fuel ratio (Vm/Vu) ≈ 11.54

(Referred from reference [1])

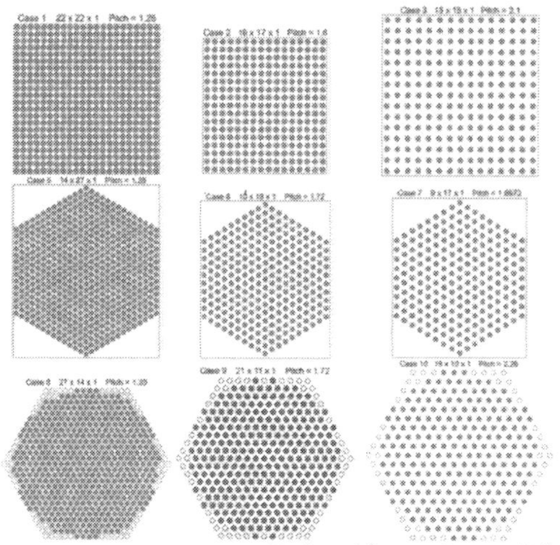

(Referred from reference [1])

Fig. 4. Arrays of fuel rods of LCT-007.

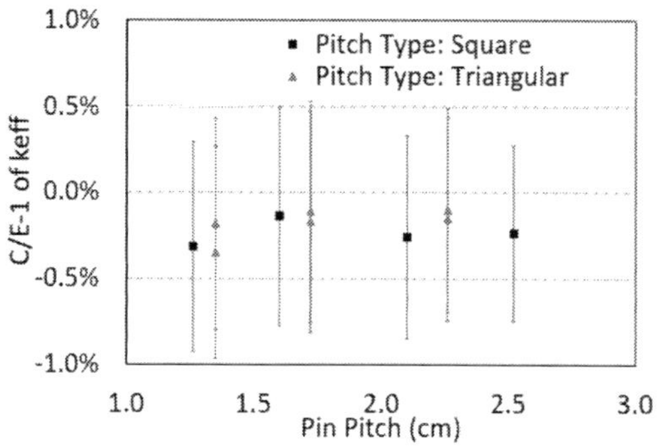

(a) C/E-1: Before reduction of cross section's uncertainty

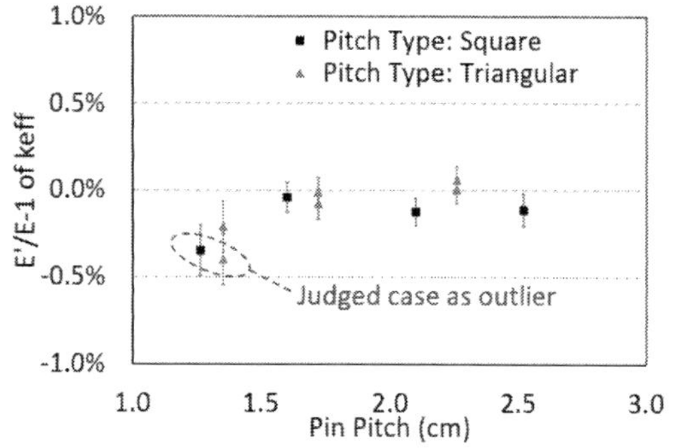

(b) E'/E-1: After reduction of cross section's uncertainty

Fig. 5. C/E-1 and E'/E-1 of keff of LCT-007.

As a trial, E' of cases 1 and 4 in Fig. 6 were estimated when cross section of H-1 and S(α,β) of H_2O were replaced to ENDF/B-VII.1 from JENDL-4.0u (The reactivities were the value evaluated in the JAEA report[3]. See gray color plots in Fig.6). Figure 6 shows that dependency on pin pitch is expected to be weaken after replacing cross section of H-1 and S(α,β) of H_2O to ENDF/B-VII.1 (After the replacing cross sections, absolute value of E'/E-1 of case 1 whose pin pitch is 1.26cm is expected to decrease rather than case 4 whose pin pitch is 2.52cm). This result implies that there is room for improvement in H-1 and S(α,β) of H_2O in JENDL-4.0u.

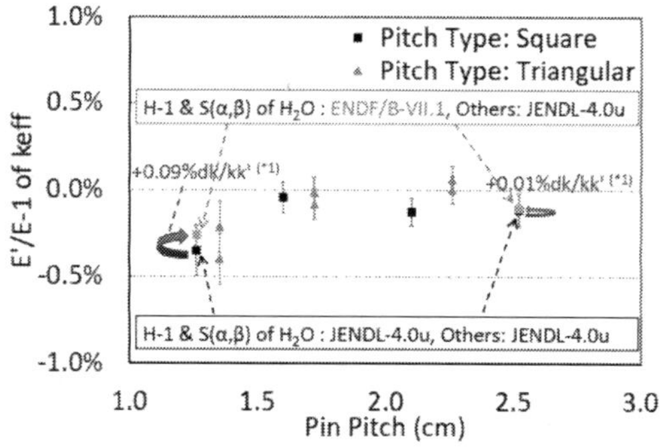

*1: The data are referred from JAEA report [3].

Fig. 6. E'/E-1 of keff of LCT-007 (added reactivities that cross section of H-1 and S(α,β) of H_2O were replaced to ENDF/B-VII.1 from JENDL-4.0u).

4.2 Outlier cause of LCT-010 cases 1 - 4

Benchmark LCT-010 (LEU-COMP-THERM-010) in ICSBEP is the experiment of water-moderated U(4.31%U-235)O_2 fuel rods reflected by lead, uranium, or steel walls[1]. The reflector of cases 1 - 4 in the experiment were lead. Table 4 and Fig.7 show array geometry of each case in the experiment.

Table 4. Array geometry of LCT-010 cases 1 - 4.

Case No.	Pitch (cm)	Cluster Dimensions (number of rods)(a) (X x Y)	Separation Between Clusters (cm)(b)	Distance Between Reflecting Walls and Fuel Clusters (cm)(c)	Reflecting Wall Material
1	2.540	13 x 8	19.495 ± .01(d)	0.0	lead
2	2.540	13 x 8	19.655 ± .02(d)	0.660 ± 0.102	lead
3	2.540	13 x 8	17.915 ± .02(d)	1.321 ± 0.076	lead
4	2.540	13 x 8	9.175 ± .02(d)	5.405 ± 0.102	lead

(Referred from reference [1])

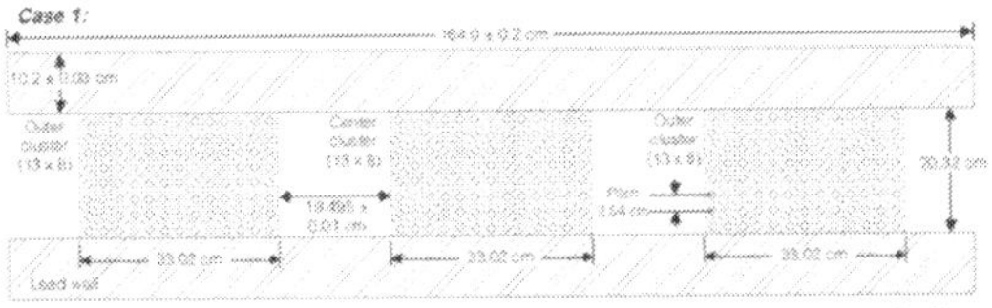

(Referred from reference [1])

Fig. 7. Arrays of fuel rods of LCT-010 case 1.

LCT-010 cases 1 - 3 are judged as outliers, and it is considered that the influence of the reflecting wall of lead is large in those cases. Therefore, Fig.8 shows elastic scattering cross sections (n,el) in Pb-206, 207 and 208. Figure 8 shows that there are about 7% relative difference in Pb-207 and about 4% in Pb-206 between ENDF/B-VIII.0 and JENDL-4.0u. On the other hand, covariance data of JENDL-4.0u for lead's nuclides in low energy are about 4E-4 that means about 2% = √(4E-4) as standard deviation. Aforementioned relative differences 7% and 4% are ~3 times of 2% in the covariance data.

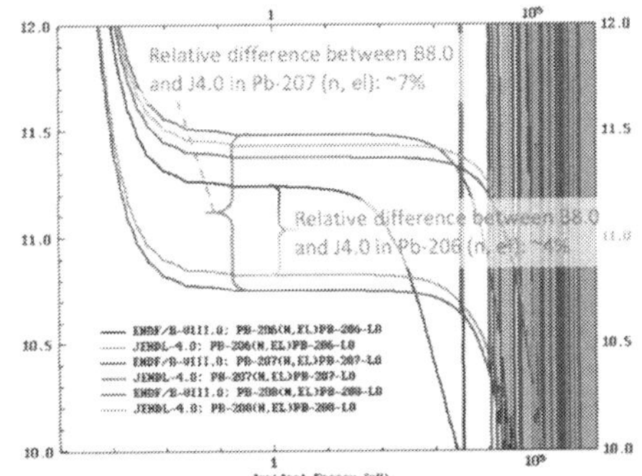

(The figure was plotted by https://www.nndc.bnl.gov/exfor/endf00.jsp)

Fig. 8. Elastic scattering cross sections of lead's nuclides.

As a trial, E'/E-1 were evaluated again with lead nuclides' covariance data (for elastic scattering cross sections) multiplied by 9 over the entire energy range, where 9 means 3 times as standard deviation aforementioned. That is, the original standard deviation 2% in covariance data was changed to 6% to cover relative differences between ENDF/B-VIII.0 and JENDL-4.0u in Fig.8. Figure 9 shows the E'/E-1. Dependency of E'/E-1 on distance from lead reflector was weaken after lead's covariance data were multiplied by 9. This result implies that there is room for improvement in lead's cross section or covariance in JENDL-4.0u.

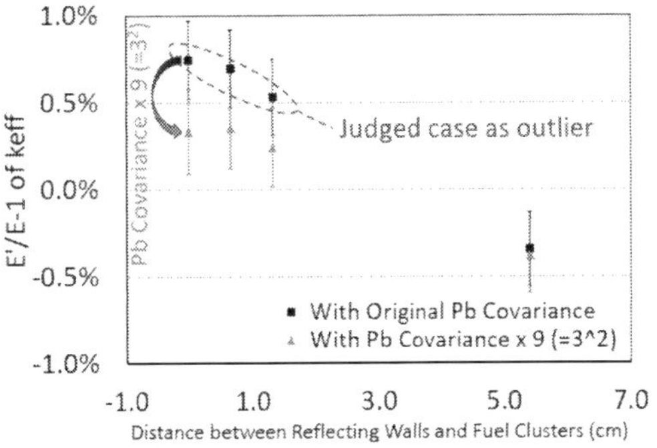

Fig. 9. E'/E-1 of keff of LCT-010 cases 1 - 4.

5 Conclusion

It was tried to get useful information feedback on cross section and covariance by application of extended bias factor method with experimental benchmarks' C/E, sensitivity coefficient and covariance data. As a result, it was succeeded to distinguish "the case unrecommended by specialists" as outlier case. And it was succeeded to get the hints that there is room for improvement in "the cross section of H-1, S(α,β) of H_2O, elastic scattering cross section or covariance of lead's nuclides" by investigating outlier cases judged as recommended case by specialists.

References

1. Nuclear Energy Agency, International Handbook of Evaluated Criticality Safety Benchmark Experiments (ICSBEP), NEA/NSC/DOC(95)03 (2019)
2. Nuclear Energy Agency, International Handbook of Evaluated Reactor Physics Benchmark Experiments (IRPhEP), NEA/NSC/DOC(2006)1 (2019)
3. Reactor Integral Test Working Group, JENDL Committee, Compilation of the Data Book on Light Water Reactor Benchmark to Develop the Next Version of JENDL - Utilization of Criticality Data in ICSBEP and IRPhEP Open Databases - , JAEA-Data/Code 2017-006 (2017) (in Japanese)
4. Y. Nagaya, K. Okumura, T. Sakurai, T. Mori, MVP/GMVP Version 3 General Purpose Monte Carlo Codes for Neutron and Photon Transport Calculations Based on Continuous Energy and Multigroup Methods, JAEA-Data/Code 2016-018 (2017)
5. C. J. Werner, J. S. Bull, C. J. Solomon, et al., MCNP version 6.2 release notes, LA-UR-18-20808, Los Alamos, National Laboratory (2018)

6. B. C. Kiedrowski, F. B. Brown, J. L. Conlin, et al., Whisper: Sensitivity/Uncertainty- Based Computational Methods and Software for Determining Baseline Upper Subcritical Limits, Nucl. Sci. Eng., **181** (2015)
7. K. Tada, Y. Nagaya, S. Kunieda, K. Suyama, T. Fukahori, Development and verification of a new nuclear data processing system FRENDY, J. Nucl. Sci. Technol., **54** (2017)
8. A. Yamamoto, K. Tada, G. Chiba, T. Endo, Multi-group neutron cross section generation capability for FRENDY nuclear data processing code, J. Nucl. Sci. Technol., **58** (2021)
9. R. E. MacFarlane, et. al., The NJOY Nuclear Data Processing System, Version 2016, LA-UR-17-20093, Los Alamos, National Laboratory (2019)
10. J. D. Drischler, The COVERX Service Module of the FORSS System, ORNL/TM-7181 (ENDF-291), Oak Ridge National Laboratory (1980)
11. W. A. Wieselquist, R. A. Lefebvre, M. A. Jessee (Eds.), SCALE Code System, ORNL/TM-2005/39, Version 6.2.4, Oak Ridge National Laboratory (2020)
12. T. Kugo, et al., Theoretical Study on New Bias Factor Methods to Effectively Use Critical Experiments for Improvement of Prediction Accuracy of Neutronic Characteristics, J. Nucl. Sci. Technol. **44** (2007)

https://doi.org/10.1051/epjconf/202328100003

Effect of correlation between cross sections and angular distributions in nuclear data of ^{63}Cu on estimation of uncertainty of neutron penetration

Naoki Yamano [1*], *Tsunenori* Inakura [1], *Chikako* Ishizuka [1], and *Satoshi* Chiba [1]

[1] Tokyo Institute of Technology, Ookayama, Meguro, Tokyo 152-8550, Japan

Abstract. Uncertainty in neutron reaction rates after penetrating a 608mm-thick copper benchmark experiment performed by the FNS facility of JAEA was estimated based on two different kinds of the Total Monte Carlo methods under random sampling methodology. 500 random nuclear data files were generated for ^{63}Cu by the T6 code system with perturbing underlying model parameters. In the first method, these files were used directly to yield processed library preserving all the correlations among different physical quantities. In the second method, the random files populated by T6 were used but the angular distribution data were kept fixed to the non-perturbed nominal ones. It was found that the two methods gave the same neutron reaction rate after the 608 mm penetration of a copper, however uncertainty of the second method was larger than that of the first method. It shows that the correlation between total cross section and angular distribution of elastic scattering at 0 degree, which stems from Wick's inequality, affects uncertainty of the calculated neutron reaction rate. The result is consistent with the case of ^{28}Si on deep penetration of thick concrete problem previously reported by the authors. It could be concluded that the uncertainty obtained by using the covariance files given in the ENDF-6 format may not give correct results for the uncertainty of neutron penetration calculation.

1 Introduction

Accurate estimation of uncertainties in the neutron transport calculation caused by nuclear data through error propagation is quite important for Verification and Validation (V&V) of design and safety analysis of nuclear systems. Usually, such an error propagation is carried out through the covariance files given in nuclear data in the ENDF-6 format [1]. However, it is also recognized that the ENDF-6 format cannot represent all the correlation of nuclear data that might be present in different kind of physical quantities. For example, the uncertainty of the angular distribution of elastic scattering is able to represent in the ENDF-6 format (MF34, MT2), however it is only processed by that of the average cosine of the angular distribution in the nuclear data processing code, which may be reasonable for application in design of nuclear reactors since the energy of neutrons involved is usually small, so the elastic scattering is quite close to be isotropic in the center-of-mass framework. However, uncertainty of the angular distribution and correlation of it to those of other quantities such as total cross section might be significantly important for systems where effects of the angular anisotropy is more important. Correlation of uncertainties of such different quantities, in this case that of angular distribution of elastic scattering and that of total cross section, is not represented properly in the ENDF-6 format, hence this correlation is normally ignored. However, we know that there is a positive correlation between these two quantities described by Wick's inequality [2], $\sigma_{el}(0) \geq \left(\frac{k}{4\pi}\sigma_t\right)^2$, where $\sigma_{el}(0)$, σ_t and k denote elastic scattering cross section at 0 degree, total cross section and wave number of incident neutron, respectively. The optical theorem of the quantum scattering theory is the underlying physics of this correlation. We have shown that the correlation of different quantities of ^{28}Si becomes significant on the deep penetration of fission neutrons through a thick concrete [3].

It is the purpose of this study to evaluate the effects of the correlation between forward elastic scattering and total cross section on the penetration of D-T neutrons through a thick copper, via Total Monte Carlo (TMC) method [4].

2 Method

We have done the following two calculations to prepare random nuclear data files.

1. In the first method, we have calculated 500 random nuclear data files for ^{63}Cu, which is one of the major constituents in natural copper, by using T6 code system [5]. In this calculation, underlying model parameters for the resolved resonance and continuous region were

* Corresponding author: yamano@zc.iir.titech.ac.jp

perturbed with probability distributions determined from a Bayesian Monte Carlo (BMC) method [6], and 500 ENDF-formatted random files were generated. These random files were used for the subsequent neutron transport calculations. This process will be explained later in more details. By this process, we could obtain the variance of the neutron reaction rates at each point of the thick copper, especially after penetrating the whole copper, and we could assign uncertainty of the neutron transport calculation from the variance. In this calculation, all the correlations of quantities present in the nuclear cross section were taken into consideration especially those between the angular distribution of elastic scattering and the total cross section.

2. In the second one, which is similar to that of the 1st method but all the angular distribution data (MF4, MT2) were kept fixed to the non-perturbed nominal ones, hence it ignores the uncertainty of the angular distribution of elastic scattering cross sections only. The 500 random files keeping all the angular distribution data to be the same were used for the subsequent neutron transport calculations.

The variances of the neutron reaction rates after complete penetration obtained by these two methods, that are the indicators of the uncertainty of neutron transport calculation, were compared, and we have elucidated the importance of the correlation of different quantities in nuclear data libraries.

The covariance of ^{63}Cu neutron cross sections, which is only evaluated for resonance parameters in ENDF/B-VIII.0 [7], was estimated using T6 as shown in the upper part of **Fig. 1**. The nominal parameters of the calculation were adjusted to reproduce neutron cross sections stored in ENDF/B-VIII.0. To estimate covariance matrices, firstly a distribution of parameters in the continuous region was calculated with the BMC method. Secondly, Monte Carlo calculation was performed based on the errors of resolved resonance parameters and the parameter distribution in the continuous region obtained from BMC as above. If the cross section errors in the resonance region were experimentally known, the uncertainties of the resonance parameters were carefully adjusted to reproduce those errors and all the resonance energies and widths were perturbed independently within the assigned errors. In the same way, all the parameters in the continuous region, which consist of more than 200 parameters including those of optical model potential, level densities, were perturbed as well as pre-equilibrium model parameters. All these perturbation calculations were conducted based on the parameter distributions which were obtained starting from a non-informative prior distribution using the BMC method, to produce 500 random files. Finally, a covariance file was generated by statistical processing of the results for these random files.

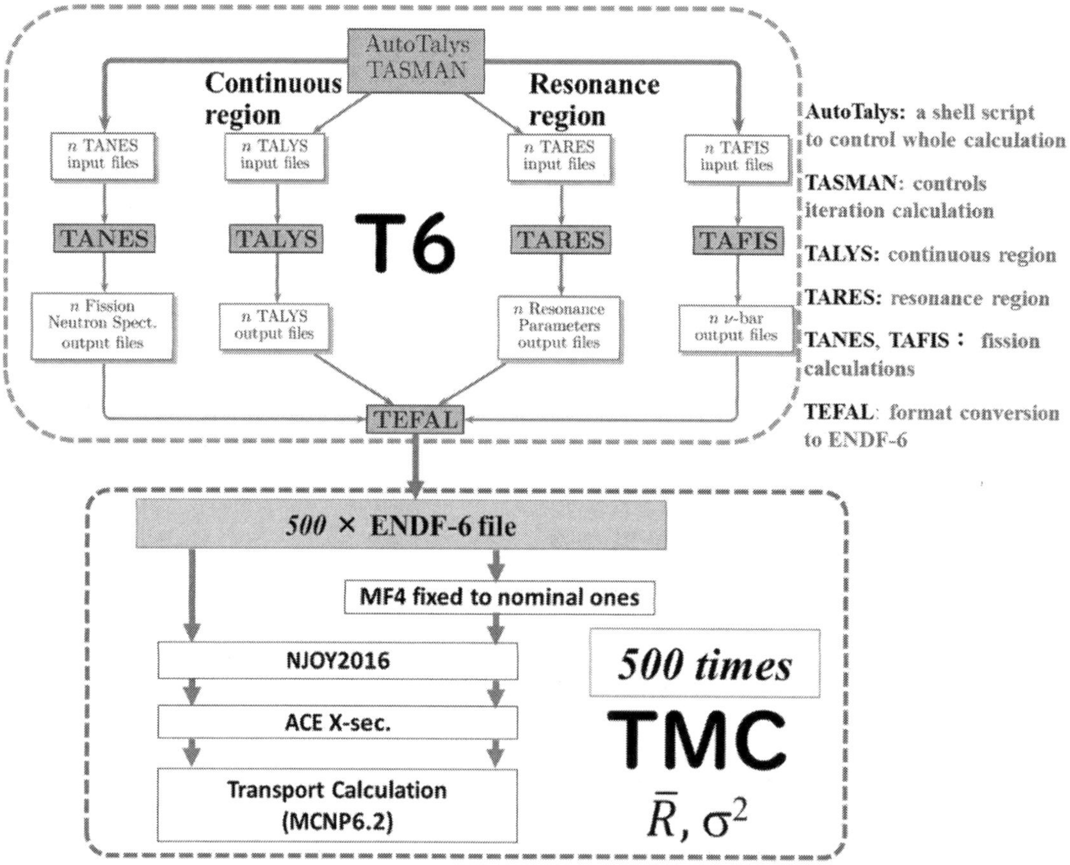

Fig. 1. Calculation scheme of cross section and associated covariance (T6, upper part), and the Total Monte Carlo (TMC) method (lower part). TANES and TAFIS in T6 were not used in the present work.

Figures 2 and **3** show the comparison of [63]Cu total cross sections and elastic cross sections between 500 random data calculated by T6 and ENDF/B-VIII.0, respectively. The cross sections in the resonance and the continuous regions calculated by T6 are consistent with those given in ENDF/B-VIII.0. The information on the correlation among the different reactions is used in the random files produced by T6.

To estimate the uncertainty of the neutron reaction rates, the TMC scheme was applied as shown in the lower part of Fig. 1. As it shows, a random file in the ENDF-6 format produced by T6 was processed using NJOY-2016 [8], and the ACE file was generated. When using NJOY, the total cross section is reconstructed from all the partial cross sections.

Then, a neutron transport calculation using MCNP6.2 [9] was performed to compute the 608 mm-thick copper benchmark experiment conducted by FNS of JAEA for which a D-T neutron source is located at 200 mm distance from the front surface of the copper slab [10]. The energy spectra of the D-T neutrons were also estimated by JAEA [10]. The two steps described above were repeated 500 times for each random file generated by T6, and the standard deviations of the calculated reaction rates were obtained by statistical processing using each of these 500 results. The computational model of the neutron transport calculation is shown in **Fig. 4** and the other conditions relating to the transport calculation are the same during repeated TMC runs, therefore, these uncertainties can be ignored.

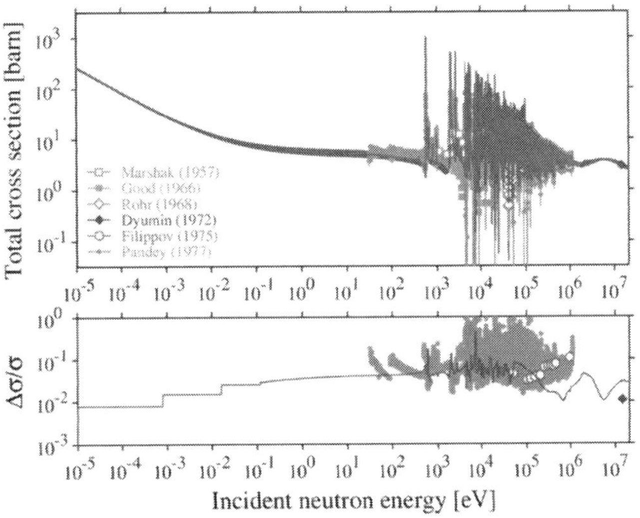

Fig. 2. [63]Cu neutron total cross sections estimated by T6 system (gray lines) compared with ENDF/B-VIII.0 (red line) and experimental data and the errors [12-17].

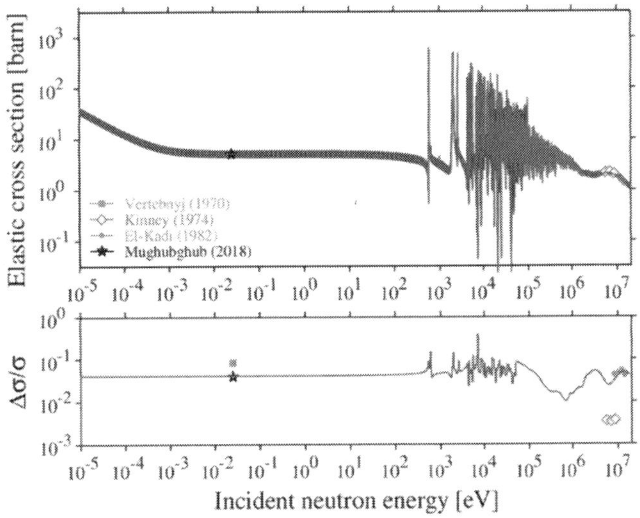

Fig. 3. [63]Cu neutron elastic cross sections estimated by T6 system (gray lines) compared with ENDF/B-VIII.0 (red line) and experimental data [18-21].

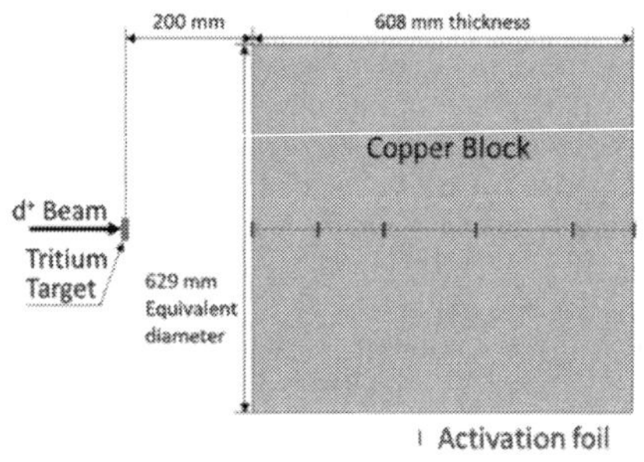

Fig. 4. Computational geometry model of FNS copper benchmark experiment.

3 Results and discussion

Neutron reaction rates for ^{58}Ni(n,2n)^{57}Ni reaction together with experimental data in the copper slab and their standard deviations are shown in in the lower panel of **Fig. 5**. The average reaction rates obtained with random files generated by T6 and those ignoring the perturbation of the angular distribution were equivalent to each other, and the calculated values were in good agreement with experimental data except for the 608 mm penetration. It was reported that scattered neutrons from concrete wall around the experimental room affected the experimental data [10]. However, the fractional standard deviation (fsd) which means the relative standard deviation to the reaction rate, as shown in the upper panels, was slightly different. The standard deviation of ^{58}Ni(n,2n)^{57}Ni reaction rate at the 608 mm copper transmission obtained by random files generated with ignoring the perturbation of the angular distribution was 18.4%, which must be compared to 14.3% obtained by those generated by T6.

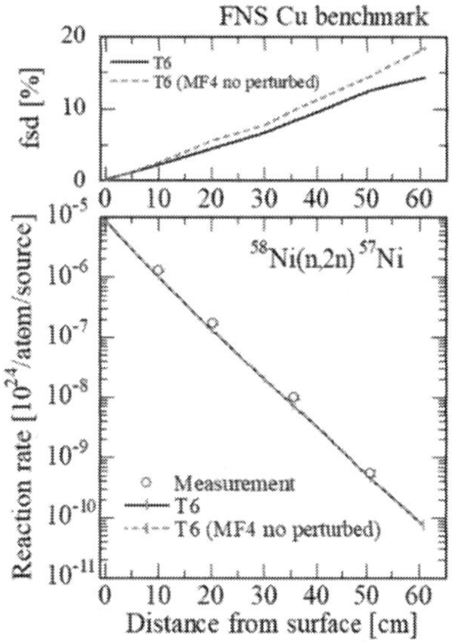

Fig. 5. Comparison of ^{58}Ni(n,2n)^{57}Ni reaction rate distribution and the standard deviation between T6 and the result ignoring perturbation of angular distribution.

The fractional standard deviation of ^{93}Nb(n,2n)^{92m}Nb reaction rate at the 608 mm copper transmission obtained by random files generated by those ignoring the perturbation of the angular distribution was 18.6%, whereas 14.5% obtained by those generated by T6 as shown in **Fig. 6**.

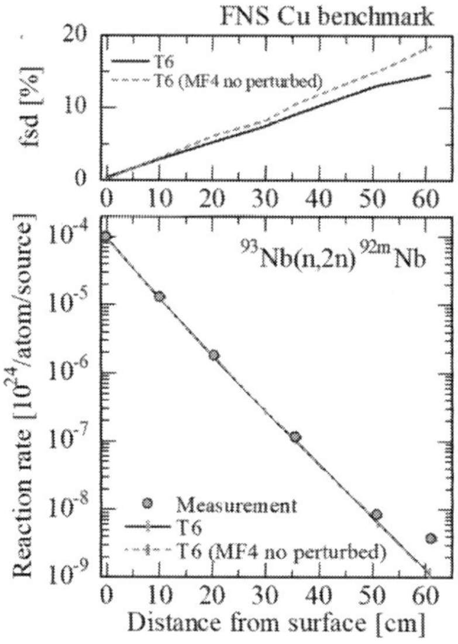

Fig. 6. Comparison of ^{93}Nb(n,2n)^{92m}Nb reaction rate distribution and the standard deviation between T6 and the result ignoring perturbation of angular distribution.

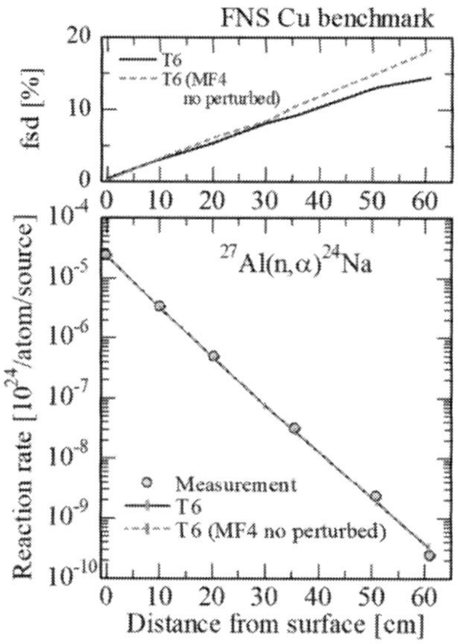

Fig. 7. Comparison of ^{27}Al(n,α)^{24}Na reaction rate distribution and the standard deviation between T6 and the result ignoring perturbation of angular distribution.

The fractional standard deviation of $^{27}Al(n,\alpha)^{24}Na$ reaction rate at the 608 mm copper transmission obtained by random files generated by those ignoring the perturbation of the angular distribution was 18.3%, whereas 14.5% obtained by those generated by T6 as shown in **Fig. 7**.

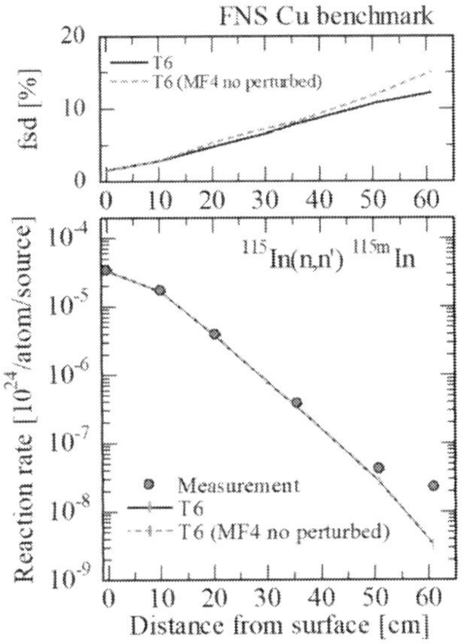

Fig. 8. Comparison of $^{115}In(n,n')^{115m}In$ reaction rate distribution and the standard deviation between T6 and the result ignoring perturbation of angular distribution.

The fractional standard deviation of $^{115}In(n,n')^{115m}In$ reaction rate at the 608 mm copper transmission obtained by random files generated by those ignoring the perturbation of the angular distribution was 15.1%, whereas 12.2% obtained by those generated by T6 as shown in **Fig. 8**.

The standard deviation of the latter, which provides for uncertainties of the angular distributions and energy distributions of secondary neutrons as well as cross sections, is smaller than the former which includes only uncertainties of the cross sections primarily. The other threshold reactions showed similar trends.

Neutron reaction rates for (n,γ) reactions in the copper slab and their standard deviations are also shown in **Figs. 9** and **10**. The fractional standard deviations for (n,γ) reactions showed different trends compared with the threshold reactions. The standard deviations for the (n,γ) reactions were almost the same between the two methods.

The random files shown as "MF4 no perturbed" were created that the angular distribution data (MF4, MT2) were kept fixed to the non-perturbed nominal ones where the perturbation of the elastic scattering angular distribution was ignored, so its correlation to the total cross section was also ignored. On the other hand, the random files generated by T6 includes perturbations of the angular distribution of secondary neutrons and correlation to other quantities.

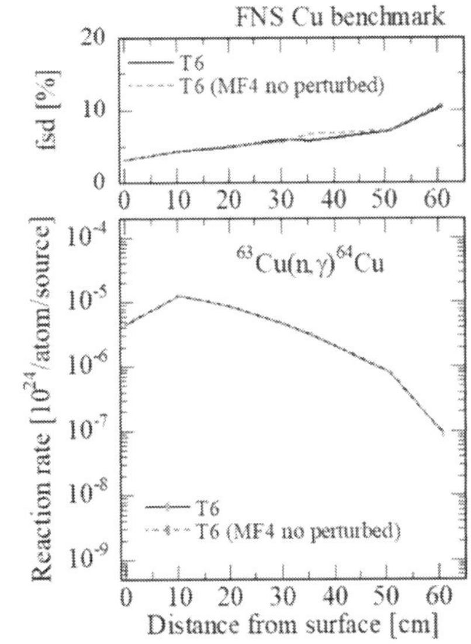

Fig. 9. Comparison of $^{63}Cu(n,\gamma)^{64}Cu$ reaction rate distribution and the standard deviation between T6 and the result ignoring perturbation of angular distribution.

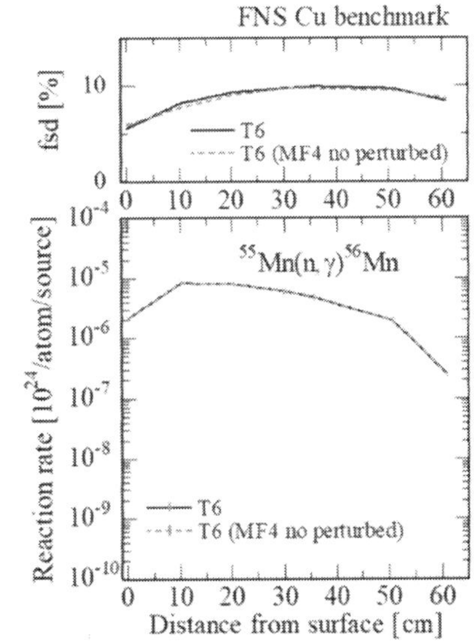

Fig. 10. Comparison of $^{55}Mn(n,\gamma)^{56}Mn$ reaction rate distribution and the standard deviation between T6 and the result ignoring perturbation of angular distribution.

The statistical error of estimated uncertainty (fsd) for the TMC method (N=500) was estimated to be about 3% based on the χ-distribution [11] for the $^{58}Ni(n,2n)^{57}Ni$ reaction rate at the 608 mm transmission.

Figures 11(a) and **(b)** show the correlations between the total cross section and the elastic scattering cross section at 0 degree in the 1,000 random files generated by T6 at the incident neutron energy of 14 MeV and 2.2 MeV, respectively.

In these figures, red points indicate that the angular distribution data (MF4, MT2) were kept fixed to the non-perturbed nominal ones. It is clear that there is a positive correlation between the total cross section and the forward elastic scattering cross section, which is consistent with Wick's inequality derived from the optical theorem.

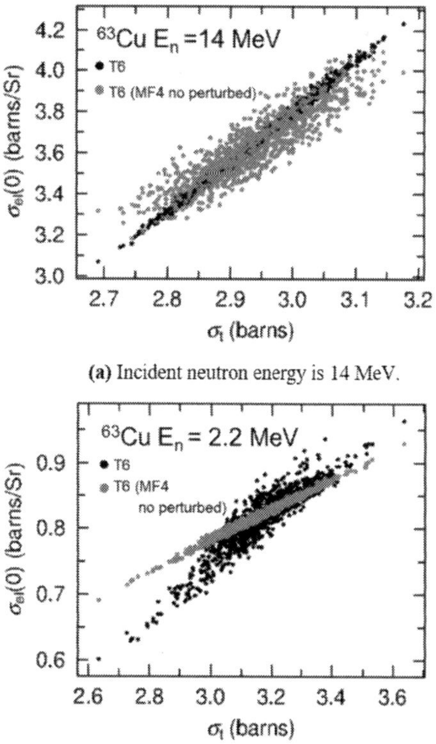

(a) Incident neutron energy is 14 MeV.

(b) Incident neutron energy is 2.2 MeV.

Fig. 11. Correlation between ^{63}Cu total cross section and elastic scattering cross section at 0 degree of 1,000 random files generated by T6.

This correlation gives rise to the cancellation of the variance of neutron threshold reaction rates of the penetrated neutrons in the following manner: in general, neutron transmission decreases as the total cross section increases. However, the forward scattering cross section, which is positively correlated with the total cross section, works to increase neutron transmission. Therefore, the forward elastic scattering cross section changes so as to cancel the change in neutron transmission caused by the change in the total cross section caused by the perturbation of the parameter. That is, the standard deviation of neutron threshold reaction rates were reduced when the correlation of the differential elastic scattering cross section to the total cross section was considered properly. This result is consistent with the ^{28}Si uncertainty on deep penetration of thick concrete problem previously reported by the authors [3]. Therefore, it was found that 1) uncertainty of the angular distribution data must be properly considered, 2) its correlation to other quantities like total cross section must also be considered properly, and 3) if the items 1) and 2) were ignored, the uncertainty of the neutron threshold reaction rates are overestimated. These conclusions give an important warning to the interpretation of the neutron transport calculations which were thoroughly based on the ENDF-formatted files, where items 1) and 2) are not fulfilled in general. It has to be noticed at the same time that these conclusions are applicable to cases where the anisotropy of neutron scattering is highly important for the results like the neutron penetration problems considered in this work.

4 Conclusion

The effect of uncertainty in nuclear data on the neutron penetration of copper slab was estimated with the TMC method based on nuclear data generated by T6. The average threshold reaction rates obtained with random files generated by T6 and those ignoring the correlation between the angular distribution of elastic scattering and the total cross sections were equivalent. However, the standard deviation of ^{58}Ni(n,2n)^{57}Ni reaction rate obtained with those ignoring the correlation between the angular distribution of elastic scattering and the total cross sections at the 608 mm copper transmission was 18.4%, which is larger than 14.3% obtained with T6 data. The cause of difference was attributed to that the correction of the angular distribution of elastic scattering and total cross section could not be considered in calculations with the ENDF-6 format. It has been concluded that the correlation of angular distribution of secondary neutrons to other quantities, especially to total cross section, needs to be dealt with correctly in the neutron deep penetration where angular anisotropy is important.

References

1. A. Trkov, M. Herman, D.A. Brown, (Eds). ENDF-6 Formats Manual, Data Formats and Procedures for the Evaluated Nuclear Data File ENDF/B-VI, ENDF/B-VII and ENDF/B-VIII. Brookhaven National Laboratory (US); 2018. (BNL report BNL-203218-2018-INRE).
2. G.C. Wick. A Theorem on Cross Sections. Phys. Rev. **75**(9), (1949) 1459.
3. N. Yamano, T. Inakura, C. Ishizuka, S. Chiba. Crucial importance of correlation between cross sections and angular distributions in nuclear data of ^{28}Si on estimation of uncertainty of neutron dose penetrating a thick concrete. J. Nucl. Sci. Technol. **59**(5), (2022) 641-646.
4. A.J. Koning, D. Rochman. Towards sustainable nuclear energy: Putting nuclear physics to work. Annals of Nuclear Energy. **35**(11), (2008) 2024-2030. DOI: 10.1016/j.anucene.2008.06.004.
5. A.J. Koning, D. Rochman. J.-Ch. Sublet, et al. TENDL: Complete Nuclear Data Library for

Innovative Nuclear Science and Technology. Nucl. Data Sheets. **155**, (2019) 1-55.

6. A.J. Koning Bayesian Monte Carlo method for nuclear data evaluation. Eur. Phys. J. A. **51**, (2015) 184. DOI: 10.1140/epja/i2015-15184-x. See also P.J. Griffin, IEEE Transactions on Nuclear Science. 66(1). Sandia National Laboratory (US); 2018. (Sandia NL report SAND-2018-11821J).
7. D.A. Brown, M.B. Chadwick, R. Capote, A.C. Kahler, A.Trkov, et al. ENDF/B-VIII.0: The 8th Major Release of the Nuclear Reaction Data Library with CIELO-project Cross Sections, New Standards and Thermal Scattering Data. Nucl. Data Sheets. **148** (2018) 1-142.
8. R.E. MacFarlane, A.C. Kahler. Methods for Processing ENDF/B-VII with NJOY. Nucl. Data Sheets. **111** (2010) 2739-2890.
9. C.J. Werner (Ed.). MCNP® USER'S MANUAL Code Version 6.2. Los Alamos National Laboratory (US); 2017. (LANL report LA-UR-17-29981).
10. F. Maekawa, Y. Oyama, C. Konno, Y. Ikeda, K. Kosako, H. Maekawa. Benchmark Experiment on a Copper Slab Assembly Bombarded by D-T Neutrons. Japan Atomic Energy Agency (Japan); 1994. (JAEA report JAERI-M 94-038).
11. B. Foad, A. Yamamoto, T. Endo. Uncertainty and regression analysis of the MSLB accident in PWR based on unscented transformation and low rank approximation. Ann. Nucl. Energy. **143** (2020) 107493.
12. H. Marshak, H.W. Newson. Neutron Resonances in the keV Region: Odd-Intermediate Nuclei. Phys. Rev. **106** (1957) 110.
13. W.M. Good, D. Paya, R. Wagner, T. Tamura. Total Cross Sections for Kilovolt Neutrons of Even-Odd Nuclei in the Region of the 3s Strength-Function Resonances. Phys. Rev. **151** (1966) 912.
14. G. Rohr, K.N. Mueller. The Total Cross-Sections of Cu-63 and Cu-65 in keV Region. Report from Euratom-countries + Euratom to EANDC, **89** (1968) 1.
15. A.I. Dyumin, D.M. Kaminker, G.N. Popova, V.A. Smolin. Total Neutron Cross Sections for Titanium, Chromium, and Copper Isotopes. Izv. Rossiiskoi Akademii Nauk, Ser. Fiz. **36** (1972) 852.
16. V.V. Filippov. Measurement of Capture and Total Cross Sections for Intermediate and Fast Neutrons by Energy Correlation Methods. 3.All Union Conf. on Neutron Phys. Vol. 2 (1975) 53, Kiev, 9-13 June 1975.
17. M.S. Pandey, J.B. Garg, J.A. Harvey. Neutron total cross sections and resonance parameters of ^{63}Cu and ^{65}Cu. Phys. Rev. C. **15** (1977) 600.
18. V.P. Vertebnyj, M.F. Vlasov, N.L. Gnidak, et al. Neutron Cross-Sections of Isotopes used as Absorbers in Atomic Reactors. Nuclear Data for Reactors Conf. Vol.1 (1970) 651, Helsinki, 1970.
19. W.E. Kinney, F.G. Perey. ^{63}Cu and ^{65}Cu neutron elastic and inelastic scattering cross sections from 5.50 to 8.50 MeV. Oak Ridge National Laboratory (US); 1974. (ORNL report ORNL-4908).
20. S.M. El-Kadi, C.E. Nelson, F.O. Purser, et al. Elastic and inelastic scattering of neutrons from ^{54,56}Fe and ^{63,65}Cu: (I). Measurements from 8 to 14 MeV and a spherical optical model analysis. Nuclear Physics, Section A. **390** (1982) 509-540.
21. S.F. Mughabghab. Atlas of neutron resonances. 6th ed. (2018) Elsevier Science; New York (NY). ISBN-9780444637697.

Development of adjusted nuclear data library for fast reactor application

Kenji Yokoyama[1,*]

[1]Japan Atomic Energy Agency, Fast Reactor Cycle System Research and Development Center, Oarai-machi, Ibaraki 311-1393, Japan

Abstract. In Japan, development of adjusted nuclear data library for fast rector application based on the cross-section adjustment method has been conducted since the early 1990s. The adjusted library is called the unified cross-section set, which is an ABBN-type group constant set with 70-group energy structure. The first version was developed in 1991 and is called ADJ91. After that, ADJ98, ADJ2000, ADJ2000R, and ADJ2010 were constantly developed. For instance, ADJ2010 was developed based on JENDL-4.0, which provides covariance data needed to apply the cross-section adjustment method, by using 488 integral experimental data acquired in typical fast reactor systems. ADJ2010 has been used as the standard cross-section set for nuclear design in the fast reactor cycle technology development project (FaCT) and the succeeding fast reactor projects. In parallel, the integral experimental data were further expanded to improve the design prediction accuracy of the core loaded with MA and/or degraded Pu. Using the additional integral experimental data, development of the next version of ADJ2017 was started in 2017. In 2022, the latest unified cross-section set AJD2017R was developed based on JENDL-4.0 by using 619 integral experimental data. An overview of the latest version with a review of previous ones will be shown.

On the other hand, the latest Japanese evaluated nuclear data library JENDL-5 was released at the end of 2021. In the development of JENDL-5, some of the integral experimental data used in ADJ2017R were explicitly utilized in the nuclear data evaluation. However, this is not reflected in the covariance data. This situation needs to be considered when developing a next version of the unified cross-section set based on JENDL-5. Preliminary adjustment calculation based on JENDL-5 is performed using C/E (calculation/experiment) values simply evaluated by a sensitivity analysis. The preliminary result of the JENDL-5-based adjustment will be also discussed.

1 Introduction

In the development of the neutronics analysis method for fast reactors in Japan, application research of the cross-section adjustment method [1] has been conducted since the early 1990s. In the core design study of an innovative nuclear reactor such as fast breeder reactors, it is important to improve the prediction accuracy of nuclear parameters from the viewpoint of both safety and economics. Several adjusted nuclear data libraries, which are called unified cross-section sets in the sense that the adjusted cross-section set unifies the differential data and the integral data, have been developed based on the cross-section adjustment theory and applied for fast reactor design work [2]. On the other hand, we in JAEA have been collecting and analyzing integral experimental data related to fast reactors for the application of the cross-section adjustment method in order to develop our own integral experimental database. Some of the integral experimental data included in the database are also registered in the International Handbook of Evaluated Reactor Physics Benchmark Experiments (IRPhEP) [3]. For instance, nine ZPPR experimental cores named ZPPR-9, -10A, -10B, -10C, -13A, -17A, -18A, -18C, and -19B, which are mock-up critical experiments for sodium-cooled MOX-fueled fast reactors in the JUPITER cooperative program between the United States and Japan [4], are registered as benchmarks in the IRPhEP handbook. In addition, the criticality, control rod worth, fuel replacement reactivity, sodium void reactivity, and burnup reactivity coefficient measured in performance tests of the experimental fast reactor JOYO MK-I in Japan are also registered as benchmark problems [5, 6].

The adjusted (unified) library is an ABBN-type group constant set with 70-group energy structure for fast reactors, which is based on the format of UFLIB [7]. The first version was developed in 1991 and is called ADJ91 [8]. After that, ADJ98, ADJ2000, ADJ2000R, and ADJ2010 were developed [9]. In 2022, the latest unified cross-section set AJD2017R [10], which is a revised version of ADJ2017 [11], was developed based on JENDL-4.0 [12] by using 619 integral experimental data. The results of ADJ2017 and ADJ2017R in detail are compiled in technical reports [10, 11]. In the present paper, an overview of the latest version with a review of previous ones will be shown.

On the other hand, the latest Japanese evaluated nuclear data library JENDL-5 [13] was released at the end of 2021. In the development of JENDL-5, some of the integral experimental data, which were also used in the development of ADJ2017R, were explicitly used in the

*e-mail: yokoyama.kenji09@jaea.go.jp

nuclear data evaluation. However, this is not reflected in the covariance data. These changes in circumstances need to be taken into account in the development of a next version of the unified cross-section set by using JENDL-5 as a base library. Preliminary adjustment calculation based on JENDL-5 is performed using C/E (calculation/experiment) values simply evaluated by a sensitivity analysis. The preliminary result of the JENDL-5-based adjustment will be also discussed.

2 Overview of unified cross-section set

2.1 Brief development history of unified cross-section set

Table 1 shows the unified cross-section sets developed in Japan. As mentioned before, the first version is ADJ91, which was developed in 1991 [8]. The base nuclear data library of ADJ91 is JENDL-2 [14]. The next version ADJ98 [15] was developed in 1998. The base library of ADJ98 was changed to JENDL-3.2 [16], which was the latest Japanese nuclear data library at that time. ADJ98 was used as a standard cross-section set for nuclear design in the project of the feasibility study on commercialized fast reactor cycle systems (FS) started from 1999 in Japan. This version still utilized only the integral experimental data of ZPPR. The further next version ADJ2000 [17] was developed in 2000, and its revised version, ADJ2000R [18], was developed in 2002. In ADJ2000, integral experimental data measured in cores other than ZPPR were used for the first time. In the development of ADJ2000, the integral experimental data measured in FCA, JOYO, and BFS are mainly added. This extension of the integral experimental data was an important milestone. The integral experimental data obtained in a single experimental facility might be biased, and it is difficult to remove the bias by the cross-section adjustment method. For this reason, it was necessary to introduce the other integral experimental data measured in independent facilities; extensive efforts were made to expand the database. ADJ2000 and ADJ2000R were also used as a standard cross-section set in FS.

Following the release of JENDL-4.0 [12] in 2011, ADJ2010 [9, 19, 20] was developed by extending the database and reanalyzing the integral experimental data with JENDL-4.0. ADJ2010 has been used as the standard cross-section set for nuclear design in the fast reactor cycle technology development project (FaCT) and the succeeding fast reactor projects. In the development of ADJ2010, a significant improvement was made to the method of evaluating uncertainty and correlation coefficient, i.e., covariance data, of the integral experiments. As a methodology for quantifying the correlation coefficient between the integral experimental data, the "error factor correlation method" was proposed [19]. This methodology is described in the OECD/NEA report [21] as a recommended method for evaluating the correlation coefficient between the integral experimental data when applying the cross-section adjustment method.

The latest adjustment cross-section set at the moment is ADJ2017R [10], which is a minor revision of ADJ2017 [11] developed in 2017. In the development of ADJ2017, integral experimental data regarding minor actinoides (MAs) and degraded plutonium (Pu) were added to improve the design accuracy of the core loaded with MA and/or degraded Pu.

2.2 Recent versions of unified cross-section set, ADJ2017 and ADJ2017R

2.2.1 Overview of ADJ2017

ADJ2017 uses JENDL-4.0 as a base library as well as the previous version ADJ2010 and is an adjusted cross-section set that intentionally introduces as many integral experimental data related to MAs and degraded Pu as possible.

In the development of the previous version ADJ2010, a total of 643 integral experimental data were analyzed, and 488 of the integral experimental data were finally selected to be used for the cross-section adjustment. In contrast, a total of 719 integral experimental data were analyzed, and 620 integral experimental data were eventually adopted to develop ADJ2017. As mentioned before, the increased integral experimental data are mainly related to MA and degraded Pu.

For the main neutronic characteristics of conventional sodium-cooled MOX-fuel fast reactors, ADJ2017 shows almost the same performance as ADJ2010. In addition, for neutronic characteristics related to MA and degraded Pu, ADJ2017 improves the C/E values of the integral experimental data sets, and reduces the uncertainty induced by the nuclear data.

2.2.2 Overview of ADJ2017R

ADJ2017R is a minor revision and is basically the same as ADJ2017, but the following two points were revised.

The first point is to unify evaluation methods of the correlation coefficients of experimental uncertainty. In a process of continuous review, it was found that different methods were used for evaluating a common uncertainty of the experimental correlation coefficient. The experimental correlation coefficients were, therefore, reevaluated for all the integral experimental data in the development of ADJ2017R.

The second point is to review the integral experimental data used for the cross-section adjustment. One of the experimental values of composition ratio after irradiation of Am-243 sample was eliminated from the cross-section adjustment because it has a very small uncertainty compared to the others, and the reason could not be clearly explained. In the development of ADJ2017R, a total of 619 integral experimental data, one less than ADJ2017, were used.

Table 2 shows an overview of the integral experimental data used for the development of ADJ2017R. The above two revisions have little effect on the cross-section adjustment results, but they are important from a quality assurance point of view.

Table 1. Features of unified (adjusted) cross-section sets developed in Japan.

	ADJ91/98 (1991/1998)	ADJ2000/2000R (2000/2001)	ADJ2010 (2010)	ADJ2017/2017R (2017/2020)
Base library	JENDL-2/3.2 (1989/1994)	JENDL-3.2 (1994)	JENDL-4.0 (2010)	JENDL-4.0 (2010)
Covariance of nuclear data	rough estimation	JENDL-3.2[1]	JENDL-4.0	JENDL-4.0
Number of energy group	18 (70)[2]	18 (70)[2]	70	70
Number of adjusted parameters	σ_∞: 32/37 χ: 2 β: 6	σ_∞: 41 χ: 2 β: 6 SSF: 1	σ_∞: 155 χ: 2 β: 11 SSF: 1 PFP: 4	σ_∞: 179 χ: 19 β: 20 SSF: 1 PFP: 4
Number of integral experiments	82/185 (ZPPR only)	237 (+ FCA, JOYO, BFS, MASURCA, and LANL)	488 (+ ZEBRA, SEFOR, MONJU(1994), JOYO(MA sample), and BFS(Np))	620/619 (+ ZPPR(D-Pu worth, F41/F25), BFS(D-Pu worth), MONJU(2010), ZEBRA(D-Pu fission RRR), FCA(MA fission RRR), JOYO(MA sample (Pu-241/Pu-240)), PFR(MA sample), and YAYOI(Np capture RRR))

[1]: The covariance data were evaluated after the release of JENDL-3.2.
[2]: The cross sections were adjusted with 18 groups and expanded to 70 groups.
σ_∞: infinite dilute cross section, χ: fission spectrum, β: delayed neutron fraction, RRR: reaction rate ratio,
SSF: self-shielding factor, PFP: pseudo fission product, D-Pu: degraded plutonium

3 Issues on development of next version

3.1 Background

As can be seen from the above explanation, the recent development of the unified cross-section set has been based on JENDL-4.0 for over a decade since JENDL has not had a major revised version since JENDL-4.0 was released in 2010. However, a new major version of JENDL-5 was finally released at the end of 2021. We will proceed with the development of the unified cross-section set based on JENDL-5.

On the other hand, the research situation on the cross-section adjustment methodology has greatly changed in the last decade. In particular, a series of activities started from 2005 in the framework of Sub-Groups (SGs) under the Working Party on International Nuclear Data Evaluation Cooperation (WPEC) in OECD/NEA. One of the important conclusions by the WPEC/SG26 [22] was that a combined use of integral experiments and differential nuclear data measurements is necessary to meet the design target accuracies because many of reactor core target accuracies are not likely to be achieved only with the current differential data. Following this suggestion, a new sub-group SG33 [21] was started in 2009. In SG33, a world-wide comparative study of theoretical formulas of the cross-section adjustment methodology and practical benchmark exercises were conducted. These activities started mainly from the viewpoint of improving the target accuracy of nuclear reactors. After that, activities were handed over to SG39 and SG46 [23, 24], and methodology for effectively utilizing the cross-section adjustment method for nuclear data evaluation and validation were also examined.

In addition, through these activities, an important issue regarding the covariance data of the nuclear data library has become widely known. In the validation of the nuclear data library, benchmark tests using integral experimental data are commonly performed. However, if the information obtained from the benchmark tests using integral experimental data is reflected only in the mean value of the nuclear data, an inconsistency between the mean value and the covariance data will occur. This is because the correlation coefficients in the covariance data should be changed because the integral experimental data are often sensitive to the other nuclear data.

3.2 Proposal from IAEA consultants meeting: general-purpose library and application library

There have been various discussions on this issue, but nevertheless, the discussions and proposals at the IAEA consultants meeting [25], which was held in 2017 soon after the issue became a hot topic, are still informative. In the summary report of the IAEA consultants meeting, the nuclear data libraries are categorized into two groups: A) general-purpose libraries and B) application libraries. The application library is a nuclear data library derived from the general-purpose library for the purpose of using it within a limited range. For simplicity, in the present paper, the general-purpose library and application library are called "A-file" and "B-file", respectively.

The summary report suggests creating covariance data of B-file if the use of integral experimental data in the

Table 2. Integral experimental data used for ADJ2017R.

Facility	Core name	Features	Parameters
ZPPR	9, 10A – 10C	600-800 MWe-class, two-region homogeneous MOX cores.	k_{eff}, RR, CRW, SVR, DR(sample)
	13A	650 MWe-class, Radially-heterogeneous MOX cores.	k_{eff}, RR, CRW, SVR, DR(sample)
	18A, 18C, 19B	1000 MWe-class, two-region homogeneous MOX cores with enriched-uranium regions.	k_{eff}, RR, CRW, SVR
ZEBRA	MZA	550 liter-sized one-region MOX core as a clean benchmark.	k_{eff}, SVR
	MZB, MZC	2,300 liter-sized two-region homogeneous MOX cores to simulate the prototype fast reactor MONJU.	k_{eff}, CWR, SVR
JOYO	MK-I	300 liter-sized 50/75MWt fast power reactor core with MOX and enriched-uranium fuel with blanket	k_{eff}, CWR, SVR, ZMRR, ITC, BRC
	MK-II	240 liter-sized 100MWt fast power reactor core with MOX and enriched-uranium fuel with blanket	MA post-irradiation test
MONJU	Start-up Tests	280MWe prototype fast breeder reactor with two-region homogeneous MOX core	k_{eff}, CRW, ITC
BFS	62-1 – 62-5, 66-1	3,400 liter-sized three or four-region enriched-uranium and/or MOX fuel cores with or without radial blankets.	k_{eff}, RR, CRW, SVR
	67, 69, 66	10kg of NpO_2 loading cores in central MOX region with various grade Pu.	k_{eff}, RR, CRW, SVR
MASURCA	ZONA-2B	380 liter-sized core in the CIRANO experiment series, which aimed at the study of Pu burner cores.	SVR, ZMRR
SEFOR	CORE-I, II	20MWt power fast power reactor core fueled with MOX and cooled with sodium	DR(whole core)
LANL	FLATTOP-Pu, -25, GODIVA, JEZEBEL, JEZEBEL-Pu240	sphere-shaped cores of approx. ten cm in diameter with metallic fuel consisted of ^{239}Pu, degraded Pu or ^{235}U.	k_{eff}
FCA	FCA IX-1 – 7	systematically changed neutron spectra cores with uranium and diluent (graphite and stainless steel).	TRU fission rate ratio
YAYOI	–	fast neutron source reactor with two irradiation holes called Glory and Grazing.	

k_{eff}: criticality, RR: reaction rate, CRW: control rod worth, SVR: sodium void reactivity, DR: Doppler reactivity, ZMRR: zone material replacement reactivity, ITC: isothermal temperature coefficient, BRC: burnup reactivity coefficient

nuclear data evaluation is not reflected in the covariance data. Since the application libraries are often in multi-group data, it is possible to create full-matrix covariance data.

According to the definition of the summary report, the B-file has the following four classifications [25]:

1) Mean values and covariances are preserved[1], only additional correlations (+/-) are added[2].

2) Mean values and standard deviations are preserved, correlations are added/updated.

3) Mean values are preserved, standard deviations and correlations are added/updated.

4) Both mean values and covariances are updated.

3.3 Discussions in working group of JENDL committee

In 2018, the following year, the JENDL Committee established a new working group aimed at promoting the use of covariance data. This working group had been active for three years and published a report [26]. There were various discussions on the use and development of covariance data in the working group; the foregoing issue was also discussed with reference to the proposals of the IAEA consultants meeting. Although there was a discussion that we should not use any integral experimental data in the nuclear data evaluation, the working group concluded that it would be impractical to abandon high-performance libraries based on the integral benchmarks. According to the conclusions, discussions were held to explore realistic countermeasures. An important suggestion derived from the discussions is that nuclear data evaluators should inform users of the use of integral experimental data in the nuclear data evaluation.

There was also a discussion about how users should handle the covariance data in their own problems when they have access to a list of integral experimental data used in the nuclear data evaluation. In this discussion, the preparation of B-file proposed by the IAEA consultants meeting was exercised. In the working group, the B-file classification of 3) proposed by the IAEA consultants meeting was adopted, and a 70-group covariance data set as a B-file was created using the JENDL-4.0-based cross-section set for fast reactors. In the exercise, the B-

[1] In other words, A-file is used as is.

[2] As an example of this case, the summary report [25] presents a study in which the same correlation factors are added to all energy groups.

file of covariance data was created with the mathematical formula of the cross-section adjustment method by tentatively using eight criticality data used in the development of JENDL/AC-2008 [27]. As a result, it was confirmed that the uncertainty evaluated from the variation of C/E values and the uncertainty evaluated from the B-file of covariance data for criticality, reaction rate, and reactivity are comparable.

The conclusions on this issue in the working group are summarized:

- It is necessary to use integral experimental data in the nuclear data evaluation in order to improve the prediction accuracy of integral parameters.
- At this moment, it is difficult to reflect the use of integral data in the covariance data of the evaluated nuclear data files.
- It is recommended for nuclear data users to reflect the use of integral data to their applications, for instance, by using a data assimilation method.
- Nuclear data evaluators should clearly explain and inform users when integral data are used in the nuclear data evaluation

3.4 Issues on development of unified cross-section set based on JENDL-5

In the working group exercise, we could not use JENDL-5 since it had not yet been released, but now we can use JENDL-5. In the first release of JENDL-5, the use of integral data in the nuclear data evaluation is not reflected in the covariance data. However, according to the suggestion of the working group, the use of the integral data used have been clarified [13]. In the nuclear data evaluation of JENDL-5, many integral experimental data of fast reactors, which were used to create ADJ2017R, were also utilized.

These circumstances raise two main issues to be considered. The first issue is the handling of covariance data. In the development of the unified cross-section set, the covariance data of the base nuclear data library, i.e., the prior covariance data, is required as input data for the cross-section adjustment calculation. In a precise sense, therefore, covariance data that is consistent with the mean values of JENDL-5 is necessary when creating an unified cross-section set using JENDL-5 as a base library. As mentioned before, JENDL-5 uses the integral experimental data to adjust the mean values of the nuclear data, but this is not reflected in the covariance data. For this reason, covariance data consistent with the mean value of JENDL-5 were prepared first for the cross-section adjustment calculation. In response to the circumstances, a B-file of JENDL-5 covariance data is created and the cross-section adjustment calculation using the B-file as the prior covariance data is examined.

The second issue is the handling of the integral experimental data that were used in the evaluation of JENDL-5. Generally speaking, it is considered better not to repeatedly use the integral experimental data in the adjustment calculation. Even if the experimental data were actually measured twice, there should be a strong correlation between the two experimental data. Ignoring such a strong correlation in the adjustment calculation would change the result, but the correlation of experimental data used in separate adjustment calculations cannot be considered. For this reason, it is better to avoid the repeated use of the experimental data used in the evaluation of JENDL-5 in our adjustment calculation for creating a unified cross-section set.

4 Preliminary adjustment calculations based on JENDL-5

4.1 Covariance B-file of JENDL-5

In the evaluation of JENDL-5, an adjustment calculation was applied by using the generalized least squares method (GLS), in which fission, capture, and elastic scattering of U-235, U-238, and Pu-239 were adjusted [13]. In addition to the GLS-based adjustment calculation, there are several descriptions on the use of integral experimental data.

We checked the descriptions on the use of integral experimental data in the evaluation of JENDL-5, and listed up the integral experimental data that were commonly used in the development of ADJ2017R. As a result, it was confirmed that there are 82 nuclear characteristics including criticality, reaction rate ratio, sodium void reactivity, and composition ratios of irradiated MA samples. In order to create a consistent B-file of covariance data, it is desirable to confirm the details of how to use these integral experimental data and reflect the use in the B-file. In the present study, however, we adopt a simple method of creating the B-file by using the posterior covariance formula of the cross-section adjustment method, which is the same method adopted in the exercise of the JENDL Committee's working group.

Thus, a B-file of 70-group covariance data were created; this B-file is used as the prior covariance for the preliminary cross-section adjustment based on JENDL-5 in the present paper. Although this method of creating a B-file of covariance data is simplified, it is considered that one of realistic methods for creating a JENDL-5-based adjusted cross-section set at the moment.

4.2 Calculation cases

Three calculation cases are considered for the purpose of comparative study. Table 3 shows a summary of the calculation cases compared to ADJ2017R.

In the first case, the original covariance data of JENDL-5 is used as the prior covariance data, and the cross-section adjustment calculation is performed using the same total set of the integral experimental data as used in the development of ADJ2017R. Note that the integral experimental data are used repeatedly in this case because JENDL-5 utilizes the 82 integral experimental data that were also used in the development of ADJ2017R. Although no explicit adjustment using integral experimental data was made in the major releases prior to JENDL-5,

Table 3. Calculation cases

	ADJ2017R	ADJJ5case1	ADJJ5case2	ADJJ5case3
Base library	JENDL-4.0	JENDL-5		
Number of integral data	619		537 (= 619 - 82[1])	
Covariance data	JENDL-4.0		JENDL-5	JENDL-5 B-file
Evaluation of C/E values	Deterministic or Monte Carlo	Sensitivity analysis		
Repeated use of integral data	—	Yes	No	No
Inconsistency of covariance data	—	Yes	Yes	No[2]

[1]: The common 82 integral experimental data were used in JENDL-5 and ADJ2017R.
[2]: The inconsisency between mean values and covarinace data was eliminated by B-file.

e.g., in JENDL-4.0, the information of the integral experimental data might be implicitly used regardless of the evaluator's intention since the nuclear data evaluators know the results of the integral benchmarks. It is considered that, therefore, the unified cross-section sets developed in the past, including ADJ2017R, are actually in the similar condition as this calculation case. This calculation case is called ADJJ5case1.

In the second case, the original covariance data of JENDL-5 is used, but the integral experimental data utilized in the evaluation of JENDL-5 are eliminated in the cross-section adjustment calculation so as to avoid the repeated use of the same integral experimental data. Note that, in this case, the cross-section adjustment calculation is performed using the covariance data that do not have any information of the use of integral experimental data. In other words, there is a possibility that the mean values may be adjusted inconsistently or overly at the stage of adjustment calculation because it is not possible to take into account the fact that the mean values have already been adjusted in the nuclear data evaluation. In practice, this calculation case should be avoided, but the importance of covariance data in the cross-section adjustment calculation can be checked by comparing with this case. This calculation case is called ADJJ5case2.

The third case is the reference case. In the third case, the B-file of covariance data is used as the prior covariance data, and the cross-section adjustment calculation is performed excluding the 82 nuclear characteristics used in the JENDL-5 evaluation. Although the method of creating the B-file of covariance data is simplified, not perfect, in this case, but at least the use of the 82 integral experimental data in the JENDL-5 evaluation is consistently taken into account and the repeated use of integral experimental data is avoided. This calculation case is called ADJJ5case3.

4.3 The other calculation conditions

In the development of ADJ2017R, the C/E values of JENDL-4.0 were evaluated by a deterministic method or a Monte Carlo method, but these evaluations are very time consuming. In the present study, the C/E values of JENDL-5 were obtained by using sensitivity analysis, in which 70-group sensitivity coefficients and differences of nuclear data between JENDL-5 and JENDL-4.0 were used. This sensitivity analysis is considered to have sufficient accuracy for this study. The other calculation conditions are basically the same as ADJ2017R.

4.4 Calculation results

Figure 1 shows C/E values of criticality. From this figure, it can be seen that ADJ2017R, ADJJ5case1, and ADJJ5case3 have similar results. In general, the same tendencies were seen for the other nuclear characteristics. On the other hand, it can be seen that ADJJ5case2 tends to be quite different. For instance, the C/E values of ZPPR and BFS have not improved much. In particular, the C/E value of GODIVA is getting worse. In ADJJ5case2, the use of the integral experimental data in the evaluation of JENDL-5 is not reflected in the prior covariance. For this reason, the restriction by the covariance data does not work correctly in the adjustment calculation, and incorrect large adjustment is made to improve the C/E values of the other integral experimental data. This result is considered to be a non-physical adjustment result. It also suggests the important role of the prior covariance data when applying the cross-section adjustment method.

Figure 2 shows C/E values of reaction rate ratio of C28/F49 (U-238 capture to Pu-239 fission). From this figure, it can be seen that all calculation cases give almost the same results for this nuclear characteristic. The differences of the calculation cases had little effect because the C28/F49 reaction rate ratios were not used in the evaluation of JENDL-5.

Figure 3 shows C/E values of reaction rate ratios of F42/F49 (Pu-242 fission to Pu-239 fission) and F64/F49 (Cm-244 fission to Pu-239 fission). These are the integral experimental data related to the degraded Pu and MA, which were introduced in the development of ADJ2017/ADJ2017R. In JENDL-4.0, F42/F49 was slightly overestimated, and F64/F49 was clearly overestimated. From this figure, it can be seen that ADJ2017R improves these overestimations. It can be seen that the C/E values of JENDL-5 are significantly improved; these benchmark results were also used in the evaluation of JENDL-5. In ADJJ5case3, where the integral experimental data utilized in the evaluation of JENDL-5 are not repeatedly used in the adjustment calculation, it is seen that the results of F64/F49 in ADJJ5case3 are almost the same as JENDL-5. In addition, the results of F64/F49 in ADJJ5case3 were not improved as much as ADJ2017R. If they had not been used in the evaluation of JENDL-5, they might have been improved to the same level as ADJ2017 by the cross-section adjustment. The results of ADJJ5case3 will be acceptable in practice, but these results suggest that there should be cases where it is better

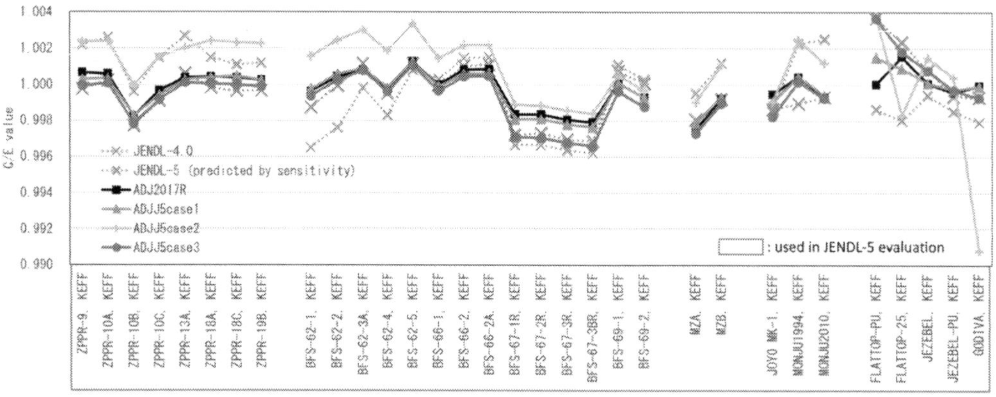

Figure 1. C/E values of criticality

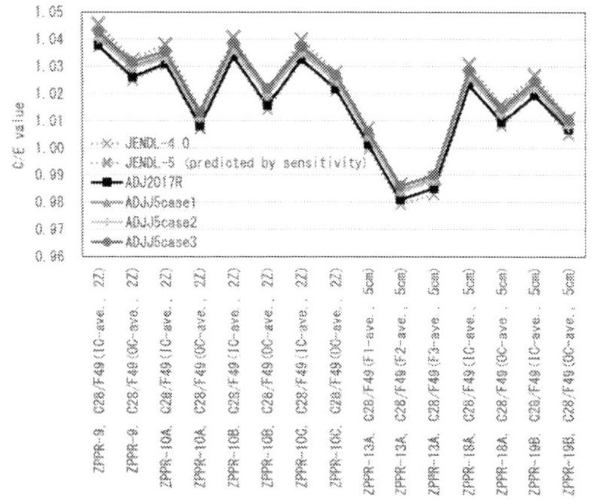

Figure 2. C/E values of reaction rate ratio of C28/F49

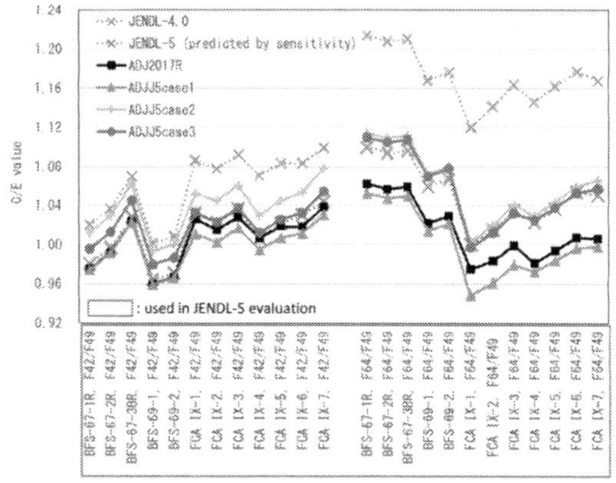

Figure 3. C/E values of reaction rate ratio of F42/F49 and F64/F49

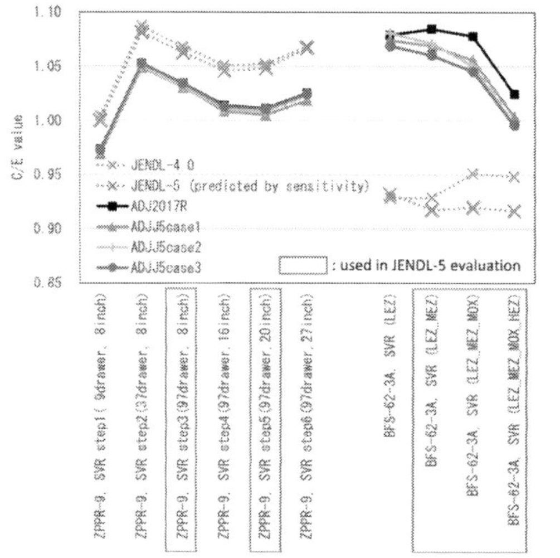

Figure 4. C/E values of sodium void reactivity

for the users to make adjustments directly. On the other hand, in ADJJ5case1, it is seen that the C/E values of F64/F49 are smaller than those of ADJ2017R. Since the same integral experimental data were used twice in this case, it is considered that the C/E values are over-adjusted.

Figure 4 shows the C/E values of sodium void reactivity. In the evaluation of JENDL-5, the sodium void reactivities of ZPPR-9 step3 and step5, BFS-62-3A LEZ_MEZ, LEZ_MEZ_MOX, and LEZ_MEZ_MOX_HEZ, were utilized. From this figure, it can be seen that the results of ADJJ5case1, ADJJ5case2 and ADJJ5case3 are almost the same. This is probably because the benchmark results of sodium void reactivity, which have relatively large uncertainty, do not have a strong impact on the evaluation of JENDL-5 compared to those of criticality and reaction rate ratio.

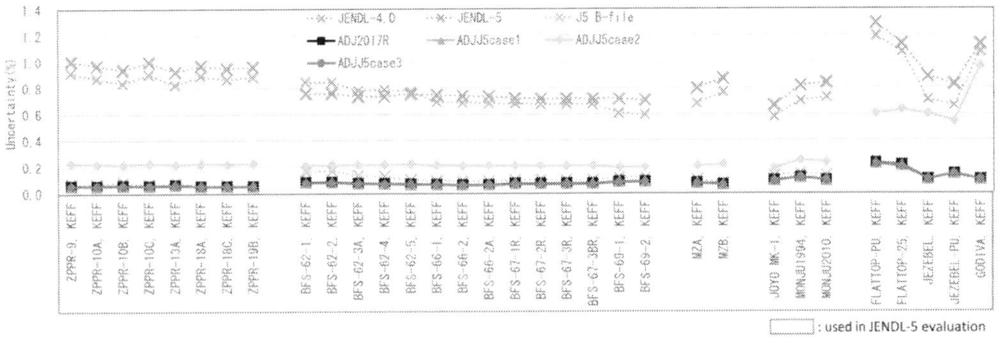

Figure 5. Nuclear-data-induced uncertainty of criticality

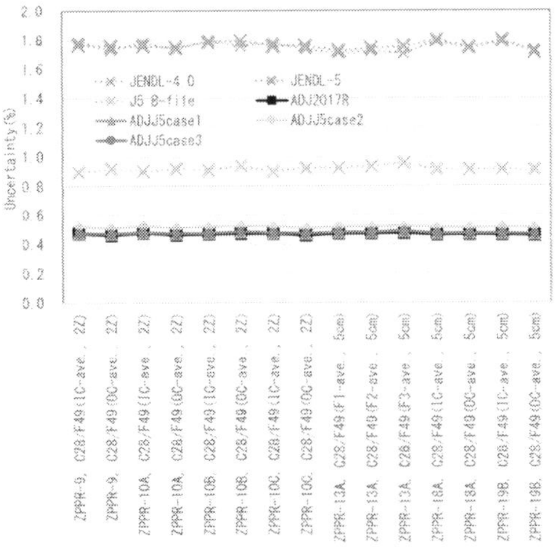

Figure 6. Nuclear-data-induced uncertainty of reaction rate ratio of C28/F49

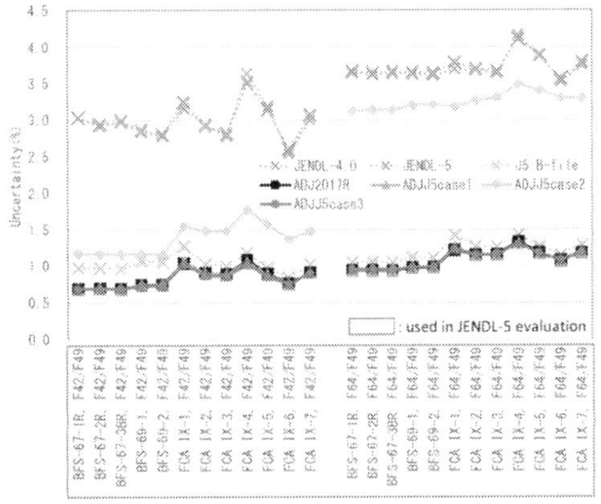

Figure 7. Nuclear-data-induced uncertainty of reaction rate ratio of F42/F49 and F64/F49

4.5 Nuclear-data-induced uncertainty

Figure 5 shows the uncertainty due to nuclear data for criticality. From this figure, it is seen that the nuclear-data-induced uncertainties evaluated with the B-file are considerably small. In the evaluation of JENDL-5, the all criticalities shown here are used, and the uncertainties evaluated with the B-file are almost the same as in AJD2017R. Although the adjustment calculation performed in the evaluation of JENDL-5 is not exactly the same as that in the creation of B-file, the true JENDL-5 covariance data are expected to give considerably smaller uncertainties with respect to criticality.

Figure 6 shows the uncertainties due to nuclear data of reaction rate ratio C28/F49. In the evaluation of JENDL-5, the integral experimental data of C28/F49 were not used, but since it is the ratio of the major reactions in fast reactors, it is considered that the information of C28/F49 is are included in the other integral experimental data such as criticality. The uncertainties of the B-file are approximately 0.9%, which are considerably smaller than 1.8% of JENDL-5. From this result, it is expected that the true JENDL-5 covariance data has correlations that reduce the uncertainty due to nuclear data of C28/F49.

Figure 7 shows the uncertainties due to nuclear data with respect to F42/F49 and F46/F49. The uncertainties of the B-file are small, and is almost the same as the uncertainties of ADJ2017R. As with the other integral experimental data, the true JENDL-5 covariance data would reduce the uncertainties of F42/F49 and F64/F49.

Figure 8 shows the uncertainties due to nuclear data of sodium void reactivity. Although the uncertainties of the B-file are smaller than those of the original JENDL-5, they are not as small as those of ADJ2017R and ADJJ5cases. From this result, it is considered that the number of integral experimental data used in the JENDL-5 evaluation was not sufficient for the sodium void reactivity compared to that used in the adjustment calculations of ADJ2017R and ADJJ5cases. This may be one of the reasons why the C/E values of the sodium void reactivity did not change significantly depending on the calculation cases.

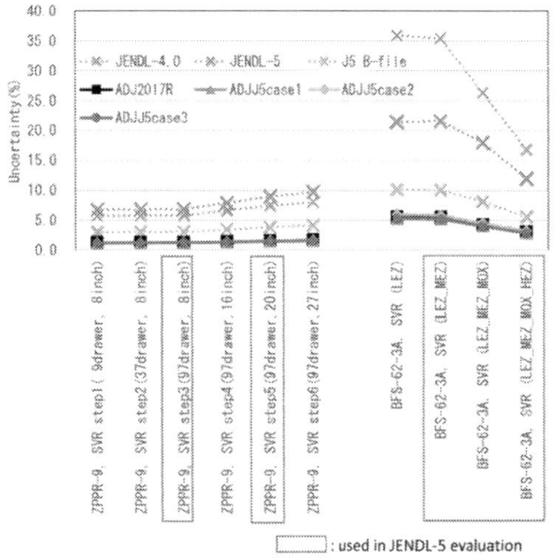

Figure 8. Nuclear-data-induced uncertainty of sodium void reactivity

5 Conclusions

This paper briefly reviews the history of the development of adjusted nuclear data libraries for fast reactors, which are called the unified cross-section sets, and provides an overview of the latest version ADJ2017R. The latest version uses JENDL-4.0 as a base library, but the next version will be developed by using JENDL-5 as the base library because JENDL-5 was released in December 2021. In the evaluation of JENDL-5, integral experimental data were intentionally used for the purpose of improving the accuracy of reactor analysis; an adjustment with the generalized least square method, which is equivalent to the cross-section adjustment method, was also adopted. The use of integral experimental data in the evaluation of JENDL-5, however, has not been reflected yet in the covariance data at the moment. Fortunately, according to the proposal based on the discussions in the working group of the JENDL Committee, the use of integral experimental data in the evaluation were clearly described. Under the present circumstances, we investigated a method of creating an adjusted cross-section set using JENDL-5 as a base library.

In the present paper, since the list of the integral experimental data used in the evaluation of JENDL-5 is available, we used the list to create 70-group covariance data that are more consistent with the mean values of JENDL-5, i.e., an application library (B-file) proposed by the IAEA consultants meeting. Furthermore, we performed preliminary cross-section adjustment calculations using the B-file as the prior covariance. By this procedure, we could create a unified cross-section set based on JENDL-5 that has almost the same performance as the latest version of the unified cross-section set ADJ2017R, which is based on JENDL-4.0. To avoid the repeated use of the integral experimental data used in the evaluation of JENDL-5, we did not use them in the reference case of the preliminary cross-section adjustment calculation. In this case, we found that the adjustment results could be a little worse than ADJ2017R. However, we would like to emphasize that the working group's proposal has shown a realistic, though not perfect, solution at the moment. The use of integral experimental data in the nuclear data evaluation is no longer inevitable because discarding the existing high-performance libraries is impractical. If so, we have no choice but to find a better and more practical solution for the use of integral experimental data in the nuclear data evaluation.

This paper includes a part of the results of "Technical development program on a common base for fast reactors" entrusted to the Japan Atomic Energy Agency (JAEA) by the Ministry of Economy, Trade and Industry (METI). The authors also wish to express their gratitude to Mr. K. Numata of NESI corporation for his valuable comments on the adjustment calculations.

References

[1] J.B. Dragt, J.W.M. Dekker, H. Guppelaar, A.J. Janssen, Nucl. Sci. Eng. **62**, 117 (1977)
[2] M. Ishikawa, K. Sugino, W. Sato, K. Numata, J. Nucl Sci Technol, Supplement 2 pp. 1073–1076 (2002)
[3] OECD Nuclear Energy Agency, *International handbook of evaluated reactor physics benchmark experiments*, NEA/NSC/DOC(2006)1 (2019)
[4] M. Ishikawa, T. Ikegami, T. Sanda, Nucl. Sci. Eng. **178**, 335 (2014)
[5] K. Yokoyama, M. Ishikawa, Nucl. Sci. Eng. **157**, 249 (2007)
[6] K. Yokoyama, M. Ishikawa, Nucl. Sci. Eng. **178**, 350 (2014)
[7] T. Hazama, G. Chiba, W. Sato, K. Numata, Tech. Rep. JAEA-Review 2009-003, Japan Atomic Energy Agency (2009)
[8] M. Ishikawa, T. Hoshi, T. Sanda, T. Kamei, T. Kawakita, *Improvement of Nuclear Design Method for Large LMFBR Cores Using the Cross-section Adjustment*, in *Proc. Joint Int. Conf. on Mathematical Methods and Supercomputing in Nuclear Applications, M&C SNA '93, 19-23, April 1993* (Kongresszentrum, Karlsruhe, Germany, 1993)
[9] K. Yokoyama, M. Ishikawa, Nucl. Data Sheet **123**, 97 (2015)
[10] K. Yokoyama, S. Maruyama, H. Taninaka, S. Ohki, Tech. Rep. JAEA-Data/Code 2021-019, Japan Atomic Energy Agency (2022), [In Japanese]
[11] K. Yokoyama, K. Sugino, M. Ishikawa, S. Maruyama, Y. Nagaya, T. Jin, Tech. Rep. JAEA-Research 2021-019, Japan Atomic Energy Agency (2018), [In Japanese]
[12] K. Shibata, O. Iwamoto, T. Nakagawa, N. Iwamoto, A. Ichihara, S. Kunieda, S. Chiba, K. Furutaka, N. Otuka, T. Ohsawa et al., J. Nucl. Sci. Technol. **48**, 1 (2011)
[13] O. Iwamoto, N. Iwamoto, S. Kunieda, F. Minato, S. Nakayama, Y. Abe, K. Tsubakihara, S. Okumura,

C. Ishizuka, T. Yoshida et al., J. Nucl. Sci. Technol., DOI:10.1080/00223131.2022.2141903 (2022)

[14] T. Nakagawa, Tech. Rep. JAERI-M 84-103, Japan Atomic Energy Research Institute (1984)

[15] K. Yokoyama, K. Numata, M. Ishikawa, Tech. Rep. JNC TN9400 99-042, Japan Nuclear Cycle Development Institute (1999), [In Japanese]

[16] T. Nakagawa, K. Shibata, S. Chiba, T. Fukahori, Y. Nakajima, Y. Kikuchi, T. Kawano, Y. Kanda, T. Ohsawa, H. Matsunobu et al., J. Nucl. Sci. Technol. **32**, 1259 (1995)

[17] M. Ishikawa, K. Numata, W. Sato, K. Sugino, Tech. Rep. JNC TN9400 2001-071, Japan Nuclear Cycle Development Institute (2001), [In Japanese]

[18] T. Hazama, G. Chiba, K. Numata, W. Sato, Tech. Rep. JNC TN9400 2002-064, Japan Nuclear Cycle Development Institute (2002), [In Japanese]

[19] K. Sugino, M. Ishikawa, K. Numata, T. Iwai, T. Jin, Y. Nagaya, T. Hazama, G. Chiba, K. Yokoyama, T. Kugo, Tech. Rep. JNC TN9400 2002-064, Japan Nuclear Cycle Development Institute (2002), [In Japanese]

[20] K. Sugino, M. Ishikawa, K. Yokoyama, Y. Nagaya, G. Chiba, T. Hazama, T. Kugo, Journal of the Korean Physical Society **59**, 1357 (2011)

[21] M. Salvatores, G. Palmiotti, G. Aliberti, R. McKnight, P. Archier, C. De Saint Jean, E. Dupont, M. Herman, M. Ishikawa, K. Sugino et al., Tech. Rep. NEA/NSC/WPEC/DOC(2013)445, OECD/NEA (2013)

[22] M. Salvatores, G. Aliberti, M. Dunn, A. Hogenbirk, A. Ignatyuk, M. Ishikawa, I. Kodeli, A.J. Koning, R. McKnight, R.W. Mills et al., Tech. Rep. Nuclear Science NEA/WPEC-26, OECD/NEA (2008)

[23] G. Palmiotti, M. Salvatores, K. Yokoyama, M. Ishikawa, Tech. Rep. NEA/NSC/R(2016)6, OECD/NEA (2017)

[24] M. Salvatores, G. Palmiotti, *Method and approaches to provide feedback from nuclear and covariance data adjustment for improvement of nuclear data files: Major findings of the NEA WPEC Subgroup 39*, in *Proc. of Int. Conf. on Nuclear Data for Science and Technology, ND2019, EPJ Web of Conferences* (2020), Vol. 239

[25] V. Radulovic, A. Trkov, Tech. Rep. Summary Report of an IAEA Consultants Meeting, IAEA Headquaters, Vienna, Austria, 14-17 November 2017, INDC(NDS)-0746, IAEA (2017)

[26] Covarinace Data Utilization and Promotion Working Group, Tech. Rep. JAEA-Review 2021-014, Japan Atomic Energy Agency (2021), [In Japanese]

[27] O. Iwamoto, T. Nakagawa, N. Otuka, K.O. S. Chiba, G. Chiba, J. Nucl. Sci. Technol. **46**, 510 (2009)

EPJ Web of Conferences **281**, 00005 (2023) https://doi.org/10.1051/epjconf/202328100005
CW2022

Covariances from model variation: Application to quantities for astrophysics

Dimitri Rochman[1,*], *Arjan* Koning[2,**], and *Stéphane* Goriely[3,***]

[1]Reactor Physics and Thermal Hydraulic Laboratory, Paul Scherrer Institut, Villigen, Switzerland
[2]Nuclear Data Section, International Atomic Energy Agency, Vienna, Austria
[3]Institut d'Astronomie et d'Astrophysique, Université Libre de Bruxelles Brussels, Belgium

Abstract. In this work, covariance matrices coming from model variations, in contrast to the usual parameter variations, are presented. The considered phenomenological and microscopic models are included in the code TALYS, and concern level densities, gamma strength functions, optical potentials and masses. A total of 288 model sets for each isotope is used to estimate both uncertainties and correlations from systematical origin. The calculated quantities are of interest for astrophysical applications, such as capture cross sections, and reaction rates. The isotopes ($3 \leq Z \leq 100$) are from the proton to neutron drip lines, covering about 8800 cases, and are included in the TENDL-2021 library.

1 Introduction

Covariance matrices are nowadays included in almost all nuclear data libraries. They mostly concern neutron-induced reactions [1, 2], with some addition of charged particles [3]. Such matrices are useful for a number of applications, ranging from nuclear energy production to astrophysics, with also utilization for medical isotopes [4–6].

The method of assessment of such covariance matrices is based on (1) estimation of experimental uncertainties and correlations, (2) estimation of theoretical uncertainties and correlations, and (3) combination of (1) and (2). In some cases, only (1) or only (2) is used, for instance when no experimental data exists, or if theoretical values are not needed. For the theoretical estimation, almost all covariance evaluations rely on the following approach: selection of a set of preferred models, adjustment of model parameters to reproduce experimental data (or simple use of values from systematics), and then vary those parameters within acceptable ranges to produce a set of varied calculations, later used to build covariance matrices.

Such methods are well justified when the reaction models are within their range of applicability, as for stable isotopes, or isotopes "*not too far*" from the stability line. For the majority of applications related to energy production and medical isotopes, the use of such covariance matrices from libraries is therefore reasonable. In the case of astrophysics, as a large number of relevant isotopes are far from the stability line, covariance matrices as presented in the TENDL libraries (for isotopes with half-lives as short as 1 second) might be less justified: the main reason is that most of the theoretical reaction models are not proven to be correct for exotic nuclei and model uncertainties are known to dominate over parameter uncertainties (see for instance Refs. [7, 8]). A number of examples can also be found in Refs. [9, 10] for the impact of various gamma-ray strength functions.

An alternative approach to the estimation of covariance matrices is proposed here, by varying reaction models and not model parameters. For instance, a capture cross section might differ with different level density models, or with different gamma-ray strength functions. Such changes can be used to assess systematical changes in cross section calculations. A similar approach was presented in Ref. [11] for the proton-induced reactions on ^{59}Co where the goal was to provide the most appropriate set of models and their parameters. All cross sections and covariance matrices were affected by the variation of models.

Details of the methods and examples are presented in the following, based on TALYS calculations applied to capture cross sections and Maxwellian Averaged (n,γ) Cross Sections (or MACS). A large number of nuclei are considered ($3 \leq Z \leq 100$), from the proton to neutron drip lines, representing about 8800 isotopes. As presented in the following, these models were not validated for such number of isotopes, and the goal of this work is to assess the impact of the available models. Quantities for nuclear reactions are currently used in astrophysics for very exotic isotopes, often based on a single choice of models. The current work will help to recognize the impact of reaction models. Most of the present results are included in the TENDL-2021 nuclear data libraries.

2 Model variation

The method applied here is conceptually relatively simple: perform different TALYS calculations with different

*e-mail: dimitri-alexandre.rochman@psi.ch
**e-mail: a.koning@iaea.org
***e-mail: stephane.goriely@ulb.be

input models, and compare calculated quantities. The outcome of such calculations is an estimation of changes due to model selections. Compared to the traditional approach of selecting one set of models and changing model parameters, an estimation of systematical effects can be performed.
In practice, the following models are considered (as implemented in TALYS):

- E1 gamma strength functions with two possibilities (TALYS keyword *strength* with values 8 or 9): either Gogny D1M HFB+QRPA, or SMLO.
- Level density (TALYS keyword *ldmodel* with values 1, 2 or 5): Constant temperature + Fermi gas model, or Back-shifted Fermi gas model, or Microscopic HFB with combinatorial model.
- Optical model potential (TALYS keyword *jlmomp* with values y or n): JLM microscopic optical model potential or KD optical model.
- Gamma strength function for M1 (TALYS keyword *strengthM1* with values 3 or 8): SMLO or Gogny D1M HFB+QRPA,
- Collective enhancement for level density (TALYS keyword *colenhance* with values y or n): yes or no,
- Width fluctuation (TALYS keyword *widthmode* with values 0, 1 or 2): no width fluctuation, Moldauer model, or Hofmann-Richert-Tepel-Weidenmueller model,
- Mass model (TALYS keyword *massmodel* with values 0, 1, 2 or 3): Duflo-Zuker formula, FRDM 2012 table, Skyrme-HFB table, or D1M Gogny-HFB table (except for known masses, where the experimental value is always used)

In total, one can combine the variation of 7 model categories; in the following, specific combinations of models are indicated with 7 digits, such as 91n3n12 (default TALYS choice), or 81n8n12. Not all combinations are considered: for instance the microscopic gamma-ray strength model (*strength=8*) is only used with *strengthM1=8* for consistency. Similarly, the collective enhancement of the level density is disabled with the microscopic level density (*ldmodel=5*). In total, 288 combinations of models are used.
Among these 288 combinations, model parameters were adjusted to fit the experimental capture cross sections in the fast energy range for 12 model combinations: 91n3n12, 81n8n12, 81n8y12, 91n3y12, 95n3n12, 95n3y12, 85n8n12, 85n8y12, 82n8n12, 82n8y12, 92n3n12 and 92n3y12. It means that for these model sets, the agreement between calculated and measured neutron capture cross sections is expected to be excellent, within typically 10 %. When no experimental data exist, the parameters of these model sets will be based on non adjusted values.

3 Examples

Many examples can be extracted from the large amount of calculated quantities. A short selection is presented below to illustrate the content of the TENDL-2021 database for astrophysics applications.

3.1 Tin neutron-capture cross sections

To illustrate the variation of cross sections from stable isotopes to exotic ones, Fig. 1 presents the variation of neutron-capture cross sections for two Sn isotopes: ^{120}Sn and ^{160}Sn. For each of them, 288 TALYS calculations

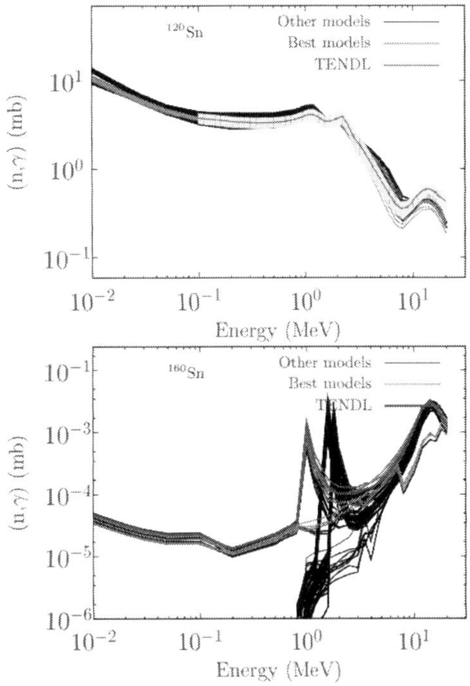

Figure 1: Example of radiative neutron-capture cross sections for two Sn isotopes (from top to bottom: 120 and 160). For each mass, 288 TALYS calculations are presented, based on different model sets.

are performed, leading to 288 different cross sections per isotope. The 12 selected models are indicated as "Best models" and the other 276 as "Other models". It can be observed that in the case of the stable isotope ^{120}Sn, the variations are smaller than for the very exotic one. In the case of ^{160}Sn, the capture cross section is small enough to be out of the range of the plot. Naturally, uncertainties and correlations will strongly differ between these two isotopes. If only model parameters for a selected set of models were varied (for instance for the default TALYS models 91n3n12), then such strong effect would not appear. For ^{120}Sn, the uncertainties due to the variation of model parameters for 91n3n12 (called TENDL in the figure) are significantly smaller than the spread of all capture cross sections from model variations. The effect on this stable isotope is not as dramatic as for ^{160}Sn. Correlations also strongly changed between both approaches, as presented in Figs. 2 and 3.

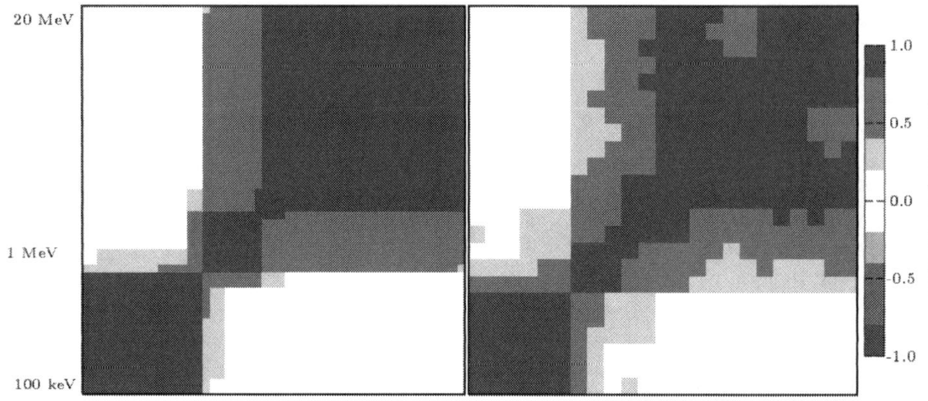

Figure 2: Correlation matrices for ^{120}Sn based on variation of model parameters (left) and models (right)

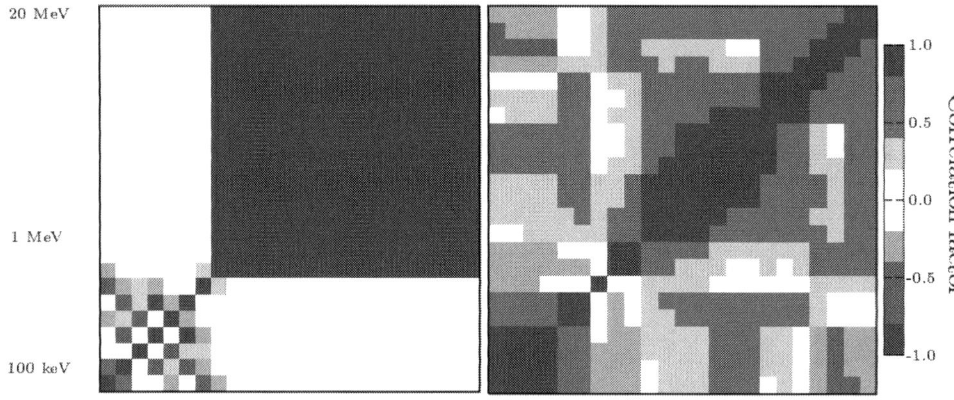

Figure 3: Same as Fig. 2, but for ^{160}Sn.

In these figures, the X- and Y-axis correspond to the incident neutron energy. For ^{120}Sn, the correlation matrix is not strongly modified between both approaches, as all calculated capture cross sections are very similar. Varying models lead to more degrees of freedom for exotic nuclei: for ^{160}Sn, correlations are strongly modified, due to more pronounced changes in cross sections.
These examples illustrate the effect of varying models; additionally, these examples show that uncertainties and correlations always depend on the underlying modeling assumptions.

3.2 30 keV neutron-capture cross sections

As an additional example, the ratio of the neutron capture cross section at 30 keV from the default TALYS models (91n3n12) over the model set 81n8n12 is presented in Fig. 4 for all the 8800 isotopes. Clearly, strong differences can be observed for exotic nuclei. As parameters for both model sets were adjusted for stable isotopes, differences are not pronounced in the stability valley. Following this approach, all model sets can be used to calculate quantities of interest for astrophysics, such as neutron capture cross sections from 0.01 eV to 20 MeV (available from the TALYS output file called *xs000000.tot*), neutron capture reaction rates for all temperatures of astrophysics relevance taking into account the thermal population of the target (from the TALYS file *astrorate.g*), the normalized partition function G(T) (from the TALYS file *astrorate.tot*) and the Laboratory (not stellar) MACS (from the TALYS file *macs.g*).

3.3 Cross isotope correlations

All the quantities presented in this work are obtained with the same approach, identical calculations and model sets. Because of these similarities, correlations among isotopes for any calculated quantities can be expected. An example is presented in the following for the MACS quantity at the neutron energy of 30 keV. For each isotope, uncertainties from the variation of models can be obtained (these values can be found in the TENDL-2021 database). Additionally, as the same 288 models are used for all isotopes, correlations among isotopes can also be calculated for the MACS, as well as for other quantities. The correlation matrix in this case is presented in Fig. 5. In this figure, the

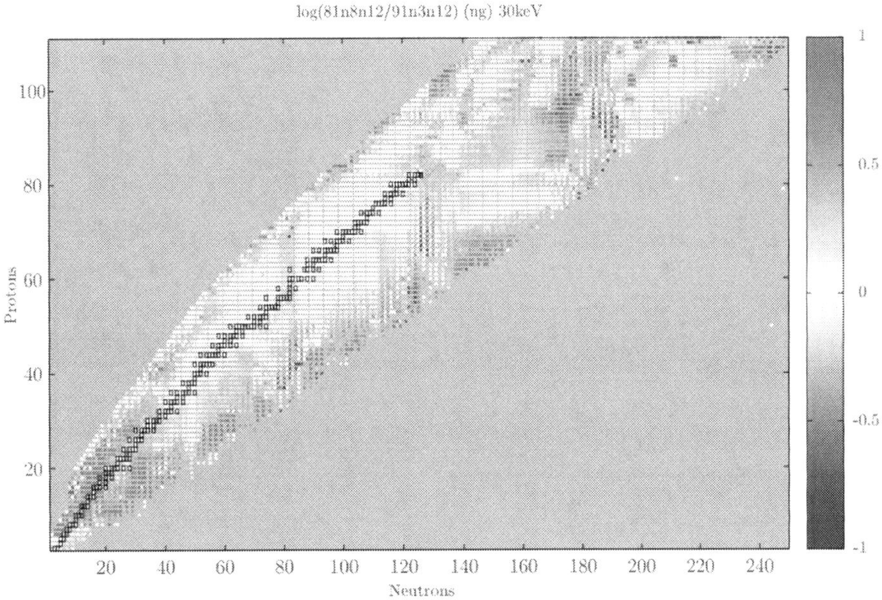

Figure 4: Ratios of model set 81n8n12 over the default TALYS model set 91n3n12 for the neutron capture cross sections at 30 keV. The color scale is the log of the ratio.

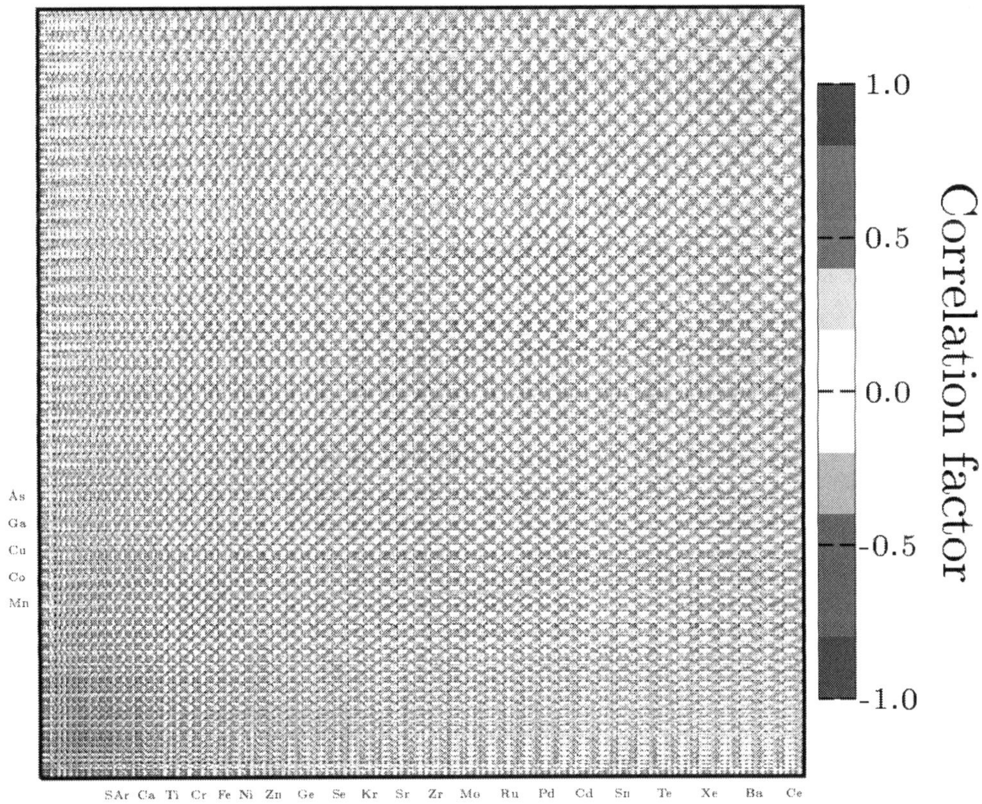

Figure 5: Correlation between nuclei for the MACS at 30 keV, taking into account 288 variations of models. Each element, included in one box, is presented with increasing mass.

 https://doi.org/10.1051/epjconf/202328100005

X- and Y-axis are presenting the element with increasing mass from the left to right (and from the bottom to the top). Isotopes of the same element are strongly correlated with each other (with some exceptions), but elements are also correlated with each other. It means that certain model sets provide systematically high (or low) MACS values, almost independently of the mass and charge of the isotopes. This indicates that a specific selection of models globally and systematically influences cross sections (and other quantities) for all isotopes and therefore subsequent calculations based on these cross sections.

4 Future developments

As mentioned, all the calculations described in the previous sections are included in the TENDL-2021 library and can be accessed online through the TENDL website (https://tendl.web.psi.ch/tendl_2021/tar_files/astro/astro.html). This is the first attempt to quantify systematically the effect of model variations. A long term goal is to update the Brussels nuclear library for astrophysics applications, called BRUSLIB [12]. Such a library contains more calculated values than the ones previously presented. It is therefore planned to update the current results with the addition of the following calculations:

- cross sections for the (n,p), (n,α), (p,α), (p,n), (p,γ), (α,n), (α,p) and (α,γ) cross sections from TALYS,
- include model variation for fission (keyword *fismodel* in TALYS) for masses greater than 209, using either empirical fission barriers or WKB approximation for fission path model,
- include model variation for the alpha optical potential (keyword *alphaomp* in TALYS), using either the double folding model from Demetriou and Goriely, or the alpha potential of Avrigeanu (default).

In total, 9 reactions will be provided, and for each of them, either 480 model variations for atomic masses lower or equal to 209, or 960 model variations for $A > 209$. An example of a typical TALYS input for ^{149}Eu, for the calculation of the MACS at 30 keV, is presented in Fig. 6. Similar types of input files are used for calculating all required quantities. This large number of calculations is currently ongoing and results planned to be released after 2022.

```
#
# TALYS input file
#
projectile n
element    Eu
mass       149
Ltarget    000
energy     energies
maxlevelstar    30
partable y
bins 40
#
# Do not use best
# parameters from database
#
best n
isomer 0.1
#
# Produce files for processing
#
endf y
endfdetail y
popeps 1.e-12
transeps 1.e-20
transpower 15
xseps 1.e-20
#
# Output of extra channels
#
channels y
filechannels y
#
# Recoils
#
recoil y
recoilaverage y
#
# parameters
#
elow 1.e-6
ngfit y
upbend y
recoil n
disctable 2
outdensity y
filedensity y
strength 9
ldmodel 2
jlmomp n
strengthm1 3
colenhance y
widthmode 2
massmodel 1
alphaomp 6
fismodel 1
endf n
outspectra n
outangle n
outbasic n
recoil n
transpower 15
transeps    1.00E-18
xseps       1.00E-17
popeps      1.00E-13
Energy 0.03
astroE 0.03
```

Figure 6: Typical TALYS input file for the calculation of the 30 keV MACS of ^{149}Eu.

5 Conclusion

We have presented in this work the effect of modifying TALYS reaction models for the calculations of various quantities related to astrophysics. In total, 288 model sets were used, varying mass models, width fluctuation corrections, level densities, gamma-ray strength functions and optical models. Simple comparisons with covariance matrices obtained from the variation of model parameters, for a fixed model, indicated strong changes. For isotopes far from the stability line, where default models are not always adequate, uncertainties and correlations determined by varying models help to understand systematic biases of models, found to be significantly larger than the ones coming from parameter variations (for a single model). All calculated values are included in the TENDL-2021 library, and it is planned to extend this database to other reaction channels, including fission and alpha captures.

References

[1] D.A. Brown *et al.*, Nuclear Data Sheets **148**, 1 (2018).
[2] A.J. Plompen *et al.*, Eur. Phys. Jour. **A 56**, 181 (2020).
[3] A. Koning, D. Rochman, J. Sublet, N. Dzysiuk, M. Fleming and S. van der Marck, Nucl. Data Sheets **155**, 1 (2019).
[4] O. Leray, H. Ferroukhi, M. Hursin, A. Vasiliev and D. Rochman, Annals of Nuclear Energy **110**, 547 (2017).
[5] C. Guerrero *et al.*, Physical Review Letters **125**, 142701 (2020).
[6] F.T. Tàrkànyi *et al.*, Journal of Radioanalytical and Nuclear Chemistry **319**, 487 (2019).
[7] S. Goriely and R. Capote, Phys. Rev. **C 89**, 054318 (2014).

[8] S. Goriely, Eur. Phys. J. **A 51**, 172 (2015).
[9] A.C. Larsen and S. Goriely, Phys. Rev. **C 82**, 014318 (2010).
[10] S. Goriely, S. Hilaire, S. Péru, and K. Sieja, Phys. Rev. **C 98**, 014327 (2018).
[11] E. Alhassan, D. Rochman, A. Vasiliev, R.M. Bergmann, M. Wohlmuther, A.J. Koning and H. Ferroukhi, EPJ Web of Conferences **239**, 13005 (2020).
[12] S. Goriely, AIP Conference Proceedings **704**, 375 (2004).

Development of a robust nuclear data adjustment method to outliers

Yuhei Fukui[1*], *Tomohiro* Endo[1], *Akio* Yamamoto[1] and *Shuhei* Maruyama[2]

[1]Nagoya University, School of Engineering, Aichi, Japan
[2]Japan Atomic Energy Agency, Oarai Research and Development Institute, Ibaraki, Japan

Abstract. We developed a new nuclear data adjustment method for experimental data containing outliers. This method mitigates the effect of outliers by applying M-estimation, a type of robust estimation, to the conventional nuclear data adjustment method using sensitivity coefficients. Based on the M-estimation, we derived a weighted nuclear data adjustment formula and developed a weight calculation method. The weighted nuclear data adjustment formula was derived by weighting the function to take the extremum of the conventional nuclear data adjustment. The weighting of each nuclear characteristic is calculated from the difference between the measured and calculated values of the nuclear characteristic. This weight calculation method can evaluate the validity of each nuclear characteristic by considering correlations between nuclear characteristics using singular value decomposition. The proposed method and the conventional method were compared and verified by twin experiments. In the twin experiments, the nuclear data were adjusted using experimental data that intentionally included outliers. As a result of twin experiments, it was confirmed that the nuclear data were adjusted robustly and appropriately even with the experimental data containing outliers.

1 Introduction

Nuclear data adjustment is a technique to reduce the uncertainty of nuclear data by unifying the measured nuclear characteristics in integral experiments and the numerically calculated nuclear characteristics using nuclear data. In conventional nuclear data adjustment methods, the experimental data (measured nuclear characteristics and their uncertainties) are assumed to be appropriate. Note, however, that the adjustment will not be appropriate if experimental data contains outliers because adjusted nuclear data are significantly affected by experimental data. To avoid this issue, the analyst removes experimental data containing outliers based on experience and engineering judgment in the practice of conventional nuclear data adjustment.

For example, in the previous study by Takeda et al. [1], removed data were determined from the difference between measured nuclear characteristic E and calculated nuclear characteristic C (C/E value). Specifically, the authors considered three cases in which nuclear characteristics were removed where the difference exceeded one, two, or three times the total uncertainty (experimental error + analytical uncertainty + nuclear data-induced uncertainty).

As another example, in the adjusted nuclear data ADJ2017 [2], experimental data were removed based on the following points:

(i) The data where the difference between measured E and calculated values C exceeds twice the total uncertainty (2σ);
(ii) The data where experimental analysis is considered inappropriate;
(iii) The data where the amount of adjustment of nuclear data exceeds its uncertainty (1σ).

In particular, the point (iii) poses an inverse problem that searches for the causal experimental data from the adjusted nuclear data, thus it is not easy to specify the experimental data to be removed. In order to accurately specify the group of experimental data that is the cause of the problem, the number of analyses would be very large because inclusion or exclusion of each experimental data will be necessary. Therefore, based on the experience and engineering judgment of the analyst, inappropriate experimental data groups are excluded.

To address the above issues, this study aims to apply a robust estimation method to the nuclear data adjustment. The least median of squares method, the random sample consensus, and the M-estimation are examples of robust estimation methods. In this study, we focus on the M-estimation due to its applicability to Bayesian inference. By introducing the M-estimation to the conventional nuclear data adjustment methods, we develop a new nuclear data adjustment method that is robust to experimental data including outliers.

2 Methodology

* Corresponding author : y-fukui@fermi.energy.nagoya-u.ac.jp

2.1 GLLS

This section describes a conventional nuclear data adjustment method called Generalized Linear Least Square (GLLS), which uses sensitivity coefficients. This method uses the following assumptions:

- The perturbation of nuclear characteristics is linear with respect to the perturbation of nuclear data.
- The probability distribution of nuclear data and nuclear characteristics follows a multivariate normal distribution.

From Bayes' theorem shown in equation (1), the nuclear data adjustment equation is derived [3].

$$P(\vec{T}|\vec{E}) = \frac{P(\vec{E}|\vec{T})P(\vec{T})}{P(\vec{E})}, \tag{1}$$

where $\vec{T}$ is the nuclear data; $\vec{E}$ is the nuclear characteristic measurement value; $P(\vec{T})$ is the probability distribution (prior probability) of the nuclear data $\vec{T}$; $P(\vec{E})$ is the probability distribution of the nuclear characteristic measurement value $\vec{E}$; $P(\vec{T}|\vec{E})$ is the probability distribution (posterior probability) of the nuclear data when the nuclear characteristic is $\vec{E}$; $P(\vec{E}|\vec{T})$ is the probability distribution (likelihood) of the nuclear characteristic measurement value when the nuclear data is $\vec{T}$. With the assumption of a multivariate normal distribution for the nuclear data and nuclear characteristics, equations (2)–(4) are derived for $P(\vec{T})$, $P(\vec{E})$, and $P(\vec{E}|\vec{T})$, respectively:

$$P(\vec{T}) \propto \exp\left[-\frac{1}{2}(\vec{T}-\vec{T}^{(0)})^{\mathrm{T}}\mathbf{M}_T^{(0)-1}(\vec{T}-\vec{T}^{(0)})\right], \tag{2}$$

$$P(\vec{E}) \propto \exp\left[-\frac{1}{2}(\vec{E}-\vec{R})^{\mathrm{T}}\mathbf{M}_E^{-1}(\vec{E}-\vec{R})\right], \tag{3}$$

$$P(\vec{E}|\vec{T}) \propto \exp\left[-\frac{1}{2}\left(\vec{E}-\vec{C}(\vec{T})\right)^{\mathrm{T}}\mathbf{M}_{E+C}^{-1}\left(\vec{E}-\vec{C}(\vec{T})\right)\right], \tag{4}$$

where $\vec{T}^{(0)} \in \mathbb{R}^{N_T}$ is the prior mean of the nuclear data; $\mathbf{M}_T^{(0)} \in \mathbb{R}^{N_T \times N_T}$ is the prior covariance matrix of the nuclear data; N_T is the total number of nuclear data; $\vec{R} \in \mathbb{R}^{N_R}$ is the true value of nuclear characteristics; $\mathbf{M}_E \in \mathbb{R}^{N_R \times N_R}$ is the measured covariance matrix; $\vec{C}(\vec{T}) \in \mathbb{R}^{N_R}$ is the calculation value of nuclear characteristics for nuclear data $\vec{T}$; $\mathbf{M}_{E+C} \in \mathbb{R}^{N_R \times N_R}$ is the covariance matrix of nuclear characteristic measurement and calculation; N_R is the total number of nuclear characteristics. The posterior probability $P(\vec{T}|\vec{E})$ is obtained by substituting equations (2)–(4) into equation (1):

$$\begin{aligned} P(\vec{T}|\vec{E}) \propto \\ \exp\Big[-\frac{1}{2}\Big\{&(\vec{T}-\vec{T}^{(0)})^{\mathrm{T}}\mathbf{M}_T^{(0)-1}(\vec{T}-\vec{T}^{(0)}) \\ &+\left(\vec{E}-\vec{C}(\vec{T})\right)^{\mathrm{T}}\mathbf{M}_{E+C}^{-1}\left(\vec{E}-\vec{C}(\vec{T})\right) \\ &-(\vec{E}-\vec{R})^{\mathrm{T}}\mathbf{M}_E^{-1}(\vec{E}-\vec{R})\Big\}\Big] \end{aligned} \tag{5}$$

$$\begin{aligned} &= \exp\Big[-\frac{1}{2}\{J(\vec{T}) \\ &-(\vec{E}-\vec{R})^{\mathrm{T}}\mathbf{M}_E^{-1}(\vec{E}-\vec{R})\Big], \end{aligned} \tag{6}$$

$$\begin{aligned} J(\vec{T}) = &(\vec{T}-\vec{T}^{(0)})^{\mathrm{T}}\mathbf{M}_T^{(0)-1}(\vec{T}-\vec{T}^{(0)}) \\ &+\left(\vec{E}-\vec{C}(\vec{T})\right)^{\mathrm{T}}\mathbf{M}_{E+C}^{-1}\left(\vec{E}-\vec{C}(\vec{T})\right), \end{aligned} \tag{7}$$

where part of the posterior probability with regard to the nuclear data $\vec{T}$ is defined as $J(\vec{T})$.

Because of the assumption of linearity between nuclear data and nuclear characteristics, the calculated nuclear characteristic value $\vec{C}(\vec{T})$ for the nuclear data $\vec{T}$ can be calculated using the sensitivity coefficient matrix $\mathbf{G} \in \mathbb{R}^{N_R \times N_T}$ as follows:

$$\vec{C}(\vec{T}) \approx \vec{C}(\vec{T}^{(0)}) + \mathbf{G}(\vec{T}-\vec{T}^{(0)}). \tag{8}$$

The nuclear data $\vec{T}$ that maximizes the posterior probability $P(\vec{T}|\vec{E})$ is the "adjusted nuclear data $\vec{T}^{(\mathrm{adj})}$" estimated by Bayes' theorem. To obtain the nuclear data $\vec{T}$ that maximizes the posterior probability, we consider the derivative of the function part $J(\vec{T})$ regard to $\vec{T}$ in the posterior probability $P(\vec{T}|\vec{E})$:

$$\frac{\partial J(\vec{T})}{\partial \vec{T}} = 0. \tag{9}$$

The nuclear data $\vec{T}$ satisfying equation (9) is the adjusted nuclear data $\vec{T}^{(\mathrm{adj})}$. In addition, by calculating the variance of $\vec{T}^{(\mathrm{adj})} - \vec{T}^{(0)}$, the covariance matrix $\mathbf{M}_T^{(\mathrm{adj})}$ of adjusted nuclear data can be obtained. From the above, the nuclear data adjustment formulae are obtained as follows:

$$\vec{T}^{(\mathrm{adj})} = \vec{T}^{(0)} + \mathbf{K}(\vec{E} - \vec{C}(\vec{T}^{(0)})), \tag{10}$$

$$\mathbf{M}_T^{(\mathrm{adj})} = \mathbf{M}_T - \mathbf{KGM}_T^{(0)}, \tag{11}$$

$$\mathbf{K} = \mathbf{M}_T^{(0)}\mathbf{G}^{\mathrm{T}}\left(\mathbf{GM}_T^{(0)}\mathbf{G}^{\mathrm{T}} + \mathbf{M}_{E+C}\right)^{-1}. \tag{12}$$

2.2 M-estimation

M-estimation [4] is one of the estimation methods that are robust to outliers. In M-estimation, the regression model is updated by setting the weight $w(d)$ of the observed values from the residual d between the observed values and the regression model. The initial regression model (before updating the weight) is determined in advance using the least squares method or other methods. In this method, the influence of outliers on the estimation results can be reduced because the weighting $w(d)$ on the values will be small when the residuals d are large. The function that calculates the weight w from the residual d is called the loss function. An example of the loss function is shown in equation (13).

$$w(d) = \begin{cases} \left(1 - \left(\frac{d}{D}\right)^2\right)^2 & |d| \le D \\ 0 & |d| > D, \end{cases} \tag{13}$$

where D is the tolerable error. The estimation using this loss function is called Tukey's Biweight method. [4]

In this section, we consider the regression problem for a linear function as an example of M-estimation. Let us consider a problem estimating the parameters a and b in a regression model $y^{(\mathrm{model})} = ax + b$ using pairs of given input and observed output data $\{x_n, y_n^{(\mathrm{obs})}\}$ $(1 \le n \le N)$.

The ordinary least squares method sets up the simultaneous equations shown in equation (14) and calculates the parameters that minimize the sum of squares of the residuals between the model and the observation:

$$\begin{cases} y_1^{(\mathrm{model})} = ax_1 + b \\ y_2^{(\mathrm{model})} = ax_2 + b \\ \vdots \\ y_N^{(\mathrm{model})} = ax_N + b, \end{cases} \tag{14}$$

$$J = \sum_{n=1}^{N} \left(y_n^{(\mathrm{model})} - y_n^{(\mathrm{obs})}\right)^2. \tag{15}$$

We describe equation (14) in matrix form. If the observed data $\vec{y}^{(\mathrm{obs})}$, the regression parameter $\vec{p} = [a \quad b]^{\mathrm{T}}$, and the input $\mathbf{X}$ are used, the following can be obtained for the ordinary least squares method using the pseudo-inverse matrix:

$$\vec{y}^{(\mathrm{obs})} \simeq \vec{y}^{(\mathrm{model})} = \mathbf{X}\vec{p}, \tag{16}$$

$$\begin{bmatrix} y_1^{(\mathrm{obs})} \\ y_2^{(\mathrm{obs})} \\ \vdots \\ y_N^{(\mathrm{obs})} \end{bmatrix} \simeq \begin{bmatrix} y_1^{(\mathrm{model})} \\ y_2^{(\mathrm{model})} \\ \vdots \\ y_N^{(\mathrm{model})} \end{bmatrix} = \begin{bmatrix} x_1 & 1 \\ x_2 & 1 \\ \vdots & \vdots \\ x_N & 1 \end{bmatrix} \begin{bmatrix} a \\ b \end{bmatrix}, \tag{17}$$

$$\begin{bmatrix} a \\ b \end{bmatrix} \simeq \begin{bmatrix} x_1 & 1 \\ x_2 & 1 \\ \vdots & \vdots \\ x_N & 1 \end{bmatrix}^{\dagger} \begin{bmatrix} y_1^{(\mathrm{obs})} \\ y_2^{(\mathrm{obs})} \\ \vdots \\ y_N^{(\mathrm{obs})} \end{bmatrix}, \tag{18}$$

$$\vec{p} \simeq \mathbf{X}^{\dagger}\vec{y}^{(\mathrm{obs})}, \tag{19}$$

where $\mathbf{X}^{\dagger}$ is the pseudo-inverse of the matrix $\mathbf{X}$.

On the other hand, in M-estimation, we set up a simultaneous equation obtained by multiplying the weights w. The function J to be minimized in this case is shown in equations (20) and (21). The weight w_n of the n th data $y_n^{(\mathrm{obs})}$ is calculated by the residual between the calculated value and the observed value by the temporarily set regression model:

$$\begin{cases} w_1 y_1 = w_1(ax_1 + b) \\ w_2 y_2 = w_2(ax_2 + b) \\ \vdots \\ w_N y_N = w_N(ax_N + b), \end{cases} \tag{20}$$

$$J = \sum_{n=1}^{N} {w_n}^2 \left(y_n^{(\mathrm{model})} - y_n^{(\mathrm{obs})}\right)^2. \tag{21}$$

Here, we describe equation (20) in matrix form using the weight matrix $\mathbf{W}$. The weight matrix $\mathbf{W}$ is a matrix that has the weight w_n in the diagonal:

$$\mathbf{W}\vec{y}^{(\mathrm{obs})} \simeq \mathbf{W}\vec{y}^{(\mathrm{model})} = \mathbf{W}\mathbf{X}\vec{p}, \tag{22}$$

$$\begin{bmatrix} w_1 & 0 & \cdots & 0 \\ 0 & w_2 & \cdots & 0 \\ \vdots & \vdots & \ddots & \vdots \\ 0 & 0 & \cdots & w_N \end{bmatrix} \begin{bmatrix} y_1^{(\mathrm{obs})} \\ y_2^{(\mathrm{obs})} \\ \vdots \\ y_N^{(\mathrm{obs})} \end{bmatrix} \simeq \begin{bmatrix} w_1 & 0 & \cdots & 0 \\ 0 & w_2 & \cdots & 0 \\ \vdots & \vdots & \ddots & \vdots \\ 0 & 0 & \cdots & w_N \end{bmatrix} \begin{bmatrix} x_1 & 1 \\ x_2 & 1 \\ \vdots & \vdots \\ x_N & 1 \end{bmatrix} \begin{bmatrix} a \\ b \end{bmatrix}, \tag{23}$$

$$\begin{bmatrix} a \\ b \end{bmatrix} \simeq \left(\begin{bmatrix} w_1 & 0 & \cdots & 0 \\ 0 & w_2 & \cdots & 0 \\ \vdots & \vdots & \ddots & \vdots \\ 0 & 0 & \cdots & w_N \end{bmatrix} \begin{bmatrix} x_1 & 1 \\ x_2 & 1 \\ \vdots & \vdots \\ x_N & 1 \end{bmatrix} \right)^{\dagger} \begin{bmatrix} w_1 & 0 & \cdots & 0 \\ 0 & w_2 & \cdots & 0 \\ \vdots & \vdots & \ddots & \vdots \\ 0 & 0 & \cdots & w_N \end{bmatrix} \begin{bmatrix} y_1^{(\mathrm{obs})} \\ y_2^{(\mathrm{obs})} \\ \vdots \\ y_N^{(\mathrm{obs})} \end{bmatrix}, \tag{24}$$

$$\vec{p} \simeq (\mathbf{W}\mathbf{X})^{\dagger}\,\mathbf{W}\vec{y}^{(\mathrm{obs})}. \tag{25}$$

The procedure for applying M-estimation to the problem of regression to a linear function is summarized below.

(i) Initial parameters (a, b) are tentatively determined using the least squares method, etc.
(ii) The weights are calculated from the regression model with the initial parameters and the observed values.
(iii) The regression parameters are updated using the weights.

The procedure for applying M-estimation to nuclear data adjustment is as follows.

(i) The initial parameters are the nuclear data mean (evaluated nuclear data).
(ii) The weights are obtained from the calculated and measured nuclear characteristics for the nuclear data mean.
(iii) The nuclear data are adjusted using weights.

The next section describes the application of M-estimation to nuclear data adjustment in more detail.

2.3 MGLLS

2.3.1 Fundamental concept

We consider a nuclear data adjustment method that introduces M-estimation. The basic concept of the proposed method is the adjustment of nuclear data with a focus on nuclear characteristics where the difference between calculated and measured values is within an acceptable uncertainty range. The key points to be developed are described below.

(i) We develop a method to calculate the weighting from the difference between calculated and measured values, taking into account correlations between nuclear characteristics. Typical M-estimation cannot handle correlations between nuclear characteristics because the weights are calculated based on the assumption that the observed data are independent. To solve this problem, we propose a weights calculation method using the singular value decomposition. From the singular value decomposition of the total nuclear characteristic covariance matrix (i.e., $\mathbf{M}_R = \mathbf{G}^T\mathbf{M}_T^{(0)}\mathbf{G} + \mathbf{M}_E + \mathbf{M}_C = \mathbf{U\Sigma U}^T$), base matrix $\mathbf{U} = [\vec{u}_1 \ \vec{u}_2 \ \cdots \ \vec{u}_N]$ characterizing the correlation direction is obtained. This operation allows nuclear characteristics distributions, including correlations, to be treated as independent distributions. The weights can be calculated by evaluating the difference between the calculated and measured values in each base direction. As an example, in the case of the two nuclear characteristics shown in Fig.1, the weighting is larger for the direction of $\vec{u}_1$ and smaller for the direction of $\vec{u}_2$ since the measured value is out of the scatter for the $\vec{u}_2$ direction while that is within the scatter for the $\vec{u}_1$ direction.

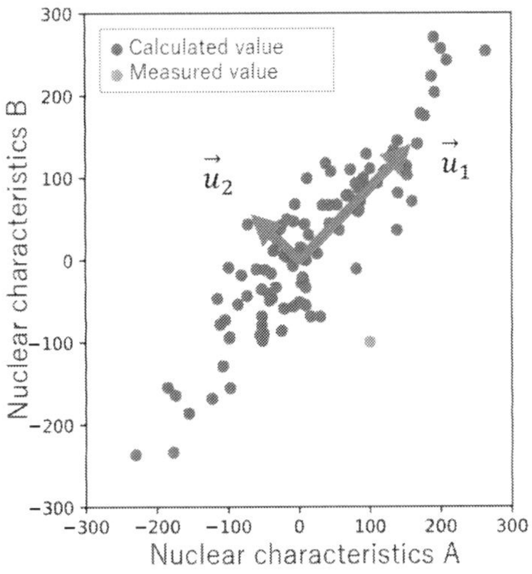

Fig.1 The concept of weights calculation

(ii) We develop a weighted nuclear data adjustment equation for each nuclear characteristics incorporating M-estimation.

2.3.2 Weights calculation

This section describes the setting of the weight matrix $\mathbf{W}$, in which the correlations between the nuclear characteristics are taken into account. In this method, the direction (i.e., base) of the correlation is determined from the singular value decomposition of the covariance matrix of the nuclear characteristics, and the weight matrix $\mathbf{W}^{(\mathbf{U})}$ is calculated based on the loss function in the obtained base direction appeared in the base matrix $\mathbf{U}$. The detailed calculation procedure is shown below.

(1) The measured nuclear characteristics $\vec{E}$, calculated nuclear characteristics $\vec{C}(\vec{T}^{(0)})$, and their covariance matrices ($\mathbf{M}_E$ and $\mathbf{M}_C$) are obtained.

(2) Obtain the total covariance matrix $\mathbf{M}_R$ of the total uncertainty from the sum of the covariance matrices:

$$\mathbf{M}_R = \mathbf{G}\mathbf{M}_T^{(0)}\mathbf{G}^T + \mathbf{M}_E + \mathbf{M}_C. \tag{26}$$

(3) We obtain the base matrix $\mathbf{U}$ of the nuclear characteristic correlations by the singular value decomposition. The covariance matrix $\mathbf{M}_R$ is decomposed into the orthonormal $\mathbf{U}$ and the diagonal matrix $\mathbf{\Sigma}$:

$$\mathbf{M}_R = \mathbf{U\Sigma U}^T, \tag{27}$$

$$\mathbf{U} = [\vec{u}_1 \ \ \vec{u}_2 \ \ \cdots \ \ \vec{u}_N], \tag{28}$$

where the column elements $\vec{u}_1 \ \vec{u}_2 \ \cdots \ \vec{u}_N$ of the base matrix $\mathbf{U}$ are the base vectors.

(4) The measured and calculated values ($\vec{E}$ and $\vec{C}(\vec{T}^{(0)})$) and the covariance matrix $\mathbf{M}_R$ are transformed (or projected) into those in the space of the base matrix $\mathbf{U}$ as expressed by $\vec{E}^{(\mathbf{U})}$, $\vec{C}^{(\mathbf{U})}$, and $\mathbf{M}_R^{(\mathbf{U})}$, respectively:

$$\vec{E}^{(\mathbf{U})} = \mathbf{U}^T\vec{E}, \tag{29}$$

$$\vec{C}^{(\mathbf{U})} = \mathbf{U}^T\vec{C}(\vec{T}^{(0)}), \tag{30}$$

$$\mathbf{M}_R^{(\mathbf{U})} = \mathbf{U}^T\mathbf{M}_R\mathbf{U}(= \mathbf{\Sigma}). \tag{31}$$

(5) Calculate the weight matrix $\mathbf{W}^{(\mathbf{U})}$ (diagonal matrix) by evaluating the difference between the transformed measured value $\vec{E}^{(\mathbf{U})}$ and the transformed calculated value $\vec{C}^{(\mathbf{U})}$ using the loss function w:

$$\mathbf{W}^{(\mathbf{U})} = \begin{bmatrix} w_1 & & & \mathbf{0} \\ & w_2 & & \\ & & \ddots & \\ \mathbf{0} & & & w_N \end{bmatrix}, \tag{32}$$

$$w_n = w\left(\vec{C}^{(\mathbf{U})}[n], \vec{E}^{(\mathbf{U})}[n], \mathbf{M}_R{}^{(\mathbf{U})}[n,n], \gamma\right), \tag{33}$$

where $[n]$ and $[n,n]$ represent the n th and the (n,n) th elements of vector and matrix, respectively. If the Gaussian function is used as the loss function, then

$$w_n = \exp\left(-\frac{\gamma}{2}\frac{\left(\vec{C}^{(\mathbf{U})}[n] - \vec{E}^{(\mathbf{U})}[n]\right)^2}{\mathbf{M}_R^{(\mathbf{U})}[n,n]}\right), \tag{34}$$

where γ is an arbitrary tuning parameter. In this paper, we set $\gamma = 1.0$.

(6) Obtain the inverse weight matrix $\mathbf{W}^{-1}$ by operating the base matrix $\mathbf{U}$ on the inverse weight matrix $\mathbf{W}^{(\mathbf{U})-1}$:

$$\mathbf{W}^{-1} = \mathbf{U}\mathbf{W}^{(\mathbf{U})-1}\mathbf{U}^{\mathrm{T}}. \quad (35)$$

When calculating the inverse of the weight matrix $\mathbf{W}^{(\mathbf{U})-1}$, the diagonal components of $\mathbf{W}^{(\mathbf{U})}$ should not contain small values very close to zero. If the diagonal components of $\mathbf{W}^{(\mathbf{U})}$ contain very small values, $\mathbf{W}$ will be a singular matrix and its inverse $\mathbf{W}^{-1}$ may not be properly calculated.

2.3.3 Weighted nuclear data adjustment

Like M estimation, the nuclear data adjustment formula is derived by assuming a weighted model equation. As an alternative of the function $J(\vec{T})$, which is the minimization target in the conventional nuclear data adjustment, let us consider the function $J_M(\vec{T})$, in which nuclear characteristics $\vec{E}$ and $\vec{C}(\vec{T})$ are weighted:

$$\begin{aligned} J_M(\vec{T}) =& \\ &(\vec{T}-\vec{T}^{(0)})^{\mathrm{T}}\mathbf{M}_T^{(0)-1}(\vec{T}-\vec{T}^{(0)}) \\ &+\left(\mathbf{W}\left(\vec{E}-\vec{C}(\vec{T})\right)\right)^{\mathrm{T}}\mathbf{M}_{E+C}^{-1}\mathbf{W}\left(\vec{E}-\vec{C}(\vec{T})\right) \\ &= (\vec{T}-\vec{T}^{(0)})^{\mathrm{T}}\mathbf{M}_T^{(0)-1} \\ &+\left(\vec{E}-\vec{C}(\vec{T})\right)^{\mathrm{T}}\mathbf{W}^{\mathrm{T}}\mathbf{M}_{E+C}^{-1}\mathbf{W}\left(\vec{E}-\vec{C}(\vec{T})\right) \\ &= (\vec{T}-\vec{T}^{(0)})^{\mathrm{T}}\mathbf{M}_T^{(0)^{-1}}(\vec{T}-\vec{T}^{(0)}) \\ &+\left(\vec{E}-\vec{C}(\vec{T})\right)^{\mathrm{T}}\mathbf{W}\mathbf{M}_{E+C}^{-1}\mathbf{W}\left(\vec{E}-\vec{C}(\vec{T})\right) \end{aligned} \quad (36)$$

Here, we find that $J_M(\vec{T})$ is obtained by replacing the nuclear characteristic uncertainty $\mathbf{M}_{E+C}^{-1}$ in the conventional nuclear data adjustment formula to $\mathbf{W}\mathbf{M}_{E+C}^{-1}\mathbf{W}$.

From this result, the conventional nuclear data adjustment formula (10)–(12) is rewritten as:

$$\vec{T}^{(\mathrm{adj})} = \vec{T}^{(0)} + \mathbf{K}\cdot(\vec{E}-\vec{C}(\vec{T}^{(0)})), \quad (37)$$

$$\mathbf{M}_T^{(\mathrm{adj})} = \mathbf{M}_T - \mathbf{K}\mathbf{G}\mathbf{M}_T^{(0)}, \quad (38)$$

$$\mathbf{K} = \mathbf{M}_T^{(0)}\mathbf{G}^{\mathrm{T}}\left(\mathbf{G}\mathbf{M}_T^{(0)}\mathbf{G}^{\mathrm{T}} + \mathbf{W}^{-1}\mathbf{M}_{E+C}\mathbf{W}^{-1}\right)^{-1}. \quad (39)$$

In this paper, we call this method MGLLS (weighted nuclear data adjustment with M estimation incorporated into GLLS).

3 Calculations

3.1 Verification conditions

We verify the proposed method using a twin experiment, which is one of the verification methods for data assimilation. The procedure for verification in this study is described below.

(1) Randomly set the reference nuclear data $\vec{T}^{(\mathrm{ref})}$ based on the prior mean $\vec{T}^{(0)}$ and the prior covariance matrix $\mathbf{M}_T^{(0)}$ .We prepared nine different $\vec{T}^{(\mathrm{ref})}$ (#1–#9) to verify for various cases.

(2) Numerically calculate the hypothetical measured nuclear characteristics $\vec{E}$ using the nuclear data is $\vec{T}^{(\mathrm{ref})}$.

(3) Intentionally change a part of the measured nuclear characteristics $\vec{E}$ to virtually generate an abnormal value (i.e., outlier).

(4) Adjust the nuclear data $\vec{T}^{(0)}$ using the measured nuclear characteristics $\vec{E}$, and obtain adjusted nuclear data $\vec{T}^{(\mathrm{adj})}$ and its covariance $\mathbf{M}_T^{(\mathrm{adj})}$.

(5) Compare $\vec{T}^{(\mathrm{adj})}$ with the reference nuclear data $\vec{T}^{(\mathrm{ref})}$. Here, $\mathbf{M}_T^{(\mathrm{adj})}$ is considered as the uncertainty range of $\vec{T}^{(\mathrm{adj})}$.

The verification conditions are summarized in Table 1. We analyzed four nuclear characteristics for each of the three experiments. Therefore, the total number of nuclear characteristics is twelve.

Table 1 Calculation conditions

Experiments	Pu-Met-Fast-001 (Jezebel) [5] Pu-Met-Fast-006 (Flattop) [5] Pu-Sol-Therm-001 [5]
Adjusted nuclear data	^{239}Pu (n,γ) cross-section
Nuclear characteristics	k_{eff} C41/F41 C49/F41 F49/F41
Outlier	C49/F41 in Pu-Met-Fast-001
Nuclear characteristics analysis	MCNP6.2 [6]
Sensitivity analysis	SCALE6.2/TSUNAMI-1D [7]
Nuclear data perturbation	FRENDY [8, 9]*
Evaluated nuclear data	ENDF/B-VII.1 [10]
Covariance data	Scale.rev08.56groupcov7.1 [7]*
Energy groups	56*
Tuning parameter of loss function γ	1.0

*Using the perturbation tool of ACE file included with FRENDY, we perturbed and adjusted the continuous energy cross-section with 56 energy group structure. Though the perturbation and adjustment are carried out in multi-group structure, the point-wise cross-sections are used for nuclear characteristics analysis by MCNP6.2).

To verify the robustness to outliers, we intentionally replaced measured nuclear characteristics $\vec{E}$ to an abnormal value. In this study, abnormal experimental data were obtained by changing the sign of the relative difference $\Delta = (E - C)/C$ for one of the nuclear characteristics, here C is a calculated nuclear characteristic for nuclear data $\vec{T}^{(0)}$. In other words, we reversed the positives and negatives of the relative difference. As an example, let us consider the abnormal measured value E' against the measured value E and the calculated value C. Because of the positive and negative reversal of its relative perturbations, we can calculate the following equation for the abnormal measured value E',

$$\frac{E' - C}{C} = -\frac{E - C}{C}, \tag{40}$$

$$E' = 2C - E. \tag{41}$$

In this study, we chose C49/F41 at PMF001 as an outlier because this nuclear characteristic is highly sensitive to the ^{239}Pu (n, γ) cross-section.

3.2 Result of adjustment

For experimental data containing outliers, the nuclear data were adjusted using the conventional method (GLLS) and the proposed method (MGLLS). We compare the adjusted nuclear data $\vec{T}^{(\mathrm{adj})}$ and its uncertainty after adjustment (i.e., square root of diagonal elements of $\mathbf{M}_T^{(\mathrm{adj})}$) to reference values $\vec{T}^{(\mathrm{ref})}$. The cases with different $\vec{T}^{(\mathrm{ref})}$s (#1–#9) are shown in Fig. 2–Fig. 10, respectively. In these figures, the red single dot chain line means the reference nuclear data, i.e., virtual true nuclear data in the twin experiments. When the reference values (red single dot chain line) are within the range of estimation uncertainty (light blue dashed lines), the adjustment can be considered appropriate.

In cases #1,#3–#9, the difference between the reference and adjusted values of nuclear data in GLLS is much larger than its standard deviation. In contrast, in the MGLLS, the difference between the reference and adjusted values of the nuclear data is more consistent with the range of uncertainty (i.e., the standard deviation σ).

The result of case #2 shows no significant difference between GLLS and MGLLS. This is because, in case #2, the perturbations of PMF-001-C49/F41 to outliers are small. In other words, since the sign of a small perturbation amount is reversed to generate the outlier of case #2, there is a small difference between the original and the perturbed (including outliers) nuclear characteristics.

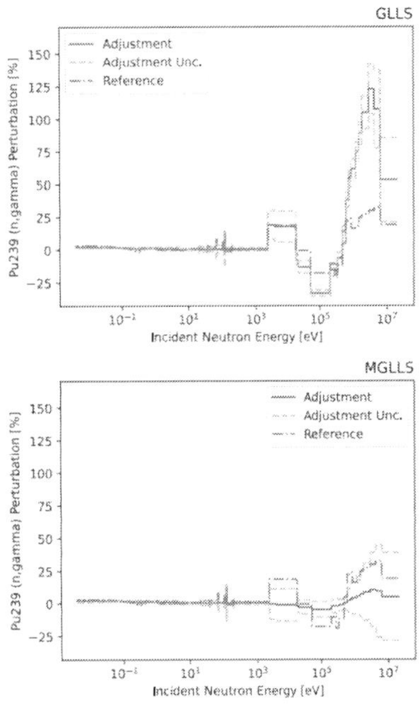

Fig. 2 Comparison of adjusted nuclear data #1

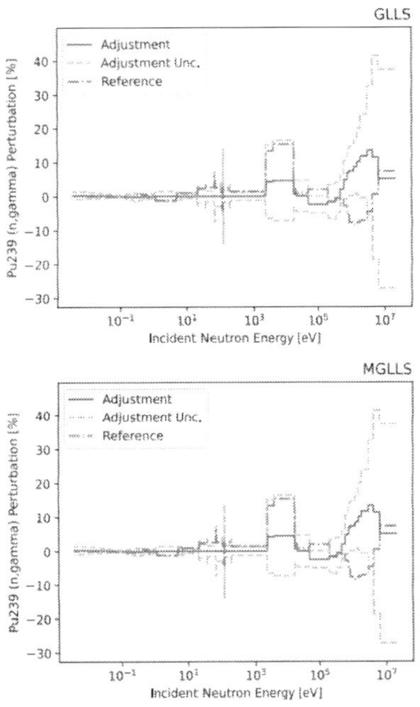

Fig. 3 Comparison of adjusted nuclear data #2

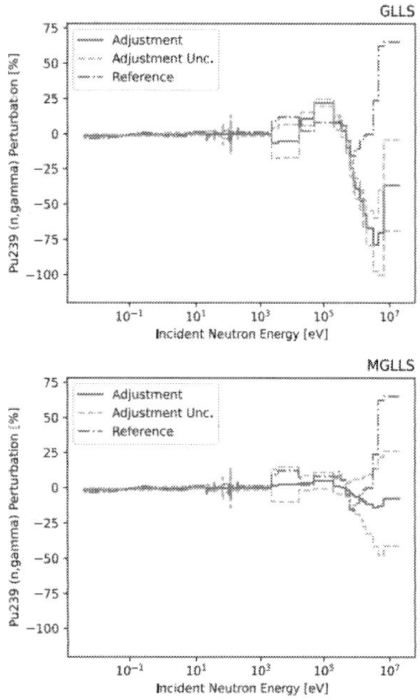

Fig. 4 Comparison of adjusted nuclear data #3

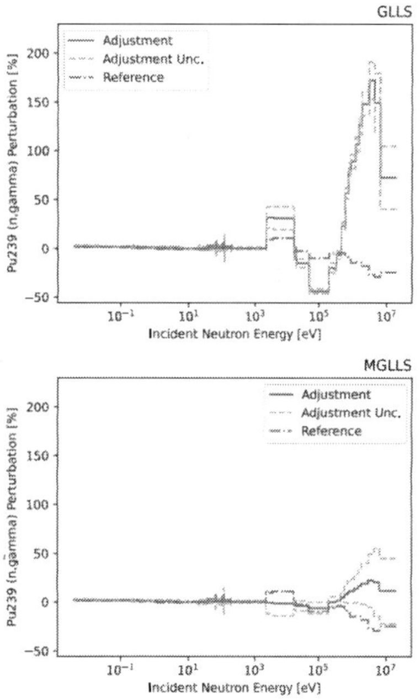

Fig. 5 Comparison of adjusted nuclear data #4

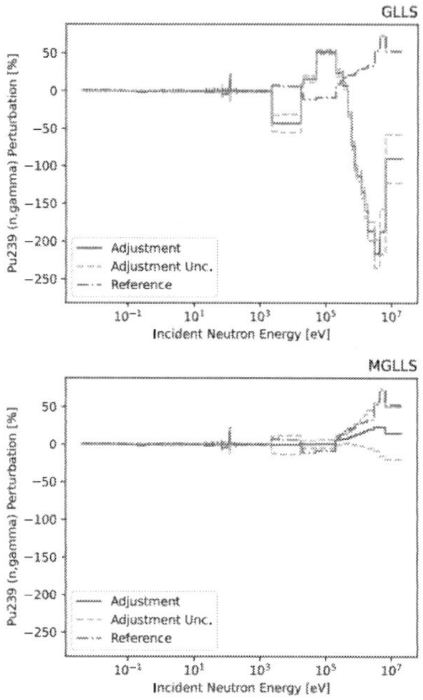

Fig. 6 Comparison of adjusted nuclear data #5

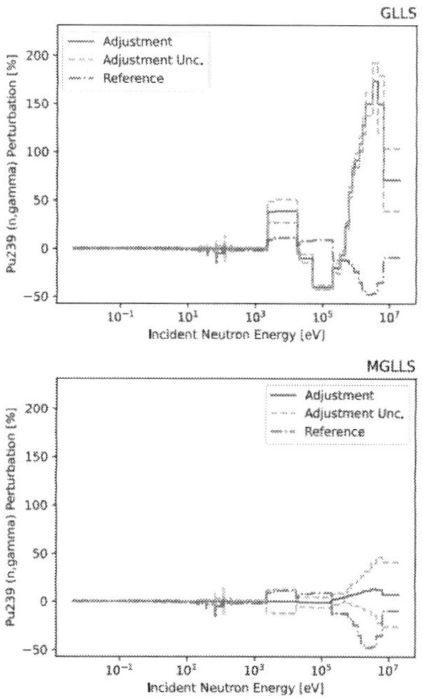

Fig. 7 Comparison of adjusted nuclear data #6

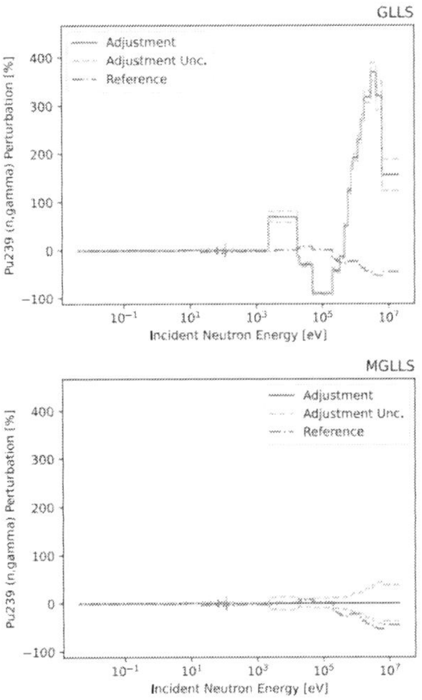

Fig. 8 Comparison of adjusted nuclear data #7

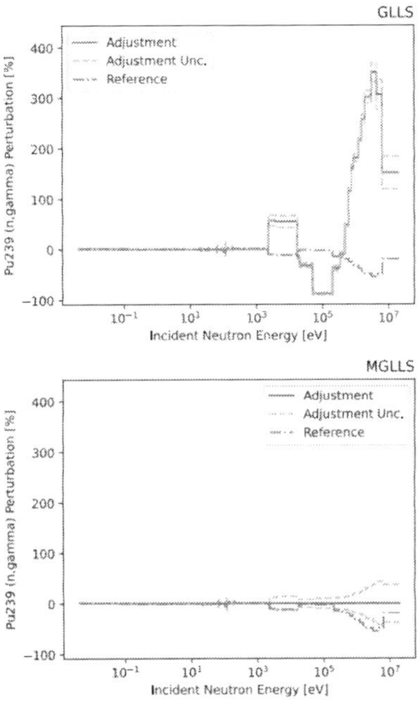

Fig. 9 Comparison of adjusted nuclear data #8

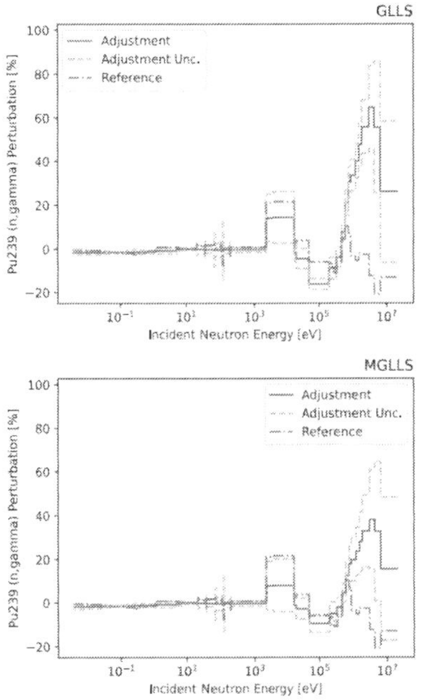

Fig. 10 Comparison of adjusted nuclear data #9

3.3 Quantitative performance evaluation of the adjustment methods

The adequacy of the adjusted nuclear data is quantitatively evaluated using the χ^2 value as an index. If the difference between the adjusted value and the reference value is within the uncertainty range of the adjustment, the order of χ^2 is comparable to that of the degrees of freedom, and the adequacy of the adjustment (or estimation) can be evaluated. The calculation method of the χ^2 value is as follows:

$$\chi^2 = \left[\vec{T}^{(\mathrm{adj})} - \vec{T}^{(\mathrm{ref})}\right]^{\mathrm{T}} \mathbf{M}_T^{(\mathrm{adj})-1} \left[\vec{T}^{(\mathrm{adj})} - \vec{T}^{(\mathrm{ref})}\right]. \tag{42}$$

If $\mathbf{M}_T^{(\mathrm{adj})}$ is a singular matrix, its inverse matrix cannot be properly calculated. Therefore, a low-rank approximation using singular value decomposition for $\mathbf{M}_T^{(\mathrm{adj})}$ is used:

$$\mathbf{M}_T^{(\mathrm{adj})} = \mathbf{U\Sigma U}^{\mathrm{T}} \approx \mathbf{U}_r \mathbf{\Sigma}_r \mathbf{U}_r^{\mathrm{T}}, \tag{43}$$

$$\mathbf{U}_r = \mathbf{U}[:,:r] = [\vec{u}_1 \quad \vec{u}_2 \quad \cdots \quad \vec{u}_r], \tag{44}$$

$$\mathbf{\Sigma}_r = \mathbf{\Sigma}[:r,:r] = \begin{bmatrix} \sigma_1 & & & \mathbf{O} \\ & \sigma_2 & & \\ & & \ddots & \\ \mathbf{O} & & & \sigma_r \end{bmatrix}, \tag{45}$$

where r is the order after the application of the low-rank approximation. And where $\mathbf{U}[:,:r]$ is slicing to extract the first to r-th column element of matrix $\mathbf{U}$, and $\mathbf{\Sigma}[:r,:r]$ is the slicing to extract the first to r-th square matrix element of matrix $\mathbf{\Sigma}$. Using the above singular value decomposition for $\mathbf{M}_T^{(\mathrm{adj})}$, we approximate the χ^2 values using the following equation:

$$\chi^2 = \left[\vec{T}^{(\mathrm{adj})} - \vec{T}^{(\mathrm{ref})}\right]^{\mathrm{T}} \mathbf{M}_T^{(\mathrm{adj})-1}\left[\vec{T}^{(\mathrm{adj})} - \vec{T}^{(\mathrm{ref})}\right], \quad (46)$$

$$\approx \left[\vec{T}^{(\mathrm{adj})} - \vec{T}^{(\mathrm{ref})}\right]^{\mathrm{T}} \mathbf{U}_r \boldsymbol{\Sigma}_r^{-1} \mathbf{U}_r^{\mathrm{T}}\left[\vec{T}^{(\mathrm{adj})} - \vec{T}^{(\mathrm{ref})}\right]. \quad (47)$$

For both adjustment methods, the evaluated χ^2 values are shown in Fig. 11. The blue circled and the orange x-shaped markers indicate χ^2 values for the conventional method (GLLS) and the proposed method (MGLLS), respectively. When the adjustment is appropriate, χ^2 takes the comparable value to the degree of freedom. Because the ^{239}Pu (n,γ) cross-section is adjusted by 56 group energy bins, the number of degrees of freedom is less or equal to 56 in the present calculation condition. In Fig. 11, the degrees of freedom are indicated by a black dashed line.

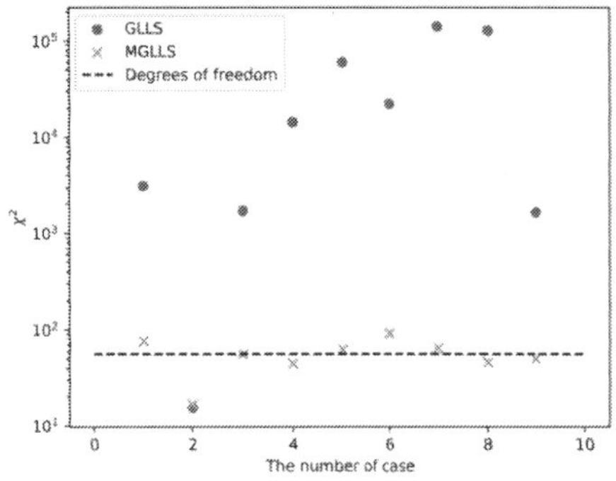

Fig. 11 Evaluated χ^2 value in each case

In cases #1, #3–#9, the χ^2 values in the GLLS case are extremely large, ranging from 1000 to 100000. On the other hand, the χ^2 values are comparable to the degrees of freedom (number of nuclear data) in the MGLLS case. The result demonstrates that the estimation results have improved using MGLLS. The reason for almost the same results using GLLS and MGGLS in case #2 is, as mentioned earlier, the small amount of perturbation to generate the outlier.

4 Conclusions

By incorporating M-estimation into the conventional nuclear data adjustment method, we proposed a robust nuclear data adjustment method for outliers, i.e., MGLLS. In this method, the "adequacy" of each nuclear characteristic is determined from the difference between the measured nuclear characteristic obtained by critical experiments and the calculated nuclear characteristic using nuclear data. Based on the "adequacy," each nuclear characteristic is weighted and the nuclear data is adjusted. Correlations between nuclear characteristics are considered in evaluating this "adequacy."

The proposed method was verified through twin experiments on critical experiments including fast and thermal systems. We quantitatively confirmed that the proposed method can robustly and appropriately adjust nuclear data even for experimental data containing inappropriate values.

A future study is an extension of the present method to the Bayesian Monte Carlo (BMC) method [11]. The BMC method uses the nuclear characteristics covariance matrix in the form of an inverse matrix when adjusting nuclear data. Therefore, the weight matrix can be used directly instead of the inverse matrix, thus we can avoid the instability of the inverse matrix calculation of the weight matrix.

This work was supported by a Japan Society for the Promotion of Science (JSPS) Grant-in-Aid for Scientific Research (C) (21K04940).

References

1. T. Takeda, K. Yokoyama, K. Sugino, Annals of Nuclear Energy **109**, 698 (2017)
2. K. Yokoyama, K. Sugino, M. Ishikawa, et al., JAEA-Research 2018-011 (2019)
3. K. Yokoyama, M. Ishikawa, T. Kugo, Journal of Nuclear Science and Technology **49**, 1165–1174 (2012)
4. R. A. Maronna, R. D. Martin, V. J. Yohai, *Robust Statistics: Theory and Methods*, (John Wiley & Sons, New York, 2006)
5. International Criticality Safety Benchmark Evaluation Project Handbook, OECD (2020)
6. C. J. Werner, editor. MCNP user's manual: code version 6.2. LA-UR-17-29981.: Los Alamos National Laboratory (2017)
7. W. A. Wieselquist, R. A. Lefebvre, M.A. Jessee, editors. SCALE code system. ORNL/TM- 2005/39 ver. 6.2.4. Oak Ridge (TN): Oak Ridge National Laboratory (2020)
8. K. Tada, Y. Nagaya, S. Kunieda, et al., Journal of Nuclear Science and Technology **54**, 806–817 (2017)
9. K. Tada, R. Kondo, T. Endo, et al. Development of ACE file perturbation tool using FRENDY, Journal of Nuclear Science and Technology, DOI:10.1080/00223131.2022.2130463
10. M. Chadwick, M. Herman, P. Oblozinsky, et.al., Nuclear Data Sheets **112**, 12, 2887–2996 (2011)
11. A. J. Koning, Eur. Phys. J. A. **51**, 184 (2015)

Understanding the impact of nuclear-data covariances on various integral responses using adjustment

D. Neudecker[1,*], *M.J.* Grosskopf[1], *J.* Alwin[1], *T.* Cutler[1], *S.* Frankle[1], *N.* Gibson[1], *W.* Haeck[1], *M.W.* Herman[1], *J.* Hutchinson[1], *N.* Kleedtke[1], *I.J.* Michaud[1], *M.E.* Rising[1], *T.* Smith[1], *N.* Thompson[1], and *S.* Vander Wiel[1]

[1]Los Alamos National Laboratory, Los Alamos, NM, 87545, USA

Abstract. The EUCLID (Experiments Underpinned by Computational Learning for Improvements in Nuclear Data) project created a library of sensitivities for nine different integral responses with respect to nuclear data. These integral responses were obtained from measurements at LLNL (Lawrence Livermore National Laboratory) pulsed spheres, critical and sub-critical assemblies. At the same time, covariances for ENDF/B-VIII.0 were processed at LANL (Los Alamos National Laboratory). The combination of these data allow us to study the impact of nuclear-data covariances on various integral responses, either by forward-propagating covariances via sensitivities, or by using nuclear data, integral responses, and sensitivities for adjustment. Here, we will present: the impact of ^{1}H, ^{9}Be, ^{12}C, ^{27}Al, ^{56}Fe, ^{235,238}U, and ^{239,240}Pu ENDF/B-VIII.0 covariances on simulated bounds of the following integral responses: LLNL pulsed-spheres neutron-leakage spectra, the effective neutron multiplication factor, reaction rates, and reactivity coefficients of ICSBEP critical assemblies. Also, adjustment results with the same nuclear-data covariances and responses will be discussed.
LA-UR-22-31309

1 Introduction

Nuclear-data covariances capture the limited knowledge we have on nuclear data due to approximations made to derive nuclear models, and limited precision of differential experimental data. These covariances are a vital ingredient to quantify economical and safety bounds on various application simulations due to imprecise nuclear data. Integral responses, such as the effective neutron-multiplication factor, k_{eff}, describe such application simulations on a small scale and validate the nuclear data for the application space the responses represent.

There has been a strong push, recently highlighted in Ref. [1], by nuclear data producer and user communities to include more integral responses than k_{eff} into the validation and adjustment of nuclear data. The reason for that request is that including more responses in the validation enables studying how nuclear data would perform for a broader application space. Also, when one uses many integral responses for validation and adjustment, some compensating errors [2] between various nuclear-data observables can be uncovered. Compensating errors are often introduced in nuclear data by calibrating several nuclear-data observables at once by hand with respect to k_{eff} [3–5]. The effective neutron multiplication factor queries nuclear-data observables in the same way. More explicitly, the ratios of sensitivities of k_{eff} to two nuclear-data observables of the same isotope are very similar in the same energy range across different assemblies. For instance, the ratios of sensitivities to fission cross sections and average prompt fission neutron multiplicity in the same energy range are very similar from one assembly to another if they are of the same class. Therefore, mistakes in one nuclear-data observable can offset mistakes in other observables. Some integral responses beyond k_{eff} were, however, shown [6–12] to have different sensitivities to the same nuclear data compared to those of k_{eff}. Given the different usage of nuclear data to simulate various responses, there is a smaller nuclear-data phase space where errors in one observable can compensate for others, while still predicting the experimental values of all responses within their uncertainties.

The EUCLID (Experiments Underpinned by Computational Learning for Improvements in Nuclear Data) LDRD-DR (Laboratory Directed Research & Development-Directed Research) project [13] explored how nuclear data is used when simulating nine different integral responses, measured from LLNL pulsed spheres, critical or sub-critical assemblies, by calculating sensitivity profiles of these responses to nuclear data [6–12]. At the same time, ENDF/B-VIII.0 mean values and covariances [14] were processed. The combination of these data—sensitivities, mean values and covariances—allowed us to quantify the bounds on the simulated integral responses due to nuclear-data covariances. We also used these data to adjust ENDF/B-VIII.0 nuclear data with respect to these integral responses.

Section 2 summarizes briefly how bounds on integral responses due to nuclear data were quantified along with the adjustment methodology. Section 3 reviews the integral responses considered (LLNL pulsed-sphere neutron-leakage spectra [15], k_{eff}, fission reaction rates and re-

*e-mail: dneudecker@lanl.gov

activity coefficients of ICSBEP critical assemblies [16]), their sensitivities and what nuclear data were used for the study. The results of assessing nuclear-data bounds on integral experiments and adjustment are discussed in Section 4 along with mentioning necessary extensions of the work. A brief summary, conclusions and outlook are provided in Section 5.

2 Methods

2.1 Forward-propagation of nuclear-data covariances to integral bounds

Nuclear-data covariances were forward-propagated to bounds on integral quantities Δ_C^{ND} with the well-known sandwich formula [17]:

$$\Delta_C^{ND} = \sqrt{\sum_{i,j} \mathbf{S}_{C,i}\mathbf{Cov}_{\mathrm{i,j}}^{\mathrm{ND}}\mathbf{S}_{C,j}^T}, \quad (1)$$

where $\mathbf{S}_{C,i}$ is the vector of sensitivities with respect to an integral values C for nuclear-data observable i. It is a vector because sensitivities are given for each nuclear-data observable i on a 51-energy grid. The variable $\mathbf{Cov}_{\mathrm{i,j}}^{\mathrm{ND}}$ denotes nuclear data covariances between observables i and j on the same energy grid as $\mathbf{S}_{C,i}$.

We consider the following observables i and j: The neutron-induced fission, elastic, inelastic, capture, (n,2n), (n,3n), (n,4n), (n,p), (n,α) and (n,t) cross section along with the prompt-fission neutron spectrum (PFNS) and the average total-fission neutron multiplicity, $\bar{\nu}$, if both $\mathbf{Cov}_{\mathrm{i,j}}^{\mathrm{ND}}$ and $\mathbf{S}_{C,i}$ are available. We consider covariances for the total cross section implicitly. I.e., we do not include the total covariances explicitly in the adjustment but rather represent it by the cross-section covariances that contribute to it. Otherwise, this would amount to double-counting uncertainties due to the sum-rule applying to the total cross section and other contributing cross sections. The same reasoning applies to why we use only covariances and sensitivities for the $\bar{\nu}$, rather than the prompt one. We chose to use covariances and sensitivities for $\bar{\nu}$ as often no covariances are supplied for the average delayed-fission neutron multiplicity. Also, covariances are rarely, if at all, supplied for the delayed-fission neutron spectrum, and never for the total-fission neutron spectrum. Hence, only PFNS covariances and sensitivities were used. The effect of missing the delayed component is expected to be small as the PFNS dominates the total-fission neutron spectrum.

If $\mathbf{S}_{C,i}$ is defined to be relative to the nuclear data as well as C, the covariances also need to be divided by the respective nuclear data. Then the resulting Δ_C^{ND} is given also relative to C.

Forward-propagating covariances to provide bounds on integral responses allows us to understand the uncertainties on predicted quantities due to nuclear data. It can help understand if a simulated value is within nuclear-data uncertainties or far beyond. If it is far beyond, this can either point to a shortcoming in experimental data, or nuclear-data mean values and covariances.

2.2 Adjustment methodology

Adjustment of nuclear data was undertaken with the generalized linear least squares technique (GLLS) as, for instance, used by the LANL Whisper [18] code. Adjusted nuclear data mean values $\boldsymbol{\sigma}^*$ and covariances $\mathbf{Cov}^*$ are obtained by:

$$\begin{aligned}\boldsymbol{\sigma}^* &= \boldsymbol{\sigma}^{ND} + (\mathbf{S}_C\mathbf{Cov}^{\mathrm{ND}})^T\boldsymbol{Q}(\boldsymbol{E}-\boldsymbol{C}),\\ \mathbf{Cov}^* &= \mathbf{Cov}^{\mathrm{ND}} - (\mathbf{S}_C\mathbf{Cov}^{\mathrm{ND}})^T\boldsymbol{Q}(\mathbf{S}_C\mathbf{Cov}^{\mathrm{ND}}),\end{aligned} \quad (2)$$

with

$$\boldsymbol{Q} = (\mathbf{S}_C\mathbf{Cov}^{\mathrm{ND}}\mathbf{S}_C^T + \mathbf{Cov}^{\mathrm{e}})^{-1}. \quad (3)$$

The variables $\boldsymbol{E}$, $\boldsymbol{C}$, and $\boldsymbol{\sigma}^{ND}$ correspond to vectors of experimental and calculated integral values as well as nuclear data. The matrix $\mathbf{Cov}^{\mathrm{e}}$ is the covariance matrix of integral experimental data. An online tool was developed by M. Grosskopf as part of EUCLID to perform adjustment [19].

3 Input data

3.1 Integral responses

The following four integral responses were considered:

- The effective neutron multiplication factor, k_{eff}, of over 1,000 ICSBEP [16] critical assemblies. More specifically, we will study here plutonium metal fast (PU-MET-FAST), metal intermediate (PUT-MET-INT), and solution thermal (PU-SOL-THERM) assemblies. k_{eff} experiments are usually considered to be the gold standard for nuclear-data validation; they are well-vetted and provide reasonable experimental uncertainties (in the range of 0.05–0.5%). These experiments are frequently used to validate nuclear data up to 5 MeV. In the case of ENDF/B-VIII.0, a sub-set of these integral data are used to tweak nuclear data by hand (within the limits of differential experimental data) to better predict k_{eff} [14].
- The neutron-leakage spectra emitted from Pu LLNL pulsed spheres. The spheres were pulsed by a 14-MeV D+T neutron source in the center of the sphere [15]. Experimental data are provided as a function of time-of-flight (TOF), and for different material thicknesses, angles, and neutron detectors. These experiments extend the energy range of validation of around 5 MeV reached by k_{eff} to 15 MeV. However, these experiments are suspected to be biased [8, 20]; their uncertainties (0.5–2%) are not rigorously quantified.
- Fission reaction rates measured in the Jezebel, Godiva, Flattop-HEU, and Flattop-Pu critical assemblies are employed as well [7]. These measurements were designed to study the PFNS and the reactions measured. Experimental uncertainties are in the range of 0.8–2%. Questions have been raised on understanding the exact measurement value and surroundings reported for these experiments [21, 22].

- Reactivity coefficients,

$$\Delta\rho = \left(1 - \frac{1}{k^{a}_{\rm eff}}\right) - \left(1 - \frac{1}{k^{b}_{\rm eff}}\right), \tag{4}$$

measured in five different locations in the Jezebel critical assembly [23] have been considered. Configuration a corresponds to a void measurement (no sample), and b with a small sample of a defined material that does not change the spectrum of the assembly. Because of the ratio type of these data, some systematic uncertainties are reduced. A blanket uncertainty of 7.5% was added nonetheless to all reported experimental uncertainties following the IRPhEP uncertainty guide [24].

3.2 Sensitivities of integral responses to nuclear data

Sensitivities for $k_{\rm eff}$ were calculated with the "ksen" card in MCNP® Code Version 6.2[1][25]. Sensitivities for LLNL pulsed-sphere neutron-leakage spectra were also calculated with MCNP but with a brute-force methodology [6, 8], while "ksen" sensitivities were used to back out sensitivities for $\Delta\rho$ [6, 11]. Reaction rate sensitivities were computed via SENSMG [26]. All sensitivites were computed on the same 51-energy grid and for a subset of relevant reactions (neutron-induced fission, elastic, inelastic, capture, (n,2n), (n,3n), (n,4n), (n,p), (n,α) and (n,t) cross section along with PFNS and $\overline{\nu}$) for all isotopes that non-negligibly appear in the experiment. Sensitivities of $k_{\rm eff}$ to up to the P5 Legendre moment for angular distributions (MF=34) were also calculated, but were not used as these could not be computed at this time for other responses either due to the prohibitive number of Monte Carlo runs required for brute-force methods or limitations in codes to provide those sensitivities.

3.3 Nuclear data mean values and covariances

ENDF/B-VIII.0 mean values and covariances [14] were processed with the code NJOY2016.67 [27] onto a 51-energy grid. PFNS covariances were processed with separate NJOY/ERRORR runs per incident-energy grid, $E_{\rm inc}$, for those $E_{\rm inc}$ given in ENDF/B-VIII.0 using the IWT=9 option (CLAW spectrum) to collapse to the 51-(outgoing) energy grid maintaining the original $E_{\rm inc}$. PFNS covariances were then expanded (assuming full correlations between the PFNS belonging originally to one $E_{\rm inc}$ bin) to the $E_{\rm inc}$ of the sensitivities. The data were processed into a JSON format for easier readability [6]. In the current analysis, we consider ENDF/B-VIII.0 covariances for ^{1}H, ^{9}Be, ^{12}C, ^{27}Al, ^{56}Fe, ^{235,238}U, and ^{239,240}Pu isotopes. In future studies, this will be extended to include covariances for all isotopes that are available in ENDF/B-VIII.0.

[1]MCNP® and Monte Carlo N-Particle® are registered trademarks owned by Triad National Security, LLC, manager and operator of Los Alamos National Laboratory. Any third party use of such registered marks should be properly attributed to Triad National Security, LLC, including the use of the designation as appropriate. For the purposes of visual clarity, the registered trademark symbol is assumed for all references to MCNP within the remainder of this paper.

4 Results and discussion

4.1 Bounds on integral data due to nuclear-data covariances

Bounds of all four studied responses are shown in Figs. 1–4 based on experimental data only as well as including in quadrature Monte-Carlo simulation and nuclear-data uncertainties. The latter were obtained by forward-propagating nuclear-data covariances via Eq. (1). One key observation is that most of the C/E values for $k_{\rm eff}$ lie within the 1-sigma nuclear data uncertainties. That is more than the expected 68% if one would assume that C/E $k_{\rm eff}$ values follow an independent Gaussian distribution. Previously, this was interpreted as that nuclear-data covariances were over-estimated with regards to the spread in C/E of $k_{\rm eff}$ [28]. However, this good agreement reflects two facts [14]:

- ENDF/B-VIII.0 mean values were tweaked by hand to better predict $k_{\rm eff}$ of a sub-set of the experimental data shown in Fig. 1. This effectively breaks the assumption that the data should follow an independent Gaussian distribution as nuclear data were manipulated to enforce agreement,
- Moreover, ENDF/B-VIII.0 covariances remained unchanged and only reflect the knowledge coming from differential experimental data and theory.

Due to that it is justified that the spread of C/E of $k_{\rm eff}$ lies well within the uncertainties encompassing nuclear-data covariances.

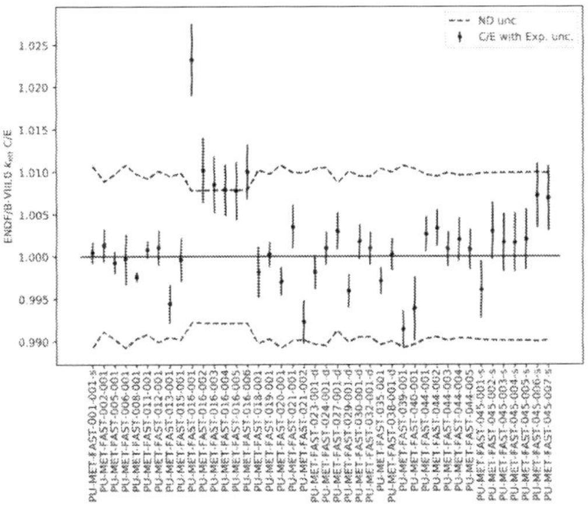

Figure 1. C/E values are shown for $k_{\rm eff}$ of PU-MET-FAST ICSBEP critical assemblies [16]. The solid, vertical lines are experimental uncertainties, while the dashed lines encompass Monte Carlo statistics, and uncertainties obtained by forward-propagating nuclear-data covariances via Eq. (1) in quadrature.

Compared to $k_{\rm eff}$, a reasonable amount of C/E values fall outside of the 1-sigma range for reaction rates (Fig. 2). ENDF/B nuclear data are usually validated with respect to these measurements [14], but most of the values are trending to be low which was already noticed in Ref. [29].

Hence, studies on these data are currently ongoing [22] to understand if either an adjustment of nuclear data is needed or whether experimental biases might be at play.

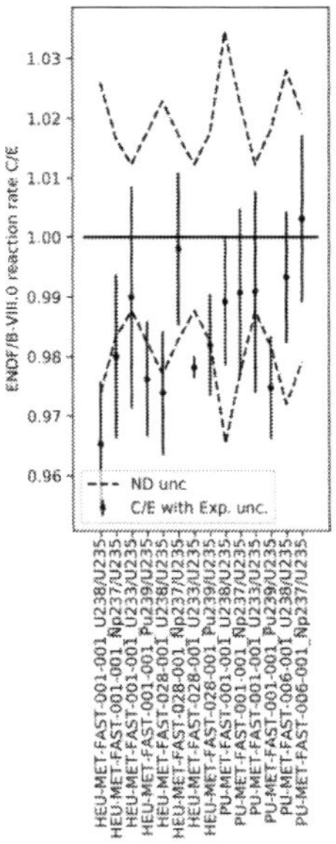

Figure 2. The same as Fig. 1 except for fission reaction rates in critical assemblies.

The C/E agreement of the Pu LLNL pulsed spheres is imperfect below 185, from 205 to 235 and above 480 ns in Fig. 3. It is difficult to model the neutron source accurately impacting possibly C/E values below 185 ns. The discrepancy in C/E from 205 to 235 ns was resolved in Refs. [30–32] by improving the ^{239}Pu nuclear data for the continuum inelastic cross section and angular distribution. Above 480 ns, it is likely that we see the impact of background or detector efficiency biasing the experimental data. It is also noteworthy that nuclear-data uncertainties on the spectrum from 173 to 187 ns seem underestimated. The spectrum in this TOF range is mostly sensitive to elastic and inelastic scattering distribution. Neither sensitivities, nor in some cases covariances, exist for these observables leading to the underestimated nuclear-data uncertainties.

Reactivity coefficients measured in Jezebel are infrequently used for validating nuclear data. Hence, it is not surprising to see C/E values outside of the 1-sigma range in Fig. 4. Still, it is obvious that there are some experimental outliers, such as for BeO or tungsten samples, while the agreement of C and E for $\Delta\rho$ using Pu samples is distinctly better. The experimental data of $\Delta\rho$ need to be further explored.

4.2 Adjustment

If we adjust nuclear-data mean values and covariances with respect to k_{eff} of Jezebel via Eq. (2), the evaluated ^{239}Pu nuclear data change minimally in Fig. 5, while the evaluated ^{239}Pu uncertainties decrease distinctly. These results illustrate again that ENDF/B-VIII.0 mean values were tweaked by hand to reproduce Jezebel k_{eff}; hence, any remaining possible adjustment of cross sections is small. ENDF/B-VIII.0 covariances, however, remained unchanged in this processing of tweaking the mean values. Therefore, adjusted uncertainties reduce significantly when the physics is constrained by Jezebel k_{eff}.

If we adjust with respect to a sub-set of PU-MET-FAST, PU-MET-INT and PU-SOLT-THERM k_{eff} values, the evaluated mean values change more distinctly (by more than a factor of 10) compared to adjustment with only Jezebel k_{eff} (Fig. 6). However, the proposed changes in evaluated data are well within the evaluated nuclear-data uncertainties of the respective reactions. Hence, here we see the effect of adjusting to k_{eff} values that were not as closely matched by hand-tuning nuclear data. The overall change of evaluated uncertainties in the fast range is comparable to adjusting to k_{eff} of Jezebel. This (EUCLID) sub-set of assemblies was chosen by identifying those that had reliable k_{eff} values and increasing uncertainties when suspected to be underestimated [33]. In a second step, we down-selected those benchmark series with multiple, strongly correlated configurations to consider only one k_{eff} value per series in the adjustment to avoid double-counting these experiments as we do not consider covariances between them at this point.

The adjustment to fission reaction rates in the Jezebel critical assembly leads to comparably modest changes in evaluated nuclear data in Fig. 7. We only adjust to reaction rates in Jezebel, while bounds for fission reaction rates in Godiva, Flattop, Flattop-Pu and Jezebel are shown in Fig. 2, because adjustment was also performed for k_{eff} and reactivity coefficients of Jezebel in Figs. 5 and 9. It is noteworthy that the fission cross section is increased by 1%, an effect seen also when including cross sections averaged over a ^{252}Cf spontaneous-fission spectrum [34] or similar adjustment work by R. Casperson [29]. The size of the change depends on what sub-set of reaction rates are used for adjustment. It is also noteworthy that neither $\overline{\nu}$ nor the capture cross section change. The reason for that is that sensitivities of reaction rates in Jezebel to ^{239}Pu $\overline{\nu}$ and the capture cross section are negligibly small (Fig. 8). Hence, fission reaction rates in the Jezebel critical assembly allow us to study the PFNS (adjusted, but not shown) and (n,f) cross sections independently from $\overline{\nu}$ as well as scattering independent from the capture cross section. Therefore, reaction rates help in disentangling compensating errors in nuclear data introduced by k_{eff}.

If we adjust to Pu LLNL pulsed-sphere-neutron leakage spectra or Pu reactivity coefficients in Jezebel, the changes in evaluated ^{239}Pu nuclear data in Fig. 9 are far beyond those seen for k_{eff}. This goes along, for the Pu pulsed sphere, with the larger C/E values seen for the simulated and experimental spectra compared to k_{eff} or reaction rates. A large adjustment of the inelastic cross section was expected given research in Refs. [30–32] that showed that the spectrum from 205 to 235 ns could be better predicted by changes in the inelastic cross sections and angular distributions. However, a change by 160% is far beyond the evaluated uncertainty of this reaction. This large adjust-

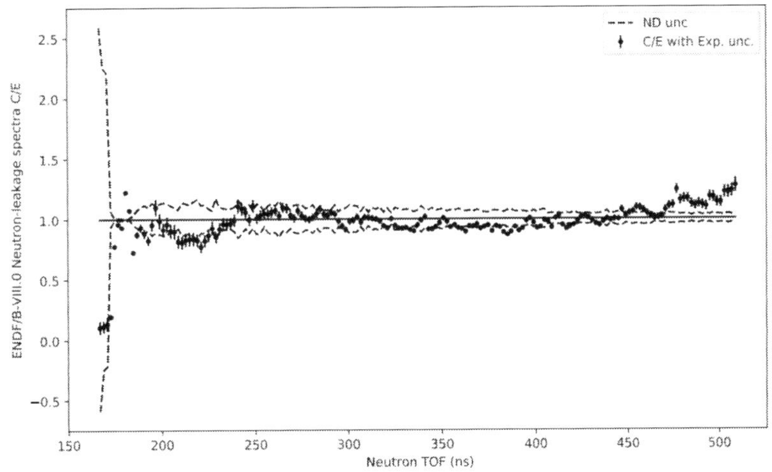

Figure 3. The same as Fig. 1 except for Pu LLNL pulsed-sphere neutron-leakage spectra measured with a sphere of 0.7 mean-free path, at 26 degree with an NE-213 detector.

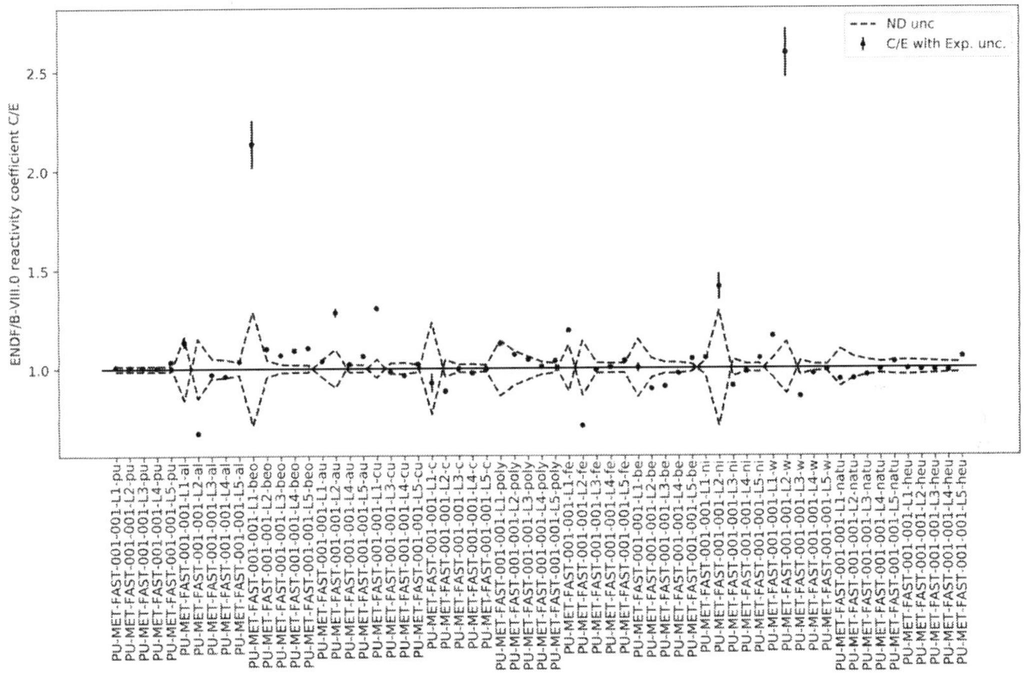

Figure 4. The same as Fig. 1 except for the reactivity coefficients in the PU-MET-FAST-001 (Jezebel) critical assembly. "Lx" specifies the location of the sample.

ment reflects that the inelastic cross section has to absorb, in this adjustment exercise, shortcomings in both the cross section and the angular distributions; the latter cannot be adjusted here due to missing sensitivities. Hence, adjustments of the inelastic cross sections are over-estimated.

Also, one should question such large adjustments in the nuclear data in the light of large discrepancies in C/E at late times that are likely caused by biases in the experimental data rather than the nuclear data. Therefore, the adjustment should account for possible bias in the experimental data by modeling this bias as part of the GLLS algorithm. In addition, one should consider that all pulsed-sphere data are implicitly assumed to have uncorrelated uncertainties as correlations are not provided. Missing, but

 https://doi.org/10.1051/epjconf/202328100007

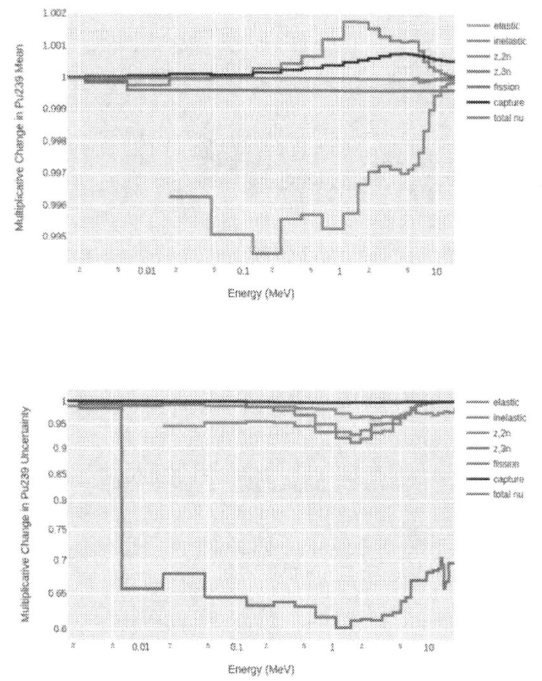

Figure 5. (Top) The % change of adjusted mean values as a ratio to ENDF/B-VIII.0 ^{239}Pu nuclear data is shown. (Bottom) The % change of adjusted uncertainties as a ratio to ENDF/B-VIII.0 ^{239}Pu nuclear-data uncertainties is shown. The adjustment was undertaken with respect to k_{eff} of Jezebel.

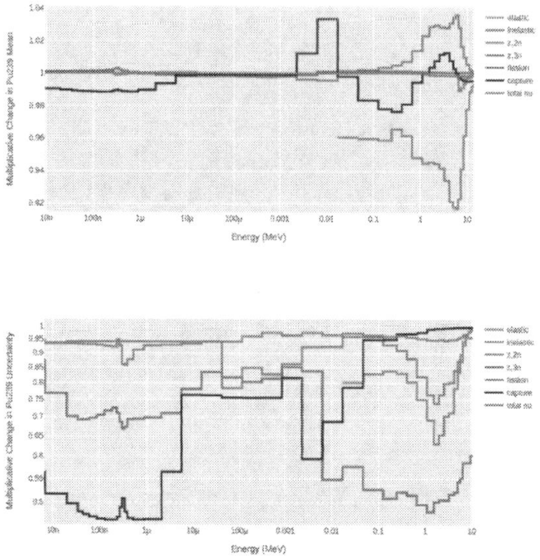

Figure 6. The same as Fig. 5 except for adjustment to selected PU-MET-FAST, PU-MET-INT and PU-MET-SOL k_{eff} values.

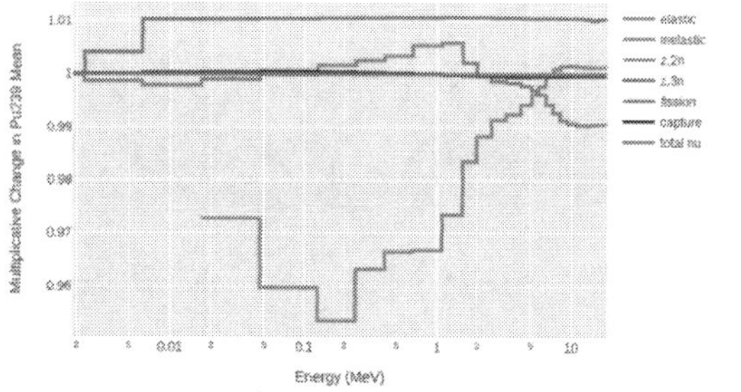

Figure 7. The same as the top-plot of Fig. 5 except for adjustment to fission reaction rates in the Jezebel critical assemblies.

plausibly non-negligible, correlations could result in distinctly different adjustment results and thus conclusions. Considering correlations for experimental uncertainties, however, requires more detailed studies of these experiments. Last but not least, it is surprising to see adjustments of nuclear data (e.g., the fission cross section) down to 10 keV despite the fact that the thin Pu sphere (0.7 mean-free path) is mostly sensitive to the fission cross section from 12–15 MeV. This observation was attributed in Ref. [29] to the strongly correlated covariance matrix of the Neutron Data Standards [35] that is currently under revision.

It is surprising to see such large changes in the elastic and inelastic ^{239}Pu cross sections (beyond their 1-sigma evaluated uncertainties) when adjusted with Pu reactivity coefficients in Jezebel despite the C/E values of these data being in reasonable agreement. In Ref. [11], it was discussed that sensitivity profiles of reactivity coefficients to the elastic cross section are hampered by large Monte Carlo uncertainties, while fission cross section sensitivities have small uncertainties. Currently, work is ongoing within EUCLID to include uncertainties in the sensitivities in the adjustment algorithm.

5 Summary, conclusion, and outlook

The EUCLID (Experiments Underpinned by Computational Learning for Improvements in Nuclear Data) LDRD-DR (Laboratory Directed Research & Development-Directed Research) project [13] computed sensitivities for various integral responses. At the same time, ENDF/B-VIII.0 covariances were processed at LANL.

Here, we use sensitivities of the effective neutron multiplication factor, k_{eff}, reaction rates, and reactivity coefficients of ICSBEP critical assemblies [16] as well as LLNL pulsed-sphere neutron-leakage spectra [15] and covariances to assess the uncertainties on simulated responses due to nuclear data. We also perform adjustment with

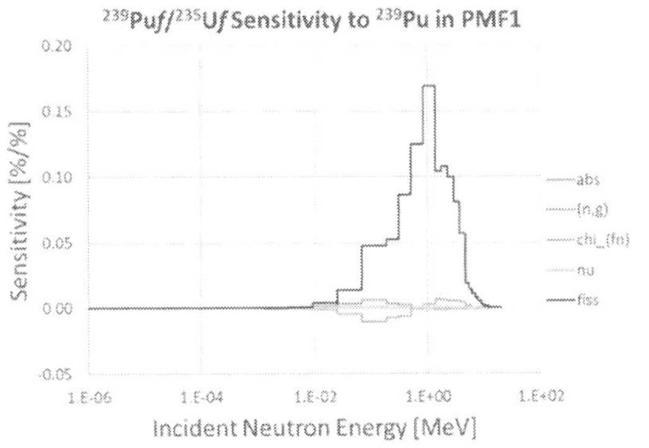

Figure 8. Sensitivities of the ^{239}Pu(n,f)/^{235}U(n,f) reaction rate in the Jezebel critical assembly to selected ^{239}Pu nuclear data are shown.

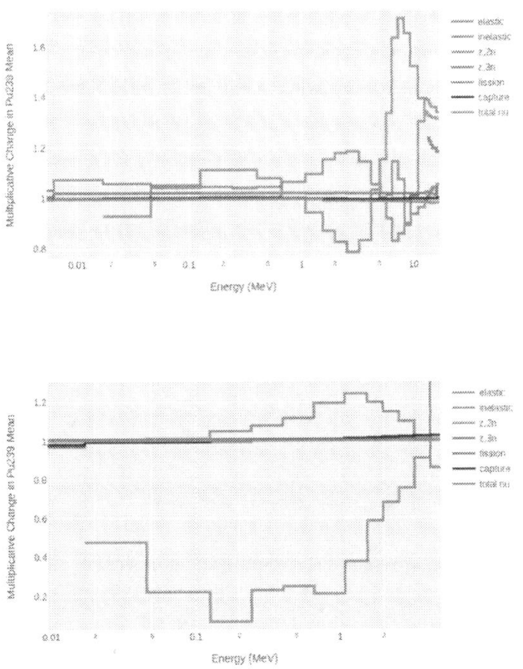

Figure 9. The same as the top-plot of Fig. 5 except for adjustment to Pu LLNL pulsed-sphere neutron-leakage spectra measured with a sphere of 0.7 mean-free path, at 26 degree with an NE-213 detector (top) and Pu reactivity coefficients measured in Jezebel (bottom).

ENDF/B-VIII.0 mean values and covariances with respect to the aforementioned integral responses.

From these studies, one can clearly observe which integral data were used for validating ENDF/B-VIII.0: The adjustment to nuclear-data mean values with k_{eff} and reaction rates is modest as nuclear data were changed by hand to better agree to some experimental data of these responses. Also, the calculated over experimental values, C/E, lie well within the spread of forward-propagated nuclear-data uncertainties. The evaluated covariances, however, change significantly with adjustment reflecting that ENDF/B-VIII.0 mean values are tweaked to better predict some integral data, while covariances remain unchanged and only reflect uncertainties from differential experiments and nuclear theory. It was also shown, with the example of adjusting ^{239}Pu nuclear data with respect to fission reaction rates in the Jezebel assembly, that one gains more insight into nuclear data by using responses beyond k_{eff}. For instance, adjustment with Pu-assembly k_{eff} would change all ^{239}Pu fission-source term along with scattering and capture nuclear data, while fission reaction rates in the Jezebel assembly lead only to adjustments in the prompt-fission neutron spectrum, fission, elastic and inelastic cross sections. Hence, using both responses allows to partially disentangle the Pu fission source term. We could fully disentangle it, if we had a response that would only be sensitive to one of the three fission-source-term observables.

LLNL pulsed-sphere neutron-leakage spectra and reactivity coefficients show a distinctly larger spread in C/E values compared to uncertainties stemming from nuclear data. That shows that nuclear data is not tweaked to the pulsed sphere data, but also that the experimental responses need to be better understood. In cases where the experimental data cannot be corrected, but biases are apparent, data needs to be either truncated, rejected, or these biases need to be modeled within the adjustment, to avoid undue impact on adjusted mean values and covariances. In addition to that, covariances and sensitivities are missing for elastic and inelastic angular distribution that would account for an important part of simulated pulsed-sphere neutron-leakage spectra uncertainties. It was also shown that large uncertainties on sensitivities can lead to unreasonably large adjustment of nuclear data. It was also noted that strong nuclear-data correlations, specifically those of the ^{239}Pu fission cross sections, can lead to adjustment of the cross section in much lower energy groups than should be affected by the measurement at hand. Lastly, no correlations are taken into account between the different experimental data as they are rarely provided. Many of the measurements could arguably have strong correlations (e.g., all data of one pulsed-sphere measurement, the different responses measured on Jezebel, k_{eff} values for multiple, slightly differing configurations on the same critical experiment), and including these correlations has the potential to significantly change adjustment results. However, estimating reliable correlations requires a detailed uncertainty analysis that is time-intensive. As a starting point, WPEC SG-33 [36] covariances could be used.

Work is currently ongoing at LANL to include modeling the bias of experimental data and uncertainties in sensitivities in the adjustment, as well as to better understand the quality of experimental data of some responses. We will also extend the adjustment studies to include more integral responses to understand whether they yield additional understanding on the nuclear data beyond k_{eff}. Last but not least, all studies presented here will be extended to include all pertinent ENDF/B-VIII.0 covariances rather than the sub-set studied here (^{1}H, ^{9}Be, ^{12}C, ^{27}Al, ^{56}Fe, ^{235,238}U, and ^{239,240}Pu).

References

[1] K. Kolos, V. Sobes, R. Vogt, C.E. Romano, M.S. Smith, L.A. Bernstein, D.A. Brown, M.T. Burkey, Y. Danon, M.A. Elsawi et al., Phys. Rev. Research **4**, 021001 (2022)

[2] D. Neudecker, et. al., EPJ Web of Conferences (2022)

[3] E. Bauge, G. Bélier, J. Cartier, A. Chatillon, J. Daugas, J. Delaroche, P. Dossantos-Uzarralde, H. Duarte, N. Dubray, M. Ducauze-Philippe et al., The European Physical Journal A **48**, 113 (2012)

[4] M. Chadwick, R. Capote, A. Trkov, M. Herman, D. Brown, G. Hale, A. Kahler, P. Talou, A. Plompen, P. Schillebeeckx et al., Nuclear Data Sheets **148**, 189 (2018), special Issue on Nuclear Reaction Data

[5] D. Neudecker, O. Cabellos, A.R. Clark, M.J. Grosskopf, W. Haeck, M.W. Herman, J. Hutchinson, T. Kawano, A.E. Lovell, I. Stetcu et al., Phys. Rev. C **104**, 034611 (2021)

[6] J. Alwin, A. Clark, T. Cutler, M. Grosskopf, W. Haeck, M. Herman, J. Hutchinson, N. Kleedtke, J. Lamproe, R. Little et al., Tech. Rep. LA-UR-22-21534, Los Alamos National Laboratory (2022)

[7] J. Alwin, J. Hutchineson, N. Kleedtke, A. Clark, T. Cutler, W. Haeck, R. Little, D. Neudecker, M. Rising, T. Smith et al., Transaction of American Nuclear Society (2022)

[8] D. Neudecker, O. Cabellos, A. Clark, W. Haeck, R. Capote, A. Trkov, M.C. White, M.E. Rising, Annals of Nuclear Energy **159**, 108345 (2021)

[9] J. Hutchinson, et. al., American Nuclear Society Winter Meeting 2021 **125**, 623 (2021)

[10] N. Kleedtke and J. Hutchinson and T. Cutler and I. Michaud and M. Rising and M. Hua and J. Alwin and M. Grosskopf and S. Vander Wiel and D. Neudecker and N. Thompson, EPJ Web of Conferences (2022)

[11] T. Cutler, J. Hutchinson, D. Neudecker, W. Haeck, A. Clark, M. Rising, ANS (2022)

[12] A. Clark, D. Neudecker, M. Grosskopf, S. Vander Wiel, EPJ Web of Conferences (2022)

[13] J. Hutchinson, J. Alwin, A. Clark, T. Cutler, M. Grosskopf, W. Haeck, M. Herman, N. Kleedtke, J. Lamproe, R. Little et al., EPJ Web of Conferences (2022)

[14] D.A. Brown, M.B. Chadwick, R. Capote, A.C. Kahler, A. Trkov, M.W. Herman, A.A. Sonzogni, Y. Danon, A.D. Carlson, M. Dunn et al., Nuclear Data Sheets **148**, 1 (2018)

[15] C. Wong, J. Anderson, P. Brown, L.F. Hansen, J.L. Kammerdiener, C. Logan, B. Pohl, LLNL UCRL-51144 Rev. 1 (1972)

[16] Tech. Rep. (NEA;7328), Paris: OECD Nuclear Energy Agency (2019)

[17] B. Rearden, M. Williams, M. Jessee, D. Mueller, D. Wiarda, Nuclear Technology **174**, 236 (2011), `https://doi.org/10.13182/NT174-236`

[18] F. Brown, M. Rising, J. Alwin, Tech. Rep. LA-UR-17-20567, Los Alamos National Laboratory (2017)

[19] M. Grosskopf, *EUCLID Adjustment Visualization*, http://euclid-adjustment-server.lanl.gov:2311/ (2022), this is only accessible internal to LANL

[20] I. Kodeli et al., EPJ Web of Conferences (2022)

[21] S. Frankle, Tech. Rep. LA-13675, Los Alamos National Laboratory (1999)

[22] N. Kleedtke, Tech. Rep. forthcoming, Los Alamos National Laboratory (2022)

[23] L. Engle, G. Hansen, H. Paxton, Nuclear Science and Engineering **8**, 543 (1960)

[24] A. Santos, A. Tsiboulia, L. Snoj, V. Radulovic, Z. Stancar, Z. Szatmary, P. Blaise, J. Bess, M. Marshall, J. Briggs, Tech. Rep. International Handbook of Evaluated Criticality Safety Benchmark Experiments, NEA/NSC/DOC/(95)03/I, Nuclear Energy Agency (2017)

[25] C. Werner, J. Armstrong, F. Brown, J. Bull, L. Casswell, L. Cox, D. Dixon, R. Forster, J. Goorley, H. Hughes et al., Tech. Rep. LA-UR-17-29981, Los Alamos National Laboratory (2017)

[26] J. Favorite, Tech. Rep. LA-UR-19-26249, Los Alamos National Laboratory (2019)

[27] R. MacFarlane, D. Muir, R. Boicourt, A. Kahler, J. Conlin, W. Haeck, Tech. Rep. LA-UR-17-20093, Los Alamos National Laboratory (2017)

[28] W. Marshall, M. Williams, B.R. D. Wiarda, M. Dunn, D. Mueller, J. Clarity, *Development and testing of Neutron cross-section covariance data for SCALE 6.2*, in *International Conference on Nuclear Criticality Safety (ICNC 2015) Charlotte, NC, USA, September 13–17, 2015* (2015), p. 1213, `arXiv:1705.07874`

[29] R. Casperson, EPJ Web of Conferences (2022)

[30] M. Mumpower, D. Neudecker, T. Kawano, M. Herman, N. Kleedtke, A.E. Lovell, I. Stetcu, P. Talou, Tech. Rep. LA-UR-22-24121, Los Alamos National Laboratory (2022)

[31] G.P. Aguilar, M. Descalle, K. Kravvaris, W. Ormand, B. Beck, C. Mattoon, I. Thompson, K. Wendt, R. Casperson, L. Snyder et al., Tech. Rep. LLNL-MI-834591, Lawrence Livermore National Laboratory (2022)

[32] R. Capote, A. Trkov, M. Pigni, D. Neudecker, M. Herman, K. Kelly, A. Lovell, J. Malec, G. Noguere, V. Pronyaev et al., EPJ Web of conferences (2022), https://www-nds.iaea.org/INDEN/

[33] D. Neudecker, J.D. Hutchinson, A.C.I. Kahler, A.R. Clark, N.W. Thompson, M.W. Herman, J.L. Alwin, M.J. Grosskopf, W. Haeck, S.A. Vander Wiel et al., Tech. Rep. LA-UR-20-25879, Los Alamos National Laboratory (2020)

[34] R. Capote, et al., EPJ Web of Conferences (2022)

[35] A. Carlson, V. Pronyaev, R. Capote, G. Hale, Z.P. Chen, I. Duran, F.J. Hambsch, S. Kunieda, W. Mannhart, B. Marcinkevicius et al., Nuclear Data Sheets **148**, 143 (2018), special issue on nuclear reaction data

[36] M. Salvatores, c.o. G. Palmiotti, Tech. Rep. NEA/NSC/WPEC/DOC(2013)445, Organization for Economic Co-operation and Development-Nuclear Energy Agency (2013)

Acknowledgments

Research reported in this publication was partially supported by the U.S. Department of Energy LDRD program at Los Alamos National Laboratory. We gratefully acknowledge partial support of the Advanced Simulation and Computing program at LANL. This work was supported by the US Department of Energy through the Los Alamos National Laboratory. Los Alamos National Laboratory is operated by Triad National Security, LLC, for the National Nuclear Security Administration of the US Department of Energy under Contract No. 89233218CNA000001.

Applicability evaluation of Akaike's Bayesian information criterion to covariance modeling in the cross-section adjustment method

Shuhei Maruyama[1*], *Tomohiro* Endo[2], and *Akio* Yamamoto[2]

[1]Japan Atomic Energy Agency, Reactor Core and Plant System Evaluation Group, Fast Reactor Cycle System Research and Development Center, 4002 Narita-cho, Oarai-machi, Higashi-Ibaraki-gun, Ibaraki, Japan
[2]Nagoya University, Department of Applied Energy, Graduate School of Engineering, Furo-cho, Chikusa-ku, Nagoya, Japan

Abstract. The applicability of Akaike's Bayesian Information Criterion (ABIC) to the covariance modeling in the cross-section adjustment method has been investigated. In the conventional cross-section adjustment method, the covariance matrices are assumed to be true. However, this assumption is not always appropriate. To improve the reliability of the cross-section adjustment method, the estimation of the covariance model using the metric ABIC has been introduced, and the performance of ABIC has been investigated through simple numerical experiments. This paper derives the formula to efficiently evaluate ABIC which is represented by a lower rank matrix to enable numerical experiments with large samples in a realistic computation time. From the results of the numerical experiments, it has been confirmed that ABIC tends to select a covariance model with fewer hyperparameters and a smaller variance for the estimation error. However, it has also been found that this desirable property of ABIC will be lost when the structure of the covariance model is far from the true one.

1 Background

Every measurement and analysis values always have uncertainties. Namely, the data follow a probability distribution with parameters such as population mean and covariance. Unfortunately, we cannot know the true parameters. Therefore, when we analyze the data, we usually estimate and/or assume these parameters. The appropriateness of the covariance matrix set (e.g., nuclear-data covariance) by analysts has been widely discussed in the framework of Sub-Groups (SGs) under the Working Party on International Nuclear Data Evaluation Cooperation (WPEC) in OECD/NEA (e.g., [1-4]). Recently, the discussion had been also done in Japan through the activity of the covariance data utilization and promotion working group organized in the JENDL committee [5].

In the cross-section adjustment method, covariance matrices are also used. One of the most important things for a reliable cross-section adjustment method is giving suitable covariance matrices close enough to the true covariance matrices. To judge the goodness of the covariance modeling, a metric is desirable. As a candidate for this metric, we focus on Akaike's Bayesian Information Criterion (ABIC) [6] which is one of the information criteria in Bayesian inference, because the cross-section adjustment method is often discussed within the framework of Bayesian inference.

In the conventional cross-section adjustment method, incorporation of the analysis method errors (errors due to the core calculation method, e.g., discretizing error in a deterministic code) as a covariance matrix still requires ad hoc treatment. In JAEA, the integral experimental database for fast reactors has been developed and the adjusted cross-section set ADJ2017 [7, 8] has been created based on this database. Many of the core characteristics in the database have been analyzed by a deterministic method. Therefore, the predicted core characteristics have non-negligible uncertainties with correlations due to some numerical approximations. However, evaluating the uncertainties and their correlations is still a challenging issue. In addition, there would be unknown uncertainties that experimenters and analysts of reactor physics experiments were not able to recognize.

In order to address the difficulties in the conventional adjustment method, we will try to incorporate ABIC into the adjustment method. ABIC is expected to work as a metric for evaluating the goodness of covariance matrix modeling related to the uncertainties. We aim to improve the reliability of the conventional cross-section adjustment method by ABIC. This paper investigates the applicability of ABIC through several numerical experiments using random sampling techniques.

We introduce ABIC and incorporate this into the conventional cross-section adjustment in the next section. The applicability is evaluated through the simple numerical experiments in section 3. The conclusion of this paper will be described in section 4.

* Corresponding author: maruyama.shuhei@jaea.go.jp

2 Methodology

Before we get to the main subject, we introduce the notations that will be used in this paper. The covariance matrix with the tilde (" ~ ") means a covariance model with hyperparameters determined later. The covariance matrix with a hat (" ^ ") means the selected covariance by determining the hyperparameters. These covariances are not always "true." The true covariance is represented by the characters without a tilde or a hat. For example,

$\mathbf{V}$: a true covariance matrix,

$\widetilde{\mathbf{V}}$: a covariance model with undetermined hyperparameters,

$\widehat{\mathbf{V}}$: the covariance selected among a covariance model $\widetilde{\mathbf{V}}$ by determining hyperparameters.

2.1 Review of conventional cross-section adjustment method

When true nuclear data set $\boldsymbol{T}$ follows a multivariate normal distribution with the mean $\boldsymbol{T}_0$ and covariance $\mathbf{M}$, the probability distribution of $\boldsymbol{T}$ is

$$\mathcal{P}(\boldsymbol{T}) = \frac{1}{(2\pi)^{\frac{n}{2}}|\mathbf{M}|^{\frac{1}{2}}}\exp\left[-\frac{1}{2}(\boldsymbol{T}-\boldsymbol{T}_0)^{\mathrm{T}} \cdot \mathbf{M}^{-1}(\boldsymbol{T}-\boldsymbol{T}_0)\right], \quad (1)$$

where n is the dimension of $\boldsymbol{T}$ and $\boldsymbol{T}_0$, $|\mathbf{X}|$ is the determinant of a matrix $\mathbf{X}$, and the superscripts -1 and T represent the inverse and transpose of the matrix respectively. When integral experimental data sets are obtained, the likelihood function is represented as follows:

$$\mathcal{P}(\boldsymbol{R}_e|\boldsymbol{T}) = \frac{1}{(2\pi)^{\frac{r}{2}}|\mathbf{V}_{\mathrm{e}}+\mathbf{V}_{\mathrm{m}}|^{\frac{1}{2}}}\exp\left[-\frac{1}{2}\left(\boldsymbol{R}_e-\boldsymbol{R}_c(\boldsymbol{T})\right)^{\mathrm{T}} \cdot (\mathbf{V}_{\mathrm{e}}+\mathbf{V}_{\mathrm{m}})^{-1}\left(\boldsymbol{R}_e-\boldsymbol{R}_c(\boldsymbol{T})\right)\right]. \quad (2)$$

$\boldsymbol{R}_e$ is a set of experimental values and $\mathbf{V}_{\mathrm{e}}$ is the experimental covariance matrix. $\boldsymbol{R}_c(\boldsymbol{T})$ is a set of calculational values if the true nuclear data set $\boldsymbol{T}$ is given, and $\mathbf{V}_{\mathrm{m}}$ is the covariance matrix due to calculation methods. Here, r denotes the dimension of $\boldsymbol{R}_e$ and $\boldsymbol{R}_c$. From Bayes' theorem,

$$\mathcal{P}(\boldsymbol{T}|\boldsymbol{R}_e) \propto \mathcal{P}(\boldsymbol{R}_e|\boldsymbol{T})\mathcal{P}(\boldsymbol{T}), \quad (3)$$

the posterior probability is represented as follows:

$$\mathcal{P}(\boldsymbol{T}|\boldsymbol{R}_e) \propto \exp\left[-\frac{1}{2}(\boldsymbol{T}-\boldsymbol{T}_0)^{\mathrm{T}}\mathbf{M}^{-1}(\boldsymbol{T}-\boldsymbol{T}_0) -\frac{1}{2}\left(\boldsymbol{R}_e-\boldsymbol{R}_c(\boldsymbol{T})\right)^{\mathrm{T}}(\mathbf{V}_{\mathrm{e}}+\mathbf{V}_{\mathrm{m}})^{-1}\left(\boldsymbol{R}_e-\boldsymbol{R}_c(\boldsymbol{T})\right)\right]. \quad (4)$$

The cross-section and its covariance after adjustment are derived from the condition that maximizes the posterior probability of Eq. (4) as follows:

$$\boldsymbol{T}' = \boldsymbol{T}_0 + \mathbf{M}\mathbf{G}^{\mathrm{T}}\mathbf{V}_{\mathrm{t}}^{-1}\left(\boldsymbol{R}_e-\boldsymbol{R}_c(\boldsymbol{T}_0)\right), \quad (5)$$

$$\mathbf{M}' = \mathbf{M} - \mathbf{M}\mathbf{G}^{\mathrm{T}}\mathbf{V}_{\mathrm{t}}^{-1}\mathbf{G}\mathbf{M}, \quad (6)$$

where

$$\mathbf{V}_{\mathrm{t}} \equiv \mathbf{G}\mathbf{M}\mathbf{G}^{\mathrm{T}} + \mathbf{V}_{\mathrm{e}} + \mathbf{V}_{\mathrm{m}}. \quad (7)$$

Here, $\mathbf{G}$ means nuclear-data sensitivity coefficients. Note that the linear approximation for $\boldsymbol{R}_c(\boldsymbol{T})$, *i.e.*,

$$\boldsymbol{R}_c(\boldsymbol{T}) = \boldsymbol{R}_c(\boldsymbol{T}_0) + \mathbf{G}(\boldsymbol{T}-\boldsymbol{T}_0), \quad (8)$$

is assumed to derive Eqs. (5) and (6). The concept of the cross-section adjustment method is shown in Fig. 1.

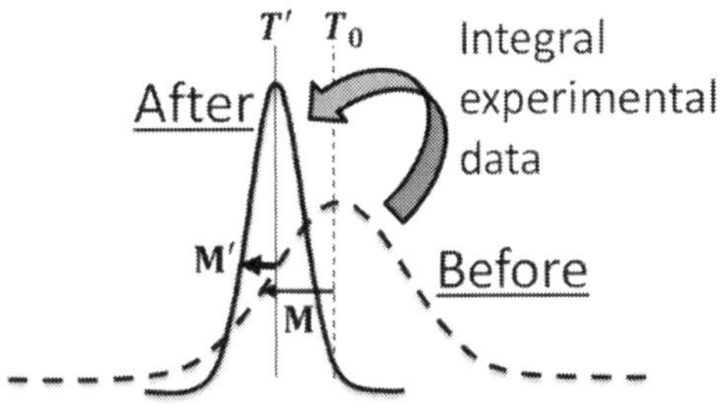

Fig. 1. Concept of cross-section adjustment

2.2 Akaike's Bayesian information criterion

In the conventional cross-section adjustment method, the covariance matrices $\mathbf{M}$, $\mathbf{V}_{\mathrm{e}}$, and $\mathbf{V}_{\mathrm{m}}$ used for the adjustment are assumed to be true. However, these are not always appropriate. Therefore, let us assume the covariance matrices have hyperparameters that are determined to suit the observed data. To find a better covariance model by tuning the hyperparameters, we propose to use ABIC in this paper. As mentioned in the previous section, since the setting of the covariance matrix $\mathbf{V}_{\mathrm{m}}$ due to a calculation method is a challenging issue in the conventional adjustment method, only $\mathbf{V}_{\mathrm{m}}$ is assumed to be unknown and to be the estimation target using ABIC in this study. In other words, $\mathbf{M}$ and $\mathbf{V}_{\mathrm{e}}$ in this study are assumed to be true.

If the covariance model $\widetilde{\mathbf{V}}_{\mathrm{m}}$ has unknown positive hyperparameters $p_1^2, p_2^2, \cdots, p_q^2$, the likelihood function of Eq. (2) is represented as follows:

$$\mathcal{P}\left(\boldsymbol{R}_e|\boldsymbol{T}; p_1^2, p_2^2, \cdots, p_q^2\right) \propto \exp\left[-\frac{1}{2}\left(\boldsymbol{R}_e-\boldsymbol{R}_c(\boldsymbol{T})\right)^{\mathrm{T}} \cdot \left(\mathbf{V}_{\mathrm{e}}+\widetilde{\mathbf{V}}_{\mathrm{m}}\left(p_1^2, p_2^2, \cdots, p_q^2\right)\right)^{-1}\left(\boldsymbol{R}_e-\boldsymbol{R}_c(\boldsymbol{T})\right)\right], \quad (9)$$

where q is the number of hyperparameters. As shown later (subsection 3.1.3), we prepare positive hyperparameters expressed as squared values to guarantee the positive definiteness of the covariance matrix $\widetilde{\mathbf{V}}_{\mathrm{m}}$. As a metric of the goodness of an inference model including the covariance modeling, we employ an information criterion called ABIC,

$$\mathrm{ABIC} = -2\ln l\left(p_1^2, p_2^2, \cdots, p_q^2\right) + 2q. \quad (10)$$

Here, l is the marginal likelihood defined as

$$l\left(p_1^2, p_2^2, \cdots, p_q^2\right) = \int d\boldsymbol{T}\ \mathcal{P}\left(\boldsymbol{R}_e|\boldsymbol{T}; p_1^2, p_2^2, \cdots, p_q^2\right)\mathcal{P}(\boldsymbol{T}). \quad (11)$$

ABIC consists of the two terms related to l and q. The marginal likelihood l is proportional to the generation probability of the observed data (integral experimental data set $\boldsymbol{R}_e$). By choosing a good inference model that leads to this probability, *i.e.*, a suitable covariance model, ABIC is decreased. On the other hand, the number of hyperparameters q works penalty in terms of the complexity of an inference model. Therefore, ABIC prefers a simpler model having fewer hyperparameters. In a previous study, a cross-section adjustment method based on the optimization of marginal likelihood l was presented [9]. Our proposed method using ABIC proposed in this paper is an extension of the previous study.

Performing the integral in Eq. (11) under the assumption of Eq. (8), the l becomes (*e.g.*, [10])

$$-2\ln l = \ln|\mathbf{M}| + \ln\left|\mathbf{V}_\mathbf{e} + \tilde{\mathbf{V}}_\mathbf{m}(p_1^2, p_2^2, \cdots, p_q^2)\right| - \ln\left|\tilde{\mathbf{M}}'(p_1^2, p_2^2, \cdots, p_q^2)\right| + \mathrm{tr}\left[\mathbf{S}\tilde{\mathbf{V}}_\mathbf{t}^{-1}(p_1^2, p_2^2, \cdots, p_q^2)\right] + r\ln(2\pi), \quad (12)$$

where

$$\tilde{\mathbf{M}}'(p_1^2, p_2^2, \cdots, p_q^2) = \left[\mathbf{M}^{-1} + \mathbf{G}^\mathrm{T}\left(\mathbf{V}_\mathbf{e} + \tilde{\mathbf{V}}_\mathbf{m}(p_1^2, p_2^2, \cdots, p_q^2)\right)^{-1}\mathbf{G}\right]^{-1}, \quad (13)$$

$$\tilde{\mathbf{V}}_\mathbf{t}(p_1^2, p_2^2, \cdots, p_q^2) \equiv \mathbf{G}\mathbf{M}\mathbf{G}^\mathrm{T} + \mathbf{V}_\mathbf{e} + \tilde{\mathbf{V}}_\mathbf{m}(p_1^2, p_2^2, \cdots, p_q^2), \quad (14)$$

$$\mathbf{S} \equiv \left(\boldsymbol{R_e} - \boldsymbol{R_c}(\boldsymbol{T_0})\right)\left(\boldsymbol{R_e} - \boldsymbol{R_c}(\boldsymbol{T_0})\right)^\mathrm{T}. \quad (15)$$

Here, $\mathrm{tr}(\mathbf{X})$ is the trace of a matrix $\mathbf{X}$. In general, the rank of $\tilde{\mathbf{M}}'$, *i.e.*, n is very large in the cross-section adjustment method. So, the evaluation of $\ln|\tilde{\mathbf{M}}'|$ becomes expensive without any treatments. In the hyperparameters tuning using ABIC, the iterative calculation due to the determination of hyperparameters is needed. Moreover, this iterative calculation will be repeated for large samples in this paper. Thus, the numerical experiments cannot be performed within a realistic computational cost. To avoid this issue, ABIC is represented as follows (see Appendix):

$$\mathrm{ABIC} = \mathcal{L} + 2q, \quad (16)$$

where

$$\mathcal{L} \equiv \ln\left|\tilde{\mathbf{V}}_\mathbf{t}(p_1^2, p_2^2, \cdots, p_q^2)\right| + \mathrm{tr}\left[\mathbf{S}\tilde{\mathbf{V}}_\mathbf{t}^{-1}(p_1^2, p_2^2, \cdots, p_q^2)\right] + r\ln(2\pi). \quad (17)$$

The equivalence of Eq. (12) and Eq. (17) was confirmed by preliminary numerical experiments.

Generally, the dimension of $\tilde{\mathbf{V}}_\mathbf{t}$, *i.e.*, r is much smaller than n. As described in Sec. 3, the total number of experimental data r is 33 in the present numerical experiments and sufficiently smaller than the total number of nuclear data $n = 14230$. Therefore, ABIC is evaluated from Eqs. (16) and (17) instead of Eqs. (10) and (12) in this study.

The covariance $\tilde{\mathbf{V}}_\mathbf{t}$ generally has a wide range of values with different orders. Then, the determinant of such matrices can easily cause a numerical underflow/overflow even if the rank of the matrix is not so high. Therefore, some treatment should be needed to stably evaluate ABIC. Various methods can be considered to avoid this numerical difficulty, but from the viewpoint of computation time, we adopted the "linalg.slogdet" method of the NumPy library [11]. This method computes a determinant via LU factorization using a LAPACK [12] routine.

The procedure of the cross-section adjustment using ABIC consists of two steps as shown in Fig. 2. The new procedure of covariance modeling is added before the conventional cross-section adjustment procedure. In this procedure, the covariance model with hyperparameters is assumed first. Then, the hyperparameters will be determined so that ABIC is minimized. As the solver of the ABIC minimization, the SciPy library [13] will be used. Finally, based on the hyperparameters determined by ABIC, the suitable covariance matrix $\hat{\mathbf{V}}_\mathbf{m}(\widehat{p_1^2}, \widehat{p_2^2}, \cdots, \widehat{p_q^2})$ is selected. The cross-section adjustment is performed using this selected covariance.

1. Covariance modeling

- Assumption of covariance model with hyperparameters p_i^2:

$$\tilde{\mathbf{V}}_\mathbf{m} = \tilde{\mathbf{V}}_\mathbf{m}(p_1^2, p_2^2, \cdots, p_q^2)$$

-Tuning of hyperparameters:

$$\widehat{p_1^2}, \widehat{p_2^2}, \cdots, \widehat{p_q^2} = \underset{p_1^2, p_2^2, \cdots, p_q^2}{\mathrm{argmin}}\ \mathrm{ABIC}(p_1^2, p_2^2, \cdots, p_q^2)$$

-Covariance model selection:

$$\hat{\mathbf{V}}_\mathbf{m} = \hat{\mathbf{V}}_\mathbf{m}\left(\widehat{p_1^2}, \widehat{p_2^2}, \cdots, \widehat{p_q^2}\right).$$

↓

2. Cross-section adjustment

$$\boldsymbol{T}' = \boldsymbol{T_0} + \mathbf{M}\mathbf{G}^\mathrm{T}\hat{\mathbf{V}}_\mathbf{t}^{-1}\left(\boldsymbol{R_e} - \boldsymbol{R_c}(\boldsymbol{T_0})\right),$$
$$\mathbf{M}' = \mathbf{M} - \mathbf{M}\mathbf{G}^\mathrm{T}\hat{\mathbf{V}}_\mathbf{t}^{-1}\mathbf{G}\mathbf{M}$$

where

$$\hat{\mathbf{V}}_\mathbf{t} \equiv \mathbf{G}\mathbf{M}\mathbf{G}^\mathrm{T} + \mathbf{V}_\mathbf{e} + \hat{\mathbf{V}}_\mathbf{m}.$$

Fig. 2. The procedure of cross-section adjustment using ABIC

3 Numerical experiments

To evaluate the applicability of ABIC to the cross-section adjustment method, some numerical experiments using the random sampling method were performed in this section. Thereby, we generated virtual nuclear data sets and integral experimental data sets based on the virtual true covariances. Using the random sampling data sets, the cross-section adjustment was performed. After that, the performance of ABIC will be discussed from the relationship between ABIC and the variance of an estimation error resulting from the cross-section adjustment method.

3.1 Conditions

3.1.1 Procedure of random sampling

First, to perform the random sampling, we virtually assumed the true covariance matrices $\mathbf{M}$, $\mathbf{V}_\mathbf{e}$, and $\mathbf{V}_\mathbf{m}$. We employed the covariance matrices in our integral experimental database [7, 8] for fast reactors as the true covariance matrices. Sensitivity coefficients $\mathbf{G}$ were also prepared from the database. As the integral experimental data for the numerical experiments, the experimental data related to criticality measurements used for the development of ADJ2017 were selected. Thirty-three experimental data listed in Table 1 were used.

It was assumed that these covariance matrices were all positive definite matrices. The singular value decomposition was performed on the original covariance matrix, and all the singular values which are smaller than a setting value Λ were replaced with Λ in advance. We set $\Lambda = 1.0 \times 10^{-8}$ here. The rank of $\mathbf{M}$ is the same as the dimension of the nuclear data set, $n =$ 14,230, and the ranks of $\mathbf{V_e}$ and $\mathbf{V_m}$ are the same as the dimension of the integral experimental data set, $r = 33$.

The procedure of the random sampling based on these prepared covariance matrices is described below. The i-th sample related to a nuclear data set $\boldsymbol{\Delta T}_i \equiv [\boldsymbol{T} - \boldsymbol{T_0}]_i$ was generated from the probability distribution of Eq. (1). Namely, the obtained sample follows the following multivariate normal distribution:

$$\boldsymbol{\Delta T}_i \sim \mathcal{N}(\mathbf{0}, \mathbf{M}), \tag{18}$$

and the ith sample related to an integral experimental data set $\boldsymbol{\Delta R}_i(\boldsymbol{T}) \equiv [\boldsymbol{R_e} - \boldsymbol{R_c}(\boldsymbol{T})]_i$ was generated from the probability distribution of Eq. (2), Namely, the obtained sample obeys the following multivariate normal distribution:

$$\boldsymbol{\Delta R}_i(\boldsymbol{T}) \sim \mathcal{N}(\mathbf{0}, \mathbf{V_e} + \mathbf{V_m}). \tag{19}$$

Here, $i = 1,2, \ldots, N$. We set sample size $N = 10{,}000$ in this paper. The sample of $\boldsymbol{\Delta R}_i(\boldsymbol{T_0}) \equiv [\boldsymbol{R_e} - \boldsymbol{R_c}(\boldsymbol{T_0})]_i$ which was used for the cross-section adjustment was obtained from $\boldsymbol{\Delta R}_i(\boldsymbol{T})$ correcting $\mathbf{G}\boldsymbol{\Delta T}_i$ based on the linear relationship of Eq. (8), *i.e.*,

$$\boldsymbol{\Delta R}_i(\boldsymbol{T_0}) = \boldsymbol{\Delta R}_i(\boldsymbol{T}) + \mathbf{G}\boldsymbol{\Delta T}_i. \tag{20}$$

Using this sample to the adjustment formula of Eq. (5), $\boldsymbol{\Delta T}'_i \equiv [\boldsymbol{T}' - \boldsymbol{T_0}]_i$ was evaluated as

$$\boldsymbol{\Delta T}'_i = \mathbf{M}\mathbf{G}^{\mathrm{T}}\widehat{\mathbf{V}}_{\mathbf{t}}^{-1}\boldsymbol{\Delta R}_i(\boldsymbol{T_0}). \tag{21}$$

Note that, since there is no way to know the true covariance matrix in the real world, $\widehat{\mathbf{V}}_{\mathbf{m}}$ was determined by minimizing ABIC and $\boldsymbol{\Delta T}'_i$ were estimated by Eq. (21) using $\widehat{\mathbf{V}}_{\mathbf{t}}$ which includes $\widehat{\mathbf{V}}_{\mathbf{m}}$.

To evaluate the performance of ABIC, the estimation error of the cross-section adjustment was defined as $\boldsymbol{\Delta T}'_i - \boldsymbol{\Delta T}_i$. Here, $\boldsymbol{\Delta T}_i$ was generated from the true covariance $\mathbf{M}$, and $\boldsymbol{\Delta T}'_i$ was evaluated from the sample $\boldsymbol{\Delta R}_i(\boldsymbol{T_0})$ generated from the true covariances $\mathbf{V_e} + \mathbf{V_m}$ and the estimated covariance $\widehat{\mathbf{V}}_{\mathbf{t}}$. To facilitate the interpretation of results, the estimation error was projected onto a sensitivity vector $\boldsymbol{g}$ as

$$\Delta r_i \equiv \boldsymbol{g}(\boldsymbol{\Delta T}'_i - \boldsymbol{\Delta T}_i). \tag{22}$$

Here, $\boldsymbol{g}$ was the sensitivity of the criticality of the JSFR core listed in Table 1. This core had been proposed as a design example for a next-generation sodium-cooled fast reactor core. Δr_i means the estimation error of the target core characteristic.

Moreover, the statistics for Δr_i were investigated. The following sample mean $\overline{\Delta r}$ and sample variance u were also calculated:

$$\overline{\Delta r} = \frac{1}{N}\sum_{i=1}^{N} \Delta r_i \tag{23}$$

$$u = \frac{1}{N-1}\sum_{i=1}^{N} (\Delta r_i - \overline{\Delta r})^2 \tag{24}$$

Table 1. Cores of integral experimental data and target core

Facility	Core name	Feature
Cores of integral experimental data		
ZPPR	9, 10A, 10B, 10C	600-800 MWe-class, two-region homogeneous MOX cores.
	13A	650 MWe-class, radially-heterogeneous MOX cores.
	18A, 18C, 19B	1000 MWe-class, two-region homogeneous MOX cores with enriched-uranium regions.
ZEBRA	MZA	550 liter-sized one-region MOX core as a clean benchmark.
	MZB, MZC	2,300 liter-sized two-region homogeneous MOX cores to simulate the prototype fast reactor MONJU
JOYO	MK-I (64, 70 assemblies)	300 liter-sized 50/75MWt fast power reactor core with MOX and enriched-uranium fuel with blanket
MONJU	1994, 2010	280MWe prototype fast breeder reactor with two-region homogeneous MOX core
BFS	62s, 66-1	3,400 liter-sized three or four-region enriched-uranium and/or MOX fuel cores with or without radial blankets.
	67s, 69s, 66-2s	10kg of NpO_2 loading cores in central MOX region with various grade Pu.
LANL	FLATTOP -Pu, -25, GODIVA, JEZEBEL, JEZEBEL -Pu240	Sphere-shaped cores of approx. 10 cm in diameter with metallic fuel consisting of Pu-239, degraded Pu or U-235.
Target core		
JSFR	U-Pu	750MWe next-generation fast reactor with two-region homogeneous MOX core

3.1.2 Method to evaluate the performance of ABIC

We describe the method to evaluate the performance of ABIC using the results obtained from random sampling here. The performance was evaluated using the property of u described below.

If the selected covariance matrix $\widehat{\mathbf{V}}_{\mathbf{m}}$ is equal to the true one, *i.e.*, if $\widehat{\mathbf{V}}_{\mathbf{m}} = \mathbf{V_m}$, it is expected that the sample Δr_i follows Gauss distribution with a mean of 0 and a variance of $v = \boldsymbol{g}\mathbf{M}'\boldsymbol{g}^{\mathrm{T}}$. Note $\mathbf{M}'$ was nuclear data covariance after adjustment evaluated from the true covariance matrix $\mathbf{V_m}$. Hereafter, we call v as "reference variance." The distribution of Δr_i is shown in Fig. 3. It is found that the distribution follows the theoretical distribution as expected. The values of variance u and v are also consistent when $\widehat{\mathbf{V}}_{\mathbf{m}} = \mathbf{V_m}$, as shown in Table 2.

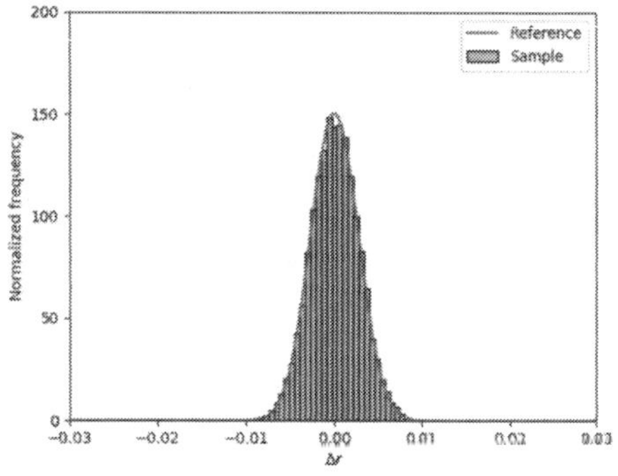

Fig. 3. Histogram of Δr (when $\widehat{\mathbf{V}}_{\mathbf{m}} = \mathbf{V}_{\mathbf{m}}$)

Table 2. Comparison example of sample variances u with the reference variance v

$\widehat{\mathbf{V}}_{\mathbf{m}}$	Sample variance u [$\times 10^{-6}$]	Reference variance v [$\times 10^{-6}$]
$\mathbf{V}_{\mathbf{m}}$	7.2	7.1
$0.1^2\mathbf{V}_{\mathbf{m}}$	24.0	
$10^2\mathbf{V}_{\mathbf{m}}$	10.7	

If we misestimate the evaluation of $\mathbf{V}_{\mathbf{m}}$, for example, when $\widehat{\mathbf{V}}_{\mathbf{m}} = 0.1^2\mathbf{V}_{\mathbf{m}}$ and $\widehat{\mathbf{V}}_{\mathbf{m}} = 10^2\mathbf{V}_{\mathbf{m}}$, the value of u becomes larger than the reference variance v as shown in Table 2 and Figs. 4 and 5. The value of u becomes minimum when $\widehat{\mathbf{V}}_{\mathbf{m}} = \mathbf{V}_{\mathbf{m}}$. It is expected that the value of u becomes larger as $\widehat{\mathbf{V}}_{\mathbf{m}}$ gets further away from the true one. Especially, the underestimation of $\mathbf{V}_{\mathbf{m}}$ has a larger impact on u than the overestimation of $\mathbf{V}_{\mathbf{m}}$.

Based on this property, we interpret the covariance model showing the smallest u (*i.e.*, closest u to the reference variance v) as the best model. We investigate the performance of ABIC based on whether the best model can be found by minimizing ABIC or not. Namely, we check for the consistency of the minimum point between ABIC and u.

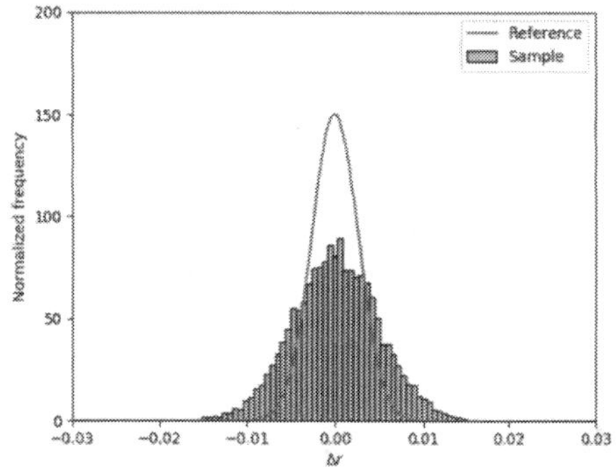

Fig. 4. Histogram of Δr (when $\widehat{\mathbf{V}}_{\mathbf{m}} = 0.1^2\mathbf{V}_{\mathbf{m}}$)

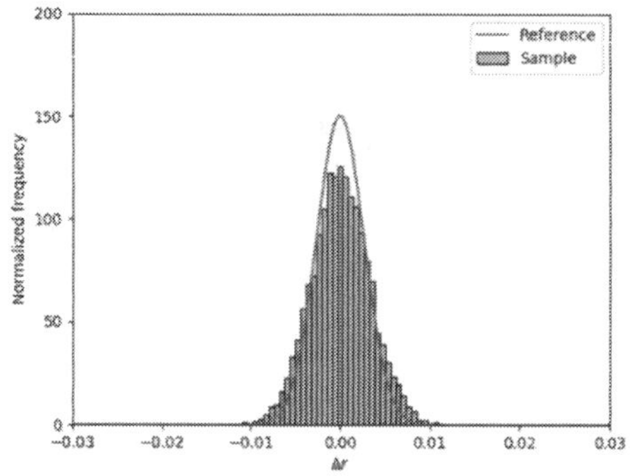

Fig. 5. Histogram of Δr when ($\widehat{\mathbf{V}}_{\mathbf{m}} = 10^2\mathbf{V}_{\mathbf{m}}$)

3.1.3 Assumption of covariance models $\widetilde{\mathbf{V}}_{m}$

The assumptions of the covariance models $\widetilde{\mathbf{V}}_{\mathbf{m}}$ corresponding to the unknown true covariance $\mathbf{V}_{\mathbf{m}}$ are listed in Table 3. The three kinds of covariance models $\widetilde{\mathbf{V}}_{\mathbf{m}}$ were prepared for the numerical experiments. Since the true covariance $\mathbf{V}_{\mathbf{m}}$ is unknown, assumptions for the covariance model $\widetilde{\mathbf{V}}_{\mathbf{m}}$ are necessary. Therefore, in the numerical tests, the covariance models $\widetilde{\mathbf{V}}_{\mathbf{m}}$ have been prepared with some modifications to the true covariance $\mathbf{V}_{\mathbf{m}}$.

Table 3. Assumptions of $\widetilde{\mathbf{V}}_{\mathbf{m}}$

$\widetilde{\mathbf{V}}_{\mathbf{m}}$	Number of hyper-parameters q	Feature of the cases
$p^2\mathbf{V}_{\mathbf{m}}$	1 (const.)	$\widetilde{\mathbf{V}}_{\mathbf{m}}$ can reproduce $\mathbf{V}_{\mathbf{m}}$ by hyperparameter tuning.
$p^2(\mathbf{V}_{\mathbf{m}} + a\mathbf{I})$ $a\mathbf{I}$: noise matrix a: positive const.	1 (const.)	$\widetilde{\mathbf{V}}_{\mathbf{m}}$ cannot reproduce $\mathbf{V}_{\mathbf{m}}$ by hyperparameter tuning.
$\sum_{k=1}^{K} p_k^2\, \boldsymbol{u}_k \boldsymbol{u}_k^{\mathrm{T}}$ $\boldsymbol{u}_k$: singular vectors of $\mathbf{V}_{\mathbf{m}}$	1 to K	Number of hyperparameters q is not constant.

The difference between the first and second cases is whether $\widetilde{\mathbf{V}}_{\mathbf{m}}$ can reproduce the true covariance $\mathbf{V}_{\mathbf{m}}$ or not by tuning the hyperparameter. In these cases, the number of hyperparameters was assumed to be constant. On the other hand, the third case had multiple hyperparameters.

The covariance structure and the singular values of the true covariance matrix $\mathbf{V}_{\mathbf{m}}$ used in the numerical experiments are shown in Figs. 6 and 7, respectively.

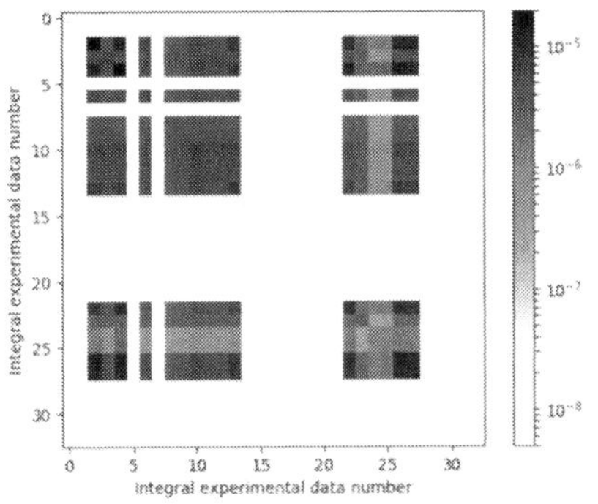

Fig. 6. Covariance structure of $\mathbf{V_m}$

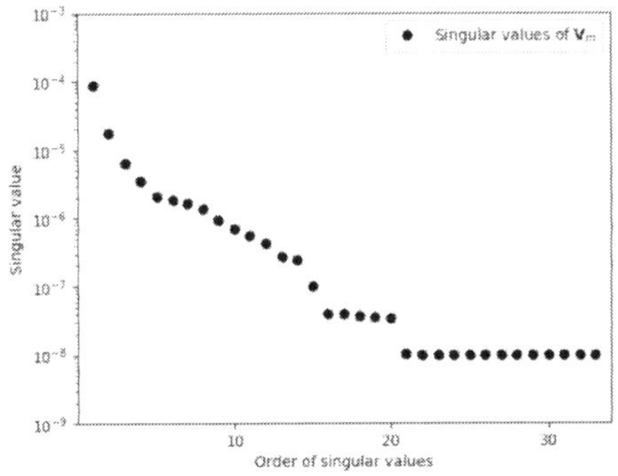

Fig. 7. Distribution for the singular values of $\mathbf{V_m}$

3.2 Results

3.2.1 The case of $\widetilde{V}_m = p^2 V_m$

As described above, in the first case, the covariance model $\widetilde{\mathbf{V}}_{\mathbf{m}}$ can reproduce the true covariance $\mathbf{V_m}$ by tuning the hyperparameter.

The histogram of the determined hyperparameter for each sample is shown in Fig. 8. The singular distribution was observed around $\widehat{p^2} = 0$. The mean of $\widehat{p^2}$ became somewhat smaller than the expected value, $p^2 = 1$. These could be caused by the assumption of a positive value as the range of the hyperparameter.

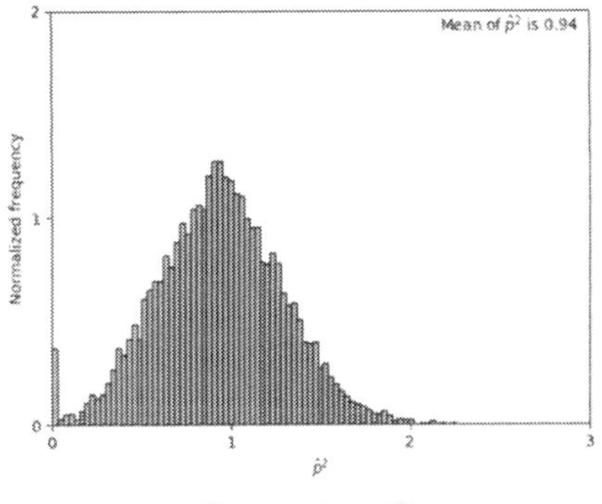

Fig. 8. Histogram of $\widehat{p^2}$ (when $\widetilde{\mathbf{V}}_{\mathbf{m}} = \widehat{p^2}\mathbf{V_m}$)

The evaluated value of the sample variance u is listed in Table 4 with the reference variance v. The histogram of Δr_i is shown in Fig. 9. The shape of the histogram of Δr_i is very close to that of the reference one. In fact, the sample variance u and the mean of the reference variance v are close. Consequently, in the first numerical experiment, ABIC showed good performance to find an appropriate covariance model.

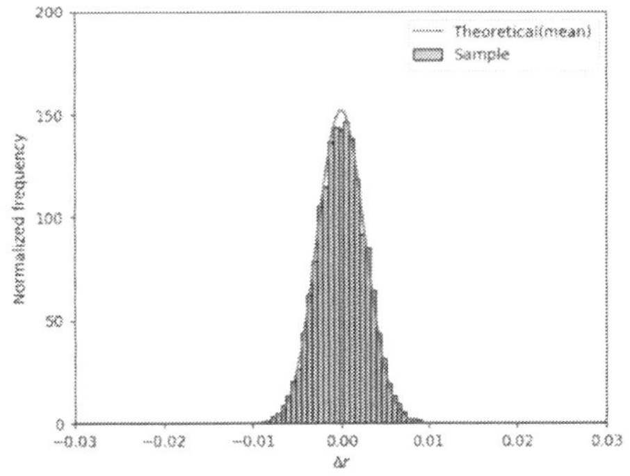

Fig. 9. Histogram of Δr (when $\widetilde{\mathbf{V}}_{\mathbf{m}} = \widehat{p^2}\mathbf{V_m}$)

Table 4. Comparison between sample variance u and the reference variance v in the first case.

$\widetilde{\mathbf{V}}_{\mathbf{m}}$	Sample variance u [$\times 10^{-6}$]	Reference variance v [$\times 10^{-6}$]
$\widehat{p^2}\mathbf{V_m}$	7.4	7.1

3.2.2 The case of $\widetilde{V}_m = p^2(V_m + aI)$

In the second case, the noise matrix $a\mathbf{I}$ was employed in the $\widetilde{\mathbf{V}}_{\mathbf{m}}$. Unlike $\widetilde{\mathbf{V}}_{\mathbf{m}}$ given in the first case, $\widetilde{\mathbf{V}}_{\mathbf{m}}$ cannot reproduce the true covariance matrix $\mathbf{V_m}$ due to the noise matrix $a\mathbf{I}$, even if the hyperparameter is tuned. Here, $\mathbf{I}$ is a unit matrix, and a is a constant value to be $a = 8.6 \times 10^{-5}$. The value of a was set equal to the maximum singular value of $\mathbf{V_m}$.

The histogram of Δr_i is shown in Fig. 10 and Table 5 shows the evaluated values of the mean of ABIC and u. The result of the first case of $\widetilde{\mathbf{V}}_{\mathbf{m}} = p^2\mathbf{V_m}$ is also shown in this table for comparison. The agreement of the histogram with the reference distribution becomes worse compared to the first case. The sample variance also becomes larger than that of the first case. As if to correspond to this phenomenon, the mean value of ABIC becomes larger. From this result, ABIC can reflect the badness of the covariance modeling of the second case compared to that of the first case. As far as this result is concerned, ABIC seems to adequately reflect the goodness of covariance modeling.

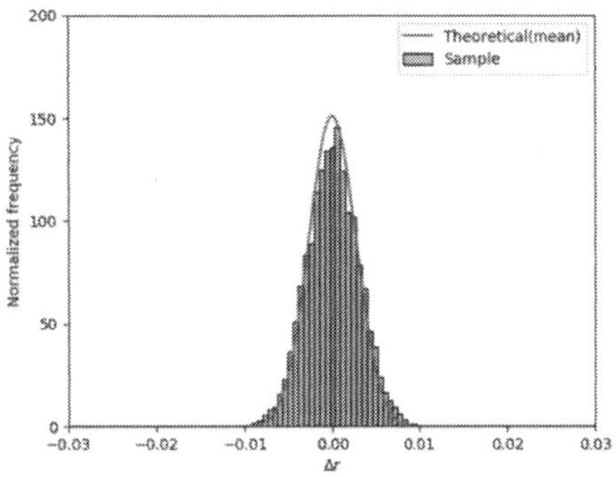

Fig. 10. Histogram of Δr (when $\hat{\mathbf{V}}_{\mathbf{m}} = \widehat{p^2}(\mathbf{V}_{\mathbf{m}} + a\mathbf{I})$)

Table 5. Comparison of the mean of ABIC and sample variance u between the first and the second cases.

$\hat{\mathbf{V}}_{\mathbf{m}}$	Mean of ABIC	Sample variance u [$\times 10^{-6}$]	Reference variance v [$\times 10^{-6}$]
$\widehat{p^2}(\mathbf{V}_{\mathbf{m}} + a\mathbf{I})$	−352	13.6	7.1
$\widehat{p^2}\mathbf{V}_{\mathbf{m}}$	−362	7.4	

However, a more detailed analysis reveals that the performance of ABIC will become poor when the covariance structure differs greatly from the true one. For the first and second cases, Figs. 11 and 12 show the changes in the mean of ABIC and the sample variance u according to the hyperparameter $\widehat{p^2}$ change, respectively. The minimum points of the mean of ABIC and the sample variance u are shown as the white and black arrows in these figures. Figure 11 represents the results when we assumed $\tilde{\mathbf{V}}_{\mathbf{m}} = p^2\mathbf{V}_{\mathbf{m}}$. Both minimum points are matching between the sample mean of ABIC and sample variance u in the first case. This matching result shows that ABIC can find the optimal hyperparameter. On the other hand, Fig. 12 represents the results when we assumed $\tilde{\mathbf{V}}_{\mathbf{m}} = p^2(\mathbf{V}_{\mathbf{m}} + a\mathbf{I})$. In contrast to the previous result in Fig. 11, the mismatching of the minimum point was observed in the second case. This mismatching represents the difficulty in finding the optimal hyperparameters using ABIC. The result of the second numerical experiment implies that the mismatch tends to increase as the covariance structure differs.

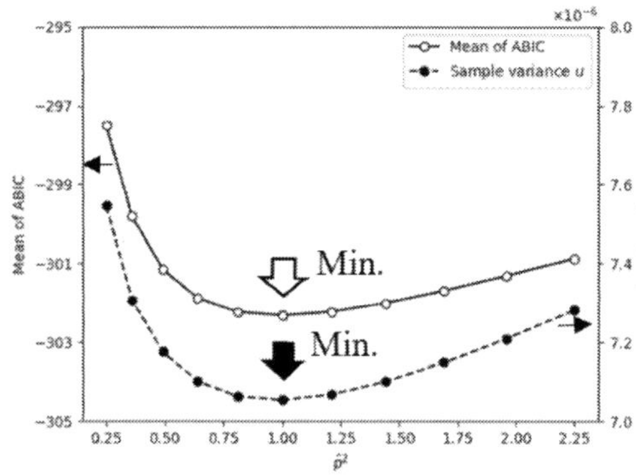

Fig. 11. Relationship between changes in the mean of ABIC and sample variance u with changes in the hyperparameters (when $\hat{\mathbf{V}}_{\mathbf{m}} = \widehat{p^2}\mathbf{V}_{\mathbf{m}}$)

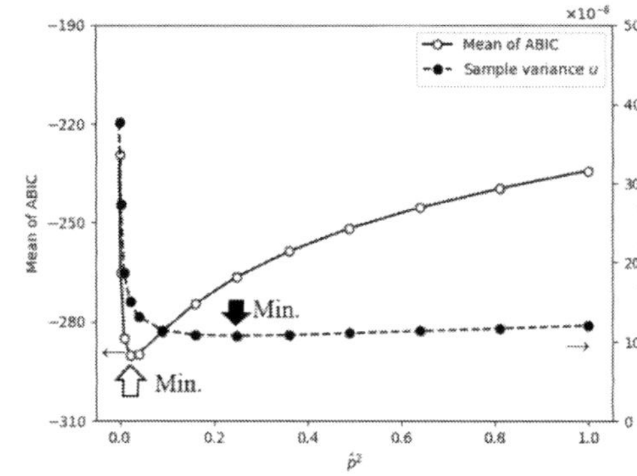

Fig. 12. Relationship between changes in the mean of ABIC and sample variance u with changes in the hyperparameters (when $\hat{\mathbf{V}}_{\mathbf{m}} = \widehat{p^2}(\mathbf{V}_{\mathbf{m}} + a\mathbf{I})$)

3.2.3 The case of $\tilde{\boldsymbol{V}}_m = \sum_{k=1}^{K} p_k^2\, \boldsymbol{u}_k \boldsymbol{u}_k^T$

In the third case, we investigated whether ABIC can choose an optimal number of hyperparameters. To clarify this, the singular vectors $\boldsymbol{u}_k$ of $\mathbf{V}_{\mathbf{m}}$ were used for $\tilde{\mathbf{V}}_{\mathbf{m}}$. Here, $\boldsymbol{u}_k$ having a smaller number of k corresponds to the larger singular value.

The results of the sample mean of ABIC and sample variance u of when $K = 1$-8 are shown in Fig. 13. In addition, the result of the case $\tilde{\mathbf{V}}_{\mathbf{m}} = \mathbf{0}$ is also shown in this figure as the case of $K = 0$. As K increases, the mean of ABIC decreases rapidly at first due to the decrease of $\mathcal{L}$ (the first log likelihood term in the right-hand side of Eq. (16)), and then increases slowly due to the increase of $2q$ (the last penalty term in the right-hand side of Eq. (16)). The mean of ABIC becomes minimum when the reduction of the value of $\mathcal{L}$ is saturated. The sample mean of ABIC becomes minimum when $K = 4$. The sample variance u becomes minimum when $K = 3$. The minimum points of the mean of ABIC and u were almost consistent. Thus, the third numerical experiment demonstrates that ABIC can appropriately suggest the number of hyperparameters. In this case, ABIC and u are larger than in the first case. This is due to the following reasons: If the covariance modeling $\widetilde{\mathbf{V}}_\mathbf{m}$ is equal to the true covariance matrix $\mathbf{V}_\mathbf{m}$, the sample variance u is consistent with the reference variance v. However, for smaller K, the true covariance matrix $\mathbf{V}_\mathbf{m}$ cannot be perfectly reproduced by $\widetilde{\mathbf{V}}_\mathbf{m}$ even if the hyperparameters are tuned. For larger K, there are too many hyperparameters to accurately estimate $\mathbf{V}_\mathbf{m}$ by $\widetilde{\mathbf{V}}_\mathbf{m}$.

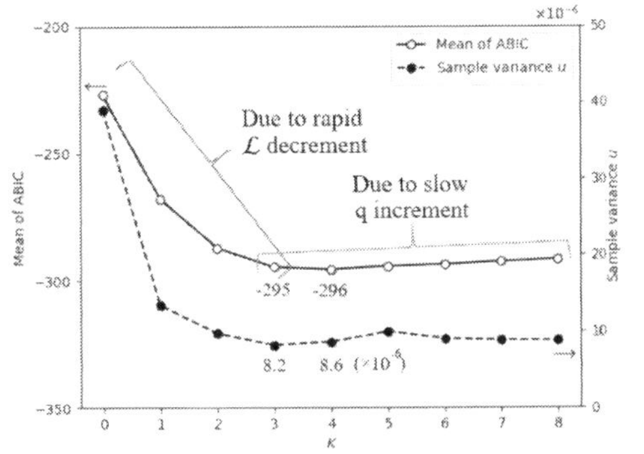

Fig. 13. Relationship between changes in the mean of ABIC and sample variance u

4 Conclusions

To improve the reliability of the cross-section adjustment method, the use of ABIC was proposed. We derived the formula to efficiently evaluate ABIC which was represented by a lower-rank matrix to enable numerical experiments with large samples in a realistic computation time.

To evaluate the applicability of ABIC to the cross-section adjustment method, some simple numerical experiments were performed in this study. Through these numerical experiments, the relationship between ABIC and u (*i.e.*, the variance of the estimation error of a core characteristic) was investigated. More specifically, the applicability of ABIC was discussed based on the consistency of the minimum points of ABIC and u. Consequently, it was confirmed that ABIC tends to select a covariance model with fewer hyperparameters and a smaller u. However, it was also found that this desirable property of ABIC will be lost if the structure of the covariance model was far from the true one. This is a future research problem to be solved when ABIC is applied to the cross-section adjustment method.

This work was supported by the Japan Society for the Promotion of Science (JSPS) Grant-in-Aid for Scientific Research (C) [Grant Number 21K04940].

References

1. M. Salvatores, G. Palmiotti, G. Aliberti, R. McKnight, P. Archier, C. De Saint Jean, E. Dupont, M. Herman, M. Ishikawa, K. Sugino, T. Ivanova, E. Ivanov, S.-J. Kim, I. Kodeli, A. Trkov, G. Manturov, S. Pelloni, C. Perfett, B.T. Rearden, A. Plompen, D. Rochman, W. Wang, H. Wu, W.-S. Yang, NEA/NSC/WPEC/DOC(2013)445, OECD/NEA (2013)
2. M. Salvatores, G. Aliberti, M. Dunn, A. Hogenbirk, A. Ignatyuk, M. Ishikawa, I. Kodeli, A.J. Koning, R. McKnight, R.W. Mills, P. Oblozinsky, G. Palmiotti, A. Plompen, G. Rimpault, Y. Rugama, P. Talou, W. S. Yang, Nuclear Science NEA/WPEC-26, OECD/NEA (2008)
3. G. Palmiotti, M. Salvatores, K. Yokoyama, M. Ishikawa, NEA/NSC/R(2016)6, OECD/NEA (2017)
4. M. Salvatores, G. Palmiotti, Proc. of Int. Conf. on Nuclear Data for Science and Technology, ND2019, EPJ Web of Conferences, **239**, 13001 (2020)
5. Covariance Data Utilization and Promotion Working Group, JAEA-Review 2021-014, Japan Atomic Energy Agency (2021) [In Japanese]
6. H. Akaike, Proceedings of the 2nd International Symposium on Information Theory, pp. 267-281, Budapest (1973)
7. K. Yokoyama, K. Sugino, M. Ishikawa, S. Maruyama, Y. Nagaya, K. Numata, T. Jin, JAEA-Research 2018-011 (2019) [in Japanese]
8. K. Yokoyama, S. Maruyama, H. Taninaka, S. Ohki, JAEA-Data/Code 2021-019 (2022) [in Japanese]
9. D. Siefman, M. Hursin, G. Schnabel, H. Sjöstrand, Annals of nuclear energy **159**, 108255 (2021)
10. C. M. Bishop, Pattern Recognition and Machine Learning (2006)
11. NumPy reference guide, release 1.19 [Internet]. NumPy.org; [updated 2020 Jun 29; cited 2022 Sep 27] Available from: https://numpy.org/doc/1.19/numpy-ref.pdf
12. LAPACK Documentation, release 3.10.1 [Internet]. Netlib.org; [updated 2022 Sep 27; cited 2022 Sep 27] Available from: https://netlib.org/lapack/explore-html/index.html
13. SciPy reference guide, release 1.4.1 [Internet]. SciPy.org; [updated 2019 Dec 19; cited 2022 Sep 27] Available from: https://docs.scipy.org/doc/scipy-1.4.1/scipy-ref-1.4.1.pdf
14. Gene H. Golub and Charles F. van Loan, Matrix Computations (1996)

Appendix

We prove the equivalence of Eqs. (12) and (17). Namely, the equivalence is represented by the following relationship:

$$\ln\left|\widetilde{\mathbf{M}}'\right| = \ln|\mathbf{M}| + \ln\left|\mathbf{V}_\mathrm{e} + \widetilde{\mathbf{V}}_\mathrm{m}\right| - \ln\left|\widetilde{\mathbf{V}}_\mathrm{t}\right|. \quad \text{(A. 1)}$$

$\ln\left|\widetilde{\mathbf{M}}'\right|$ in Eq. (12) is represented as

$$\ln\left|\widetilde{\mathbf{M}}'\right| = \ln\left|\left[\mathbf{M}^{-1} + \mathbf{G}^\mathrm{T}\left(\mathbf{V}_\mathrm{e} + \widetilde{\mathbf{V}}_\mathrm{m}\right)^{-1}\mathbf{G}\right]^{-1}\right|, \quad \text{(A. 2)}$$

using Eq. (13). From the Woodbury identity [8]:

$$(\mathbf{X} + \mathbf{YZY}^\mathrm{T})^{-1} = \mathbf{X}^{-1}$$

$$-\mathbf{X}^{-1}\mathbf{Y}(\mathbf{Z}^{-1}+\mathbf{Y}^{\mathrm{T}}\mathbf{X}^{-1}\mathbf{Y})^{-1}\mathbf{Y}^{\mathrm{T}}\mathbf{X}^{-1}, \quad (\mathrm{A.3})$$

Eq. (A.2) becomes

$$\ln\left|\widetilde{\mathbf{M}}'\right| = \ln\left|\mathbf{M}-\mathbf{M}\mathbf{G}^{\mathrm{T}}\widetilde{\mathbf{V}}_{\mathrm{t}}^{-1}\mathbf{G}\mathbf{M}\right| = \ln|\mathbf{M}| + \ln\left|\left(\mathbf{I}-\mathbf{G}^{\mathrm{T}}\widetilde{\mathbf{V}}_{\mathrm{t}}^{-1}\mathbf{G}\mathbf{M}\right)\right|, \quad (\mathrm{A.4})$$

where $\mathbf{I}$ is the unit matrix. Using the following relationship:

$$\ln|\mathbf{X}| = \mathrm{tr}[\ln(\mathbf{X})], \quad (\mathrm{A.5})$$

to the second term of Eq. (A.4), Eq. (A.4) becomes

$$\ln\left|\widetilde{\mathbf{M}}'\right| = \ln|\mathbf{M}| + \mathrm{tr}\left[\ln\left(\mathbf{I}-\mathbf{G}^{\mathrm{T}}\widetilde{\mathbf{V}}_{\mathrm{t}}^{-1}\mathbf{G}\mathbf{M}\right)\right]. \quad (\mathrm{A.6})$$

Furthermore, using the relationships:

$$\ln(\mathbf{I}-\mathbf{X}) = \sum_{m=1}^{\infty}\frac{1}{m}\mathbf{X}^{m}, \quad (\mathrm{A.7})$$

and

$$\mathrm{tr}(\mathbf{XYZ}) = \mathrm{tr}(\mathbf{YZX}) = \mathrm{tr}(\mathbf{ZXY}), \quad (\mathrm{A.8})$$

Eq. (A.6) is represented as

$$\ln\left|\widetilde{\mathbf{M}}'\right| = \ln|\mathbf{M}| + \mathrm{tr}\left[\sum_{m=1}^{\infty}\frac{1}{m}\left(\widetilde{\mathbf{V}}_{\mathrm{t}}^{-1}\mathbf{G}\mathbf{M}\mathbf{G}^{\mathrm{T}}\right)^{m}\right]. \quad (\mathrm{A.9})$$

Again, using the relationship of Eqs. (A.7) and (A.8) to Eq. (A.9), we obtain

$$\ln\left|\widetilde{\mathbf{M}}'\right| = \ln|\mathbf{M}| + \ln\left|\mathbf{I}-\widetilde{\mathbf{V}}_{\mathrm{t}}^{-1}\mathbf{G}\mathbf{M}\mathbf{G}^{\mathrm{T}}\right| = \ln|\mathbf{M}| + \ln\left|\mathbf{V}_{\mathrm{e}}+\widetilde{\mathbf{V}}_{\mathrm{m}}\right| - \ln\left|\widetilde{\mathbf{V}}_{\mathrm{t}}\right|. \quad (\mathrm{A.10})$$

EPJ Web of Conferences **281**, 00009 (2023) https://doi.org/10.1051/epjconf/202328100009
CW2022

Evaluation of covariance data in JENDL

Osamu Iwamoto[1*]

[1]Nuclear Data Center, Japan Atomic Energy Agency, 2-4 Shirakata, Tokai-mura, Naka-gun, Ibaraki 319-1195, Japan

Abstract. Evaluation of covariance for JENDL was virtually started after the release of JENDL-3.2. The covariance data were obtained for 16 nuclides and compiled to the JENDL-3.2 Covariance File. At the time of the JENDL-4.0 development, covariances were much enhanced especially for actinides; covariance data were given for 99 nuclides in total. The latest version JENDL-5 includes covariance data for 105 nuclides by adding new evaluations for light nuclides and structure materials. An overview of the covariance evaluation for JENDL is presented.

1 Introduction

Covariance of evaluated nuclear data libraries provides information about the uncertainties of the evaluated data including correlations. The covariance data enable users to estimate uncertainties originated from nuclear data. Therefore, it is important to provide the covariance data in the evaluated nuclear data libraries for nuclear data applications.

Concerning the covariance of JENDL, discussions were started as a next scope of JENDL-2 [1]. Those discussions were performed from both development and application points of view, including topics of evaluation methods on theoretical calculations, uncertainty in measurement, covariances in least square fitting, sensitivity analyses for reactivity analyses, decay heat estimations, and dosimetry utilizations. The first covariance data were provided in JENDL-3 released in 1989. However, the covariance data were given only for ^{55}Mn. To provide enough covariance data for applications, the Covariance Evaluation WG was established in 1993 under the Sigma Committee [2] and covariance data for 16 nuclides were evaluated for JENDL-3.2. The evaluation of covariance data has been continued for JENDL-3.3, JENDL-4.0 and JENDL-5 to increase both of the amounts and qualities.

At the beginning, major targets of the covariance data of JENDL were about fast reactors [3]. The covariance was used to increase the accuracy of the calculation for the fast reactors by means of nuclear data adjustment methods [4, 5]. Recently, the covariance data were also utilized to estimate the uncertainties for an innovative reactor, e.g., for accelerator-driven system [6]. In addition, to promote the use of the covariance data, a couple of working groups: Covariance Utilization WG (2011-2017), and Covariance Data Utilization and Promotion WG (2018-2020) were organized under the JENDL Committee (one of successors of the former Sigma Committee).

An overview of the evaluation and current status of covariance data in JENDL is presented, mainly for actinides.

2 Progress of covariance evaluation for JENDL

Figure 1 shows the number of nuclides with covariance data in the released libraries as JENDL so far. As shown in the figure, the evaluation of the covariance data was virtually started after the release of JENDL-3.2. For JENDL-3.2, the covariance data for 16 nuclides were evaluated and compiled as the JENDL-3.2 Covariance File. This covariance evaluation was performed with a motivation to give data for the nuclear data adjustment for fast reactor applications. The library included the data for the isotopes of H, B, O, Na, U and Pu as well as some of structural materials. Most of them were taken over to JENDL-3.3 with some new evaluations; the number of nuclides with covariance increased to 20 for JENDL-3.3. At the time of the JENDL-4.0 development, much efforts were made for covariance evaluation especially for actinides. Finally, the covariances of JENDL-4.0 were provided for the cross sections in the resonance to fast neutron energy regions, angular distributions of elastic scattering, and fission neutron related data. The number of nuclides with covariance increased to 99 (24% of all nuclides) in the updated file JENDL-4.0u. The latest version JENDL-5 provides covariance data for 105 nuclides by adding new evaluations for light nuclides and structure materials. However, due to the large increase of the total number of nuclides in JENDL-5, the percentage of isotopes with covariance data was down to 13% as shown in Fig. 1.

Regarding the evaluation methods, eye-estimation with experimental data was used to obtain the first covariance data for ^{55}Mn in JENDL-3.1. After that, for the JENDL-3.2 development, a covariance evaluation system KALMAN was developed [7]. KALMAN can be

* Corresponding author: iwamoto.osamu@jaea.go.jp

applied to various data to which nuclear model calculations were applied. In other case, the evaluation was sometimes performed with the least square method, which provided the most probable value with uncertainty and the covariance could be adopted to the libraries. The simultaneous evaluation of fission cross sections was one of those examples, the least square code SOK [8] was developed and applied to the simultaneous evaluations. It also provided the cross correlations between different nuclides. A new resonance analysis code AMUR [9] based on the R-matrix theory was developed for JENDL-5. It was mainly used for the evaluation of the light nuclides; the covariance of the cross sections of ^{16}O and ^{15}N were given based on the AMUR analyses in JENDL-5.

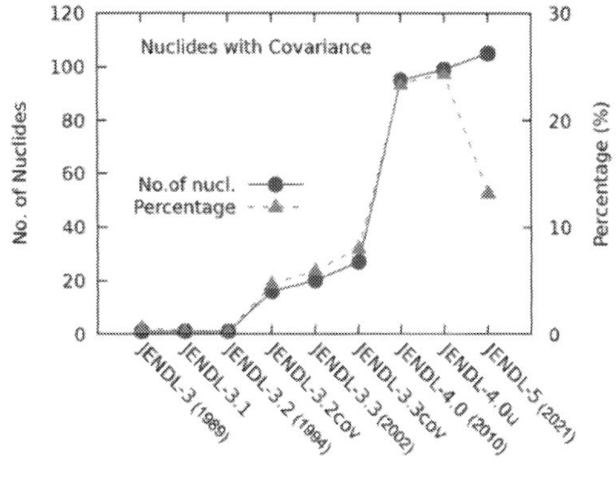

Fig. 1. The number of nuclides with covariance in JENDL

3 Covariance evaluation

The covariance evaluation was essentially based on the least square method with available experimental data. The KALMAN code developed for JENDL-3.2 was widely applied to the covariance evaluations in combination with the nuclear model codes used in the cross-section evaluations. By using the model codes depending on the evaluated data, the sensitivities of model parameters to the cross sections were calculated and applied to the least square fitting to obtain the model parameter covariance. While the nuclear reaction modelling codes were changed from the JENDL-3.2 evaluation to those of JENDL-4.0 or JENDL-5, the KALMAN code has been continuously used.

For non-model evaluations such as for fission cross sections, the values at certain energy grid points were estimated by the least square fitting with interpolation to the given energy grid; for those evaluations, the SOK [8] and GMA [10, 11] codes were applied. Both codes have similar capabilities, but SOK is better in use for the simultaneous evaluation. Covariance data obtained at the same time were adopted in JENDL with or without modifications. The least square method with a large number of data points would often give unrealistically small uncertainties which might be due to ignoring correlations and averaging the data points with a simplified model. In this case, the obtained covariance was sometimes modified to become more reasonable uncertainties by comparing with experimental data using expert judgement.

3.1 Resonance region

The resonance parameters of JENDL were sometimes taken from the published literatures, which usually gave only the resonance parameter uncertainties without correlations. The covariance data of those resonance parameters have been adopted for JENDL. However, the uncertainty of spectrum averaged cross-sections would be underestimated if the correlations were not given. To mitigate this underestimation, the background covariance was given as cross-section uncertainty.

The newly developed R-matrix resonance analysis code AMUR gives the cross sections with covariance data. For JENDL-5, AMUR was used for the evaluation of the resonance cross sections for the light nuclides of ^{16}O and ^{15}N, and the covariance data were also adopted.

Regarding the major actinides whose resonance parameters were taken from the other libraries, the covariance data of resonance region for $^{235,238}U$ and ^{239}Pu were also taken from the same library as the resonance parameter i.e., ENDF/B-VIII.0.

3.2 Simultaneous evaluation of fission cross section

The simultaneous evaluation was developed to estimate the fission cross sections of major actinides by taking into account the absolute and ratio measurements simultaneously for JENDL-3 for the first time. The covariance data deduced by the simultaneous evaluation were included in the JENDL-3.2 Covariance File. A new simultaneous evaluation code SOK was developed and applied to the evaluations for JENDL-3.3, JENDL-4.0 and JENDL-5. Figure 2 shows the uncertainties of the fission cross sections based on the simultaneous evaluation and the other evaluations in the IAEA Neutron Data Standards 2017 (IAEA-2017), ENDF/B-VIII.0 and JEFF-3.3. In general, those evaluated uncertainties have a relatively large differences each other, excepting between IAEA-2017 and ENDF/B-VIII.0 that adopts the IAEA data. Even for the JENDL evaluations that are derived by almost the same method,

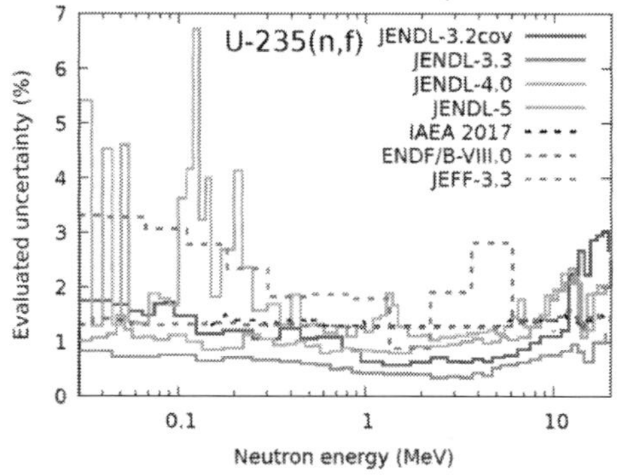

Fig. 2. Uncertainty of ^{235}U fission cross sections in JENDL and other evaluations

the differences are significant. The newest version JENDL-5 provides the largest uncertainties in spite of the accumulation of the experimental data. This situation may indicate difficulty of the evaluation of covariance. The evaluated results highly depend not only on the selection of the experimental data but also on the treatment of the experimental uncertainties. In addition, the least square method often gives unrealistically small uncertainties for fitting with a large number of the data points; to avoid this, the deduced uncertainty by the simultaneous evaluation was multiplied by a factor of 2 in the JENDL-4.0 case, where the factor was determined by the comparison with the experimental data by the evaluator.

3.3 Fission cross sections of minor actinides

The fission cross sections of minor actinides for fast neutrons were evaluated by GMA for JENDL-4.0 and most of them were carried over to JENDL-5. In the evaluation, statistical and systematic experimental uncertainties were carefully taken into account. Figure 3 shows the uncertainty and correlation of the ^{241}Am fission cross section deduced by GMA. Strong positive correlations are observed in a wide energy range; these correlations reflect the systematic uncertainties of the experimental data used for the evaluation.

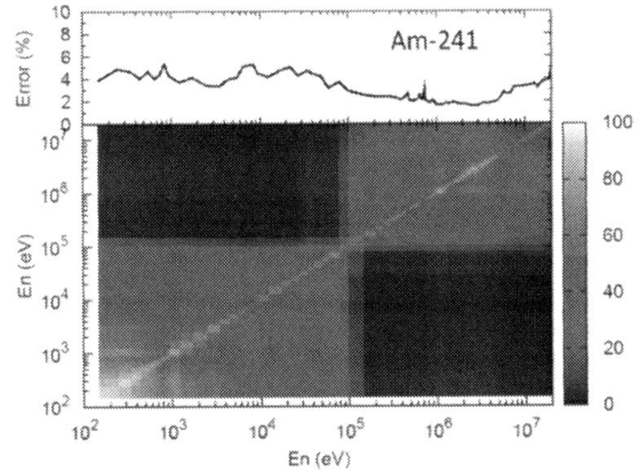

Fig. 3. Uncertainty and correlation of ^{241}Am fission cross section for fast neutrons

3.4 Capture cross section

A large part of the capture cross sections of the actinides of JENDL-4.0 was evaluated by CCONE and were taken over to JENDL-5. The KALMAN code system was applied to the evaluation of the covariance for them. The KALMAN calculation needed the sensitivities of the cross section to the model parameters. About 50 parameters such as optical model potential, level density, gamma-ray strength function and fission barrier were used.

The χ^2 value of the fitting can be a measure of the adequacy of the fitted results and the reduced chi-square may become close to 1 in an ideal case. However, in an actual case, due to unknown uncertainties of the experimental data and deficits of the nuclear modelling, large χ^2 values often come out. For the capture cross section, to compensate those deficiencies, the uncertainties of the experimental data were modified so as to become $\chi^2 = 1$ via fitting to each experimental data independently.

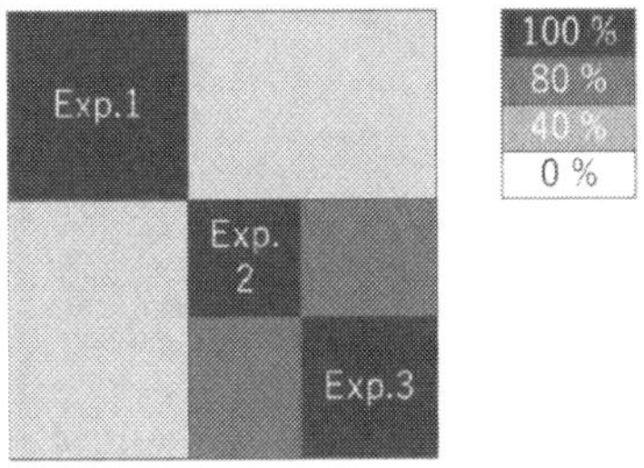

Fig. 4. Assumed correlation between experimental data sets

A generalized least square method used in KALMAN is usually applied to each experimental data by sequentially updating the covariances. However, this sequential update neglects the correlations between the experimental data sets. Because there are similarities between the experimental methods and analysis among the data sets, neglected correlations might exist. For the JENDL-4.0 evaluation, the correlations between the experimental data sets were introduced. Due to the difficulty for estimating the correlations, those correlations were assumed to be 0.8 for the data sets (e.g., Exp.2 and Exp. 3) by the same author, and 0.4 for different authors (e.g., Exp. 1 and Exp.2) with expert judgement as shown in Fig. 4.

Because the posterior cross section estimated by KALMAN would be changed from the original evaluated data, the covariance matrix was deduced as

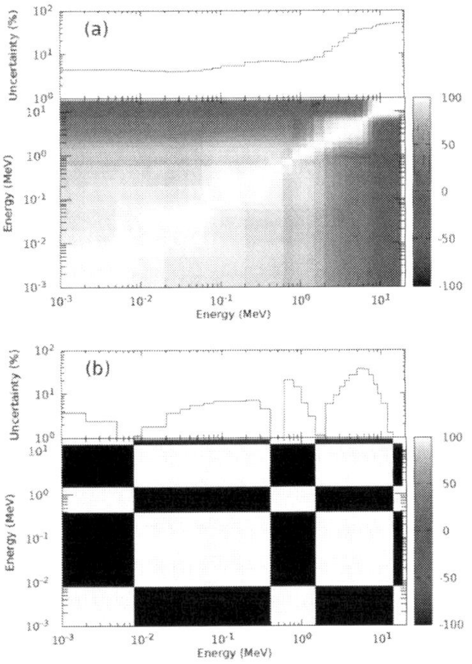

Fig. 5. Covariances of (a) KALMAN and (b) difference between posterior and original value for ^{237}Np capture cross section

covariance with respect to the original evaluated value σ_0^i, i.e.

$$C^{i,j} = \langle(\sigma^i - \sigma_0^i)(\sigma^j - \sigma_0^j)\rangle$$
$$= C_K^{i,j} + \Delta^i\Delta^j,$$

where i and j indicate the energy indexes, $C_K^{i,j}$ is a covariance deduced by KALMAN, Δ^i is difference between the posterior cross section and original value. Figure 5 shows covariance data for neutron capture cross section for ^{237}Np. Strong positive correlations are observed for covariance deduced by KALMAN, while the strong positive and negative correlations alternately appear in the component of the cross-section difference.

3.5 Other nuclear data evaluated with CCONE

For the nuclear data evaluated with CCONE, the covariances other than of the capture cross sections were deduced by a rather simple way. The KALMAN fitting was applied with the representative data that was evaluated to have the original value with the uncertainties to be consistent with the uncertainty and/or scatter of the experimental data points. With this method, posterior cross sections have the same values with the priors. Therefore, it was not needed to consider the change of the cross sections unlike for the capture cross section.

Regarding the angular distribution of elastic scattering of ^{238}U, the uncertainty of P_1 component at around 144 keV, where the P1 dominated the distribution, was estimated by comparing the experimental data. Figure 6 shows the uncertainties of the angular distributions calculated by assuming 5, 10 and 20 % uncertainties for P_1 coefficient. From the comparison the P_1 coefficient uncertainty was evaluated to be about 10 % and applied to KALMAN as a representative data to obtain the energy dependent covariances.

In the case of the nuclear data for which the experimental data were not available, the covariance was estimated with the global parameter uncertainties that were deduced by comparing the calculated results with the default model parameters and the experimental data for non-actinide nuclides, due to the difficulty in the prediction for fission cross sections with default parameters.

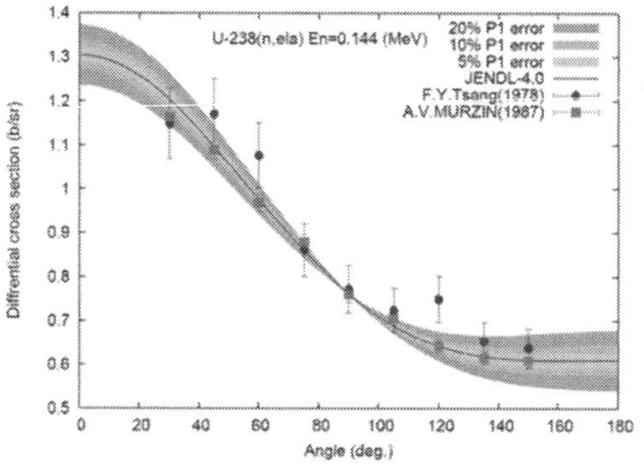

Fig. 6. Angular distribution of elastic scattering for ^{238}U. The shaged areas show uncertainties by assuming the uncertainties of the P_1 coefficient.

4 Conclusions

Overview of the covariance evaluation for JENDL was described. The covariance evaluation was started after JENDL-3. The latest version JENDL-5 provides the covariance data for 105 nuclides including a large part of the actinides that are needed for reactor applications. Evaluation methods and results for the covariance evaluation were briefly shown. Because the covariance data of JENDL are still limited in amount and quality in a viewpoint of utilization for various applications, they will be revised and enhanced in future JENDL releases.

The author acknowledges all of the contributors for JENDL covariance developments.

References

1. Sigma Committee, Proc. 1980 Seminar on Nuclear Data, JAERI-M 9523 (1981)
2. Y. Kanda, Nuclear Data News, **49**, 111 (1994), in Japanese
3. M. Ishikawa, Nuclear Data Sheets, **109**, 2778 (2008)
4. M. Ishikawa et al., J. Nucl. Sci. Technol., **39**: sup2, 1073 (2002)
5. K. Yokoyama et al., JAEA-Research 2018-011 (2019)
6. H. Iwamoto et al., J. Nucl. Sci. Technol., **50**, 856 (2013)
7. T. Kawano, K. Shibata, JAERI-Data/Code 97-037 (1997)
8. T. Kawano et al., J. Nucl. Sci. Technol., **37**, 327 (2000)
9. S. Kunieda et al, Nucl. Data Sheets, **118**, 250 (2014)
10. W. P. Poenitz, Proc. Conf. Nuclear Data Evaluation Methodsand Procedures, BNL-NCS-51363, Vol. I, 249 (1981)
11. S. Chiba, D. L. Smith, ANL/NDM-121 (1991)

A proposed method for addressing large unphysical uncertainties in mubar

D. Kent Parsons [1]*

[1]Nuclear Data Team, Group XCP-5, Los Alamos National Laboratory, Los Alamos NM, USA

Abstract. Mubar uncertainties in Section MF 34 MT 2 of ENDF/B-VIII.0 data are sometimes too large. Physical limitations of the bounds of mubar limit the maximum uncertainty to be < 1.0 and usually << 1.0. Two ad hoc methods are proposed to artificially constrain the randomly sampled values of mubar to allowable values. The application of NJOY's mubar covariance matrix to the "sandwich rule" is also discussed and a consistent way to apply sensitivity coefficients is described. Mubar uncertainties for the Jezebel k_{eff} were found to be 156 pcm – a value comparable to other evaluated cross section uncertainties.

1 Introduction

Elastic scattering mubar covariances are found in the MF 34 MT 2 section of the evaluated nuclear data files [1]. These uncertainties are given for mubar, the cosine of the lab scattering angle, from an elastic neutron scattering event. As calculated by NJOY [2], values are given in a groups by 1 vector for the Legendre P_1 out-scattering divided by the Legendre P_0 out-scattering for a given incident energy group. (Higher order mubar terms are not presently available from NJOY.) As such, the values of mubar are bounded between -1 (backward scattering) and +1 (forward scattering).

These physical limits on mubar also imply a constraint on the allowable values of the uncertainty of mubar. When the data found in the evaluation files for mubar uncertainty is too large, then the generation of correlated random samples of mubar becomes impossible without encountering unphysical (outside the -1 to 1 interval) values of mubar. A sampling technique used in NORTA [3] and Copula [4] methods is proposed as an ad hoc fix for the problems of large unphysical uncertainties in mubar. A second technique is also presented which uses a zscale-like transformation.

Additionally, a consistent application of NJOY produced mubar covariance data is presented for usage in the "sandwich rule" for the propagation of uncertainty. This is necessary, since NJOY produces a groups by 1 vector of relative variance values, while the relative sensitivity values produced by codes like SENSMG [5] and the "ksen" option of MCNP [6] are usually a matrix dimensioned as "groups by groups".

2 The constraints on the variance and standard deviation of mubar

For a distribution defined only between -1 and +1, the largest possible variance or standard deviation is 1.0. This occurs when ½ of the sampled points are at -1.0 and the other ½ are at 1.0. This configuration gives the largest possible spreading out of the points.

However, if a more realistic uniform (i.e., flat) mubar distribution is assumed, then the variance and the standard deviation are reduced further. For a uniform distribution between -1 and +1, the variance = 1/3 and the standard deviation = $1/\sqrt{3}$.

If a truncated normal distribution is forced into the interval, say at a 3 or 4 sigma truncation, then the standard deviation of mubar is even less.

Finally, as the scattering becomes very forward, then the allowable variance and standard deviations approach 0.0. Mubar = 1.0 implies that all mu values = 1.0. These uncertainty limits are shown in Figure 1.

*Corresponding author : dkp@lanl.gov

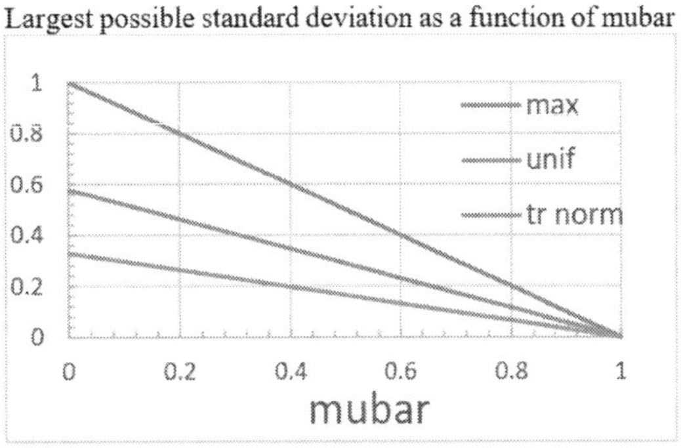

Fig 1: Constraints on mubar uncertainty due to the distribution and value of mubar

3 Proposed sampling techniques to limit the uncertainty on mubar

A version of a known statistical sampling method (NORTA – NORmal To Anything) has recently been employed to address these large uncertainties. The randomly sampled mubar points (which are initially drawn in a normal distribution) from the groups with large uncertainties are transformed into a uniform (i.e., flat) distribution which is constrained to fit inside of the allowable -1.0 to 1.0 interval in such a way as to preserve the average mubar. A CDF (cumulative distribution function) statistical function is used to transform the normal random samples to a uniform (i.e., flat) distribution (ranging initially from 0.0 to 1.0). This transformation imposes a hard limit on the variance of the group value of mubar of <1/3 and thus a limit on the standard deviation of <1/sqrt(3). For mubar = 0.0, the uniform CDF produced points between 0.0 and 1.0 are then scaled linearly to -1.0 to +1.0.

Using an EXCEL® function:

$$\text{uniform points} = \text{NORM.S.DIST}(z,\text{'true'}) \quad (1)$$

Where z = the standard scores for the normal random points

Using a MATLAB® function:

$$\text{uniform points} = \text{normcdf}(z) \quad (2)$$

The uniform points are distributed between 0 and 1 (based on the area under the normal curve measured from the left hand side). A large negative score for the normal random point is transformed to a value close to 0, while a large positive score is transformed to a value close to 1. The uniform points can then be scaled linearly to a range of -1 to 1 (for mubar = 0) or more generally to a subset of -1 to 1.

The basic transformation is illustrated in Figure 2 below. Notice how the pattern of distances between the adjacent points on the normal curve is mostly preserved in the uniform curve.

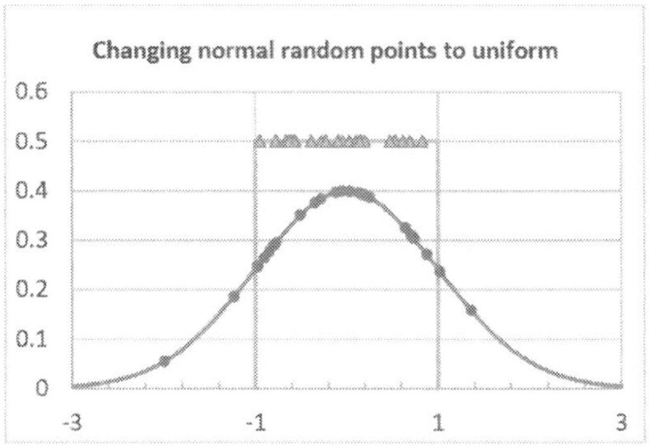

Fig. 2: Transformation of normal random points to a uniform distribution

If the value of mubar is not 0.0, then the limits of the uniform distribution are linearly scaled to a symmetric (i.e., about mubar) sub-interval that preserves mubar. For example, if mubar = 0.2, the uniform limits are -0.6 and 1.0. These results are illustrated in Figure 3.

Consistent with Figure 1, the standard deviation of the uniform distribution with mubar = 0.2 is smaller (0.46188) than the mubar = 0.0 distribution (0.57735). These values come from the formula for the standard deviation of uniform distributions.

$$\sigma = (b-a) / 2\sqrt{3} \quad (3)$$

where b = the right hand limit.
a = the left hand limit.

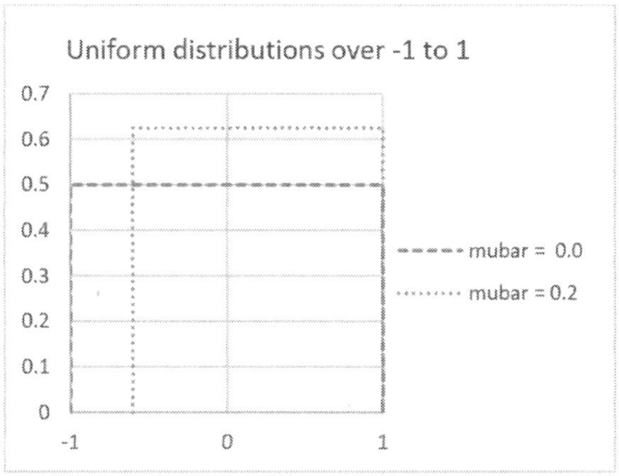

Fig 3: Comparison of mubar = 0 and 0.2 distributions

Fortunately, this transformation preserves the rank correlation (e.g., the largest and smallest points are still at their same locations in the vector) of the randomly sampled points – though it does not exactly preserve the linear correlation between the mubar data samples. However, the transformed samples all fit inside the allowed interval and the average group value of mubar is preserved.

Furthermore, iterative improvement of the linear correlation properties of the random samples is often possible. See the recently published tutorial on random sampling for nuclear data [7].

Another ad hoc sampling technique which could be used is to take the normal distribution random samples (which are spread out beyond the -1.0 to 1.0 interval) and reduce their standard deviation in a zscale-like fashion so that all the sampled points fit in between -1.0 and 1.0. Even though the normal distribution is infinite in extent, the random samples will very seldom be found beyond 3 or 4 sigma's away from the mean.

This zscale-like procedure preserves the mubar mean and the rank correlations, and also preserves the linear correlations of the original random samples. The drawback is that only very small values of the standard deviation can be accommodated. See Figure 1. The maximum value of the standard deviation would be (assuming mubar > 0) (1.0 – mubar) / (3 or 4) for 3 or 4 sigma's, respectively. These values are smaller than the uncertainty values accommodated by a uniform distribution.

4 A consistent application of mubar covariances in the sandwich rule

As produced by NJOY, mubar variances include values for each incoming group but with all outgoing groups summed together. Group to group covariance data is also given for the incident groups. Covariances may be produced in absolute or in relative form or even in correlation form.

Application of this mubar covariance data (traditionally used in relative form) in the "sandwich rule" requires elastic scattering P_1 sensitivities which are also calculated for each incoming group and with all the outgoing groups summed together. However, SENSMG [5] only produces P_1 elastic sensitivities in group-by-group form. The "ksen" option of MCNP [6] can produce groupwise P_1 elastic scattering sensitivities with or without group-by-group binning of the output groups. Sensitivities are also usually produced in relative form to be consistent with the relative covariance data.

If the sensitivities are given in group-by-group (i.e., doubly differential) form, then the sum of the P_1 elastic scattering outgoing group sensitivities for each incoming group should be used in the sandwich rule.

A thirty group ENDF/B-VIII.0 analysis of Jezebel k_{eff} uncertainty due to mubar uncertainty was carried out using MF 34 MT 2 elastic scattering mubar uncertainties for Pu-239. P_1 elastic sensitivities from SENSMG were group summed over the output groups for each input group, and the result of the sandwich rule was an uncertainty of 156 pcm. This is certainly comparable to the other Jezebel uncertainties studied in the CIELO project [see Table VIII of Reference 8].

For the simple summing of output group elastic P_1 sensitivities for each input group to be used consistently, the individual values of the P_1 out-scatter cross sections used in Sn calculations must each be changed by the same ratio of new mubar / old mubar. This consistency has been established by Jeff Favorite and verified by comparison with second order accurate central differencing estimates of the sensitivity coefficients [9]. (This has not yet been officially published.) He has also shown that if the change in mubar is allocated in other ways to the P_1 elastic scattering cross sections, then the sum of the output group sensitivities must be modified accordingly – usually in the form of a weighted sum.

It should also be noted that changing the elastic scattering P_1 cross sections in a S_n code library can be somewhat laborious. The elastic scattering cross sections are usually mixed in with inelastic, n2n, n3n, etc. cross sections in the combined Legendre scattering matrix.

5 Summary

Two ad hoc workarounds for limiting the bad effects of unphysical ENDF/B-VIII.0 mubar uncertainties in the context of random sampling have been proposed. The methods preserve mubar and ensure that all sampled points are within the desired valid interval or sub-interval. The variances and standard deviations of mubar are limited to physically valid values while preserving rank correlation. Linear correlation of the randomly sampled points is approximately preserved for the uniform method and exactly preserved for the zscale-like approach.

Hopefully, the physical constraints on the size of mubar uncertainties discussed here will be respected in future MF 34 MT 2 elastic scattering mubar evaluated data.

In connection with this proposal and these constraints, a consistent way to use the NJOY produced mubar covariance data in the sandwich rule with group-by-group sensitivities has been described. Further publications will follow.

References

1. A. Trkov, M. Herman, and D.A. Brove, "The ENDF-6 Formats Manual", BNL-90365-2009 Rev. 2, or CSEWG ENDF-102, Brookhaven National Laboratory, (2012).
2. R.E. MacFarlane, "The NJOY Nuclear Data Processing System, Version 2016", LA-UR-17-20093, Los Alamos National Laboratory, (2017).
3. M.C. Cairo and B.L. Nelson, "Modelling and generating random vectors with arbitrary marginal distributions and correlation matrix", Technique Report, Department of Industrial Engineering and Management Science, Northwestern University, Evanston, IL, 1997.

4. R. B. Nelson, **An Introduction to Copulas**, 2nd Edition, Springer Science+Business Media, Inc. 2006.
5. J.A. Favorite, "SENSMG: First-Order Sensitivities of Neutron Reaction Rates, Reaction-Rate Ratios, Leakage, keff, and α Using PARTISN," Nuclear Science and Engineering, **192**, 1, 80–114 (2018); https://doi.org/10.1080/00295639.2018.1471296. For the most current SENSMG information, see J.A. Favorite, "(U) SENSMG: First-Order Sensitivities of Neutron Reaction Rates, Reaction-Rate Ratios, Leakage, k_{eff}, α, and Subcritical Multiplication Using PARTISN," XCP-7:22–006(U), LA–UR–22–28919, August 22, 2022.
6. C.J. Werner (editor), "MCNP Users Manual - Code Version 6.2", LA-UR-17-29981 (2017).
7. D.K. Parsons, "A Tutorial for Generating Correlated Random Samples in the Context of Replica Cross Section Data Used in the Propagation of Uncertainty", LA-UR-22-28781, Los Alamos National Laboratory, (2022)
8. M. B. Chadwick, et al, « CIELO Collaboration Summary Results : International Evaluations of Neutron Reactions on Uranium, Plutonium, Iron, Oxygen, and Hydrogen », Nuclear Data Sheets, **184**, pp. 189-213, (2018).
9. J. A. Favorite, personal communication, (2022).

Discussion on variety of cross-section candidate obtained by cross-section adjustment method

Sohei Odani[1*], *Satoshi* Takeda[1], and *Takanori* Kitada[1]

[1] Osaka University, 2-1 Yamadaoka, Suita, Osaka 565-0871, Japan

Abstract. The cross-section adjustment method gives only one best-estimated cross-section set based on Bayesian theory. In this study, cross-section sets other than the best-estimated set were discussed. The cross-section sets were obtained by the random sampling method using a covariance matrix adjusted by the cross-section adjustment method. 1,000,000 cross-section sets were obtained by the random sampling method. From these cross-section sets, significant cross-section sets other than the best-estimated set were considered. 593,340 cross-section sets that accurately evaluate nuclear characteristics were obtained. The result of the cross-section adjustment method shows that the change due to U235 χ is the largest positive. However, k_{eff} is not changed to the positive side by U235 χ in 179,994 of the 593,340 cross-section sets obtained by the random sampling method. Although the best-estimated cross-section set is obtained by the cross-section adjustment method, the result suggests there are various candidates for the adjusted cross-section set.

1 Introduction

Accurate evaluation of nuclear characteristics is essential for the safe and efficient operation of nuclear reactors. Reduction of the uncertainties is important for higher evaluation accuracy. Among the uncertainties that lead to the uncertainty of nuclear characteristics, the uncertainty arising from the nuclear data is a major factor.

The cross-section adjustment method [1, 2] is one of the methods for reduction of the uncertainty arising from nuclear data. The cross-section adjustment method provides a best-estimate cross-section set whose calculated values reproduce experimental values well. However, it is pointed out that some of the results of the cross-section adjustment method have no nuclear physics justification [3]. In addition, it has been reported that there are various cross-section sets whose C/E (calculation/experiment) values are consistent with the results of the cross-section adjustment method due to the compensation effect [4, 5]. Some of these various cross-section sets may be more reliable than the result of the cross-section adjustment method. However, the significance of these various cross-section sets has not been assessed. Therefore, in this study, the significance of the various cross-section sets was discussed.

The random sampling method can generate cross-section sets based on covariance data and is used in the community of reactor physics [6-8]. In this study, the random sampling method was used to generate cross-section sets that accurately evaluate nuclear characteristics. Then, the significance of the cross-section set was assessed by comparison. The random sampling method is summarized in Sec. 2. A comparison of the cross-section sets obtained by the random sampling method with that by the cross-section adjustment method is shown in Sec. 3. Finally, the conclusion is summarized in Sec. 4.

2 Random sampling using cross-section adjustment method

For random sampling, an adjusted cross-section set and an adjusted covariance matrix obtained by the cross-section adjustment method were used. Assuming that experimental error, calculation error, and cross-section error are independent of each other, the adjusted cross-section set is obtained by

$$\boldsymbol{t}_{adj} = \boldsymbol{t}_0 + \mathbf{MG}^{\mathrm{T}}(\mathbf{GMG}^{\mathrm{T}} + \mathbf{V}_{\mathrm{e}} + \mathbf{V}_{\mathrm{m}})^{-1}(\mathbf{R}_{\mathrm{e}} - \mathbf{R}_{\mathrm{c}}(\boldsymbol{t}_0)), \quad (1)$$

where $\boldsymbol{t}_{adj}$ is the adjusted cross-section set, $\boldsymbol{t}_0$ is the original cross-section set, $\mathbf{M}$ is the covariance matrix, $\mathbf{G}$ is the sensitivity coefficient of the cross-section set, $\mathbf{V}_{\mathrm{e}}$ is the covariance matrix of the experimental error, $\mathbf{V}_{\mathrm{m}}$ is the covariance matrix of the calculation error, $\mathbf{R}_{\mathrm{e}}$ is the experimental value and $\mathbf{R}_{\mathrm{c}}$ is the calculated value. The adjusted covariance matrix $\mathbf{M}_{\mathrm{adj}}$ is obtained by

$$\mathbf{M}_{\mathrm{adj}} = \mathbf{M} - \mathbf{MG}^{\mathrm{T}}(\mathbf{GMG}^{\mathrm{T}} + \mathbf{V}_{\mathrm{e}} + \mathbf{V}_{\mathrm{m}})^{-1}\mathbf{GM}. \quad (2)$$

The multivariate normal random number according to the adjusted covariance matrix is obtained by

$$\boldsymbol{z} = \mathbf{A}\boldsymbol{x} + \boldsymbol{\mu}, \quad (3)$$

* Corresponding author: s-odani@ne.see.eng.osaka-u.ac.jp

where $\boldsymbol{z}$ is the vector whose elements are multivariate normal random numbers, $\boldsymbol{x}$ is the vector whose elements are standard normal random numbers, and $\boldsymbol{\mu}$ is the mean vector. When the adjusted covariance matrix is a positive semidefinite matrix, the adjusted covariance matrix is expanded using $\mathbf{A}$ as

$$\mathbf{M}_{\mathrm{adj}} = \mathbf{A}\mathbf{A}^{\mathrm{T}}. \quad (4)$$

To satisfy equation (4), $\mathbf{M}_{\mathrm{adj}}$ must be a positive semidefinite matrix. However, it has been reported that the covariance matrix may not be positive semidefinite [8-10]. This problem is caused by the distortion of eigenvalues due to rounding. In this study, when the covariance matrix is not positive semidefinite, the nearest covariance matrix, which is positive semidefinite matrix obtained by Higham's method [11], was used. In this case, the nearest covariance matrix $\mathbf{M}_{\mathrm{near}}$ is expanded using $\mathbf{B}$ as

$$\mathbf{M}_{\mathrm{near}} = \mathbf{B}\mathbf{B}^{\mathrm{T}}. \quad (5)$$

When using the nearest covariance matrix, the random number is obtained by

$$\boldsymbol{z} = \mathbf{B}\boldsymbol{x} + \boldsymbol{\mu}. \quad (6)$$

3 Random sampling using two uranium benchmarks

3.1 Calculation condition

The cross-section adjustment method was performed using the criticality of light water-moderated uranium benchmarks summarized in ICSBEP [12]. The benchmark problems are HEU-SOL-THERM-001-001 (HST1-1) and LEU-COMP-THERM-001-001 (LCT1-1). HST1-1 is high-enriched uranium fuel system and LCT1-1 is low-enriched uranium fuel system. In addition, both systems do not contain plutonium. Sensitivity coefficients and covariance data were used from a package included in the MCNP6.2 [13] packages, WHISPER [14]. Sensitivity coefficients included in WHISPER were calculated with ENDF/B-VII.1 [15]. The calculations used 100,000 neutrons per cycle, 100 skip cycles, and 600 active cycles. Covariance data in WHISPER is the same data that is distributed with SCALE 6.1 and is 44-group energy structure [16]. Then, the two benchmark problems are assumed to be independent, $\mathbf{V}_{\mathbf{e}}$ and $\mathbf{V}_{\mathbf{m}}$ have zero covariance elements, and the diagonal elements are each variance.

3.2 Selection of target nuclide reactions

Due to computer memory utilization issues, not all cross-sections of nuclides included in the benchmark problem were adjusted using the cross-section adjustment method. The adjusted cross-section was selected by the contribution to the uncertainty of criticality. The uncertainty of criticality due to a particular reaction is obtained by

$$\sigma_{\alpha}^{2} = \mathbf{G}_{\alpha}\mathbf{M}_{\alpha}\mathbf{G}_{\alpha}^{\mathrm{T}}, \quad (7)$$

where σ_{α} is the standard deviation of criticality due to a particular reaction, $\mathbf{G}_{\alpha}$ is the sensitivity coefficient of the reaction, and $\mathbf{M}_{\alpha}$ is the covariance matrix of the reaction.

In this study, the threshold was set at 1/100 of the uncertainty of criticality due to all reactions. Therefore, the cross-section with $\sigma_{\alpha} > 0.005$ % were adjusted by the cross-section adjustment method. The target nuclides for the cross-section adjustment method are summarized in Table 1.

Table 1. Target nuclides for the cross-section adjustment method

Nuclide	Reaction or parameter
U234	(n,γ)
U235	Elastic scattering, Inelastic scattering, (n,2n), (n,γ), ν, χ
U238	Elastic scattering, Inelastic scattering, (n,2n), (n,γ), ν, χ
H1	Elastic scattering, (n,γ)
N14	(n,p)
O16	Elastic scattering, Inelastic scattering
Al27	Elastic scattering, Inelastic scattering, (n,γ)
Cr52	Elastic scattering
Mn55	Elastic scattering
Fe56	Elastic scattering

3.3 Result and discussion

3.3.1 Results of cross-section adjustment method

The calculated and experimental values before the cross-section adjustment method are summarized in Table 2.

Table 2. The calculated and experimental values before the cross-section adjustment method

	Calculated value	Experimental value
HST1-1	0.9983 ± 0.0002	1.0000 ± 0.0025
LCT1-1	0.9996 ± 0.0001	0.9998 ± 0.0031

The changes in criticality for the two benchmark problems due to the cross-section adjustment method are summarized in Table 3. The amount of change in criticality due to each nuclide reaction in the adjusted cross-section set is summarized in Figure 1. Then, the amount of change in criticality due to each nuclide reaction $\Delta k/k$ is obtained by

$$\frac{\Delta k}{k} = \sum_{g} \mathbf{G}_g \frac{\Delta \boldsymbol{t}_g}{\boldsymbol{t}_g} \tag{8}$$

where g is a certain energy group, $\mathbf{G}_g$ is the sensitivity coefficient at a certain energy group, and $\Delta \boldsymbol{t}_g/\boldsymbol{t}_g$ is the change of cross-section at a certain energy group. As shown in Table 3, the difference between the adjusted and experimental values is small. Figure 1 shows that the C/E value is improved mainly by the adjustment of U235 χ.

Table 3. Changes in C/E-1 values of criticality with cross-section adjustment method

	Before adjustment C/E-1 [%]	After adjustment C/E-1 [%]
HST1-1	-0.172	-0.016
LCT1-1	-0.021	0.016

C/E : Calculation/Experiment

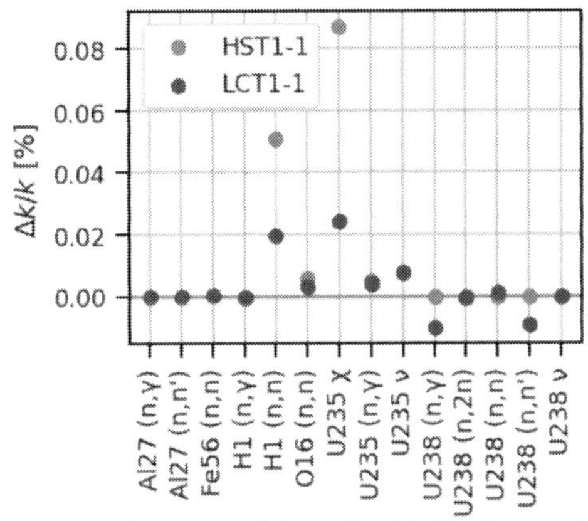

Figure 1. The amount of change in criticality due to each nuclide reaction by the cross-section adjustment method ($\Delta k/k > 1.0 \times 10^{-5}$ %)

As shown in Figure 1, the negative adjustments in U238 (n,γ) and (n,n') compensate the positive adjustments in H1 (n,n) and U235 χ. The compensation effects occurred in U238 (n,γ) and (n,n') because these reactions are sensitive only in LCT1-1, as shown in Figure 2.

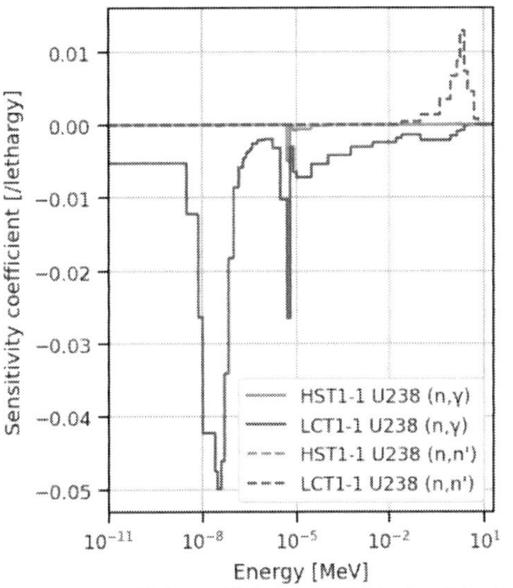

Figure 2. Sensitivity coefficients of U238 (n,γ) and (n,n')

3.3.2 Effect of using the nearest covariance matrix

As mentioned in Sec. 2, the nearest covariance matrix was used to generate multivariate normal random number. To evaluate the effect of using the nearest covariance matrix, the uncertainty of criticality arising from cross-section was compared.

First, the nearest covariance matrix was compared to the original covariance matrix by the uncertainty of the criticality of each benchmark problem. The difference in standard deviations of the criticality in each benchmark problem $\Delta\sigma_b$ is obtained by

$$\Delta\sigma_b = \sqrt{\mathbf{G}\mathbf{M}_{\mathrm{near}}\mathbf{G}^{\mathrm{T}}} - \sqrt{\mathbf{G}\mathbf{M}_{\mathrm{org}}\mathbf{G}^{\mathrm{T}}}, \tag{9}$$

where $\mathbf{M}_{\mathrm{org}}$ is the original covariance matrix. The differences in each benchmark problem are summarized in Table 4. As shown in Table 4, the difference between the nearest covariance matrix and the original covariance matrix is small, so we concluded the effect of using the nearest covariance matrix is small.

Table 4. The relative uncertainty differences for each benchmark problem

Benchmark	Difference [%]
HST1-1	5.84×10^{-4}
LCT1-1	2.40×10^{-3}

Even if each matrix element in the nearest covariance matrix is significantly different from the original covariance matrix, the differences shown in Table 4 could be underestimated due to cancellation of positive and negative values. Therefore, the focus was on the variance of the each cross-section in all energy groups for all nuclide reactions. The difference due to each cross-section $\Delta\sigma_{ig}$ is obtained by

$$\Delta\sigma_{ig} = \sqrt{\mathbf{G}_{ig}^2 \mathbf{V}_{ig}^{near}} - \sqrt{\mathbf{G}_{ig}^2 \mathbf{V}_{ig}^{org}}, \qquad (10)$$

where i is the nuclide reaction, g is the energy group, $\mathbf{V}_{ig}^{near}$ is the variance of the nearest covariance matrix, and $\mathbf{V}_{ig}^{org}$ is the original variance. The maximum difference due to each cross-section in each benchmark problem is summarized in Table 5. As shown in Table 5, the difference due to each cross-section is also small. Considering the above, the effect of using nearest covariance matrices was determined to be small.

Table 5. The maximum difference in standard deviations due to each cross-section

Benchmark	Maximum difference [%]
HST1-1	4.75×10^{-15}
LCT1-1	8.20×10^{-15}

3.3.3 Result of random sampling and discussion

The nearest covariance matrix was used to generate 1,000,000 random cross-section sets. From these random cross-section sets, the focus was on cross-section sets that accurately evaluate criticality. In this study, accurate evaluation of criticality means that the adjusted criticality is in the range shown in Table 6. This range was determined by the uncertainty of the experimental and calculated values, as shown in Table 2. At this time, statistical uncertainty and experimental uncertainty are assumed to be independent. As a result, 593,340 cross-section sets were in this range.

Table 6. The range of C values that accurately evaluate criticality

Benchmark	Range of C value
HST1-1	$0.9973 \leq C \leq 1.0027$
LCT1-1	$0.9966 \leq C \leq 1.0030$

The 593,340 cross-section sets were compared to the adjusted cross-section set obtained by the cross-section adjustment method. As shown in Figure 1, the cross-section adjustment method shows the largest positive change in criticality by U235 χ. The frequency of the change in criticality by U235 χ in the 593,340 cross-section sets in HST1-1 is shown in Figure 3. Figure 3 shows that the results of the cross-section adjustment method are included in the most frequent bin. However, in 179,994 of the 593,340 cross-section sets, criticality is not changed to the positive side by U235 χ. This result indicates that there are significant candidates for the cross-section set that accurately evaluate criticality, other than the results of the cross-section adjustment method.

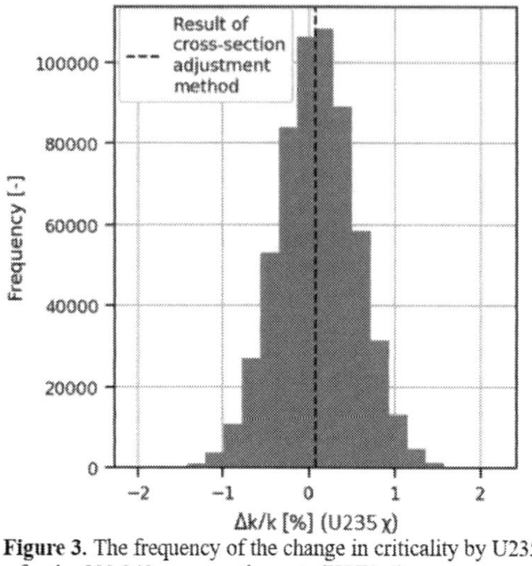

Figure 3. The frequency of the change in criticality by U235 χ for the 593,340 cross-section sets (HST1-1)

The 179,994 cross-section sets whose criticality is negatively adjusted by U235 χ were investigated. Focusing on the fourth most frequent bin in Figure 3, the amount of adjustment due to reactions other than U235 χ was investigated. The average amount of change in criticality due to all reactions is shown in Figure 4. The fourth most frequent bin is in the range of -0.35 % to -0.14 % of criticality change due to U235 χ. Then, the results shown in Figure 4 are the average of the criticality adjustments for the 84,120 cross-section sets in this range. Figure 4 shows that the negative adjustment due to U235 χ is mainly compensated by H1(n,n).

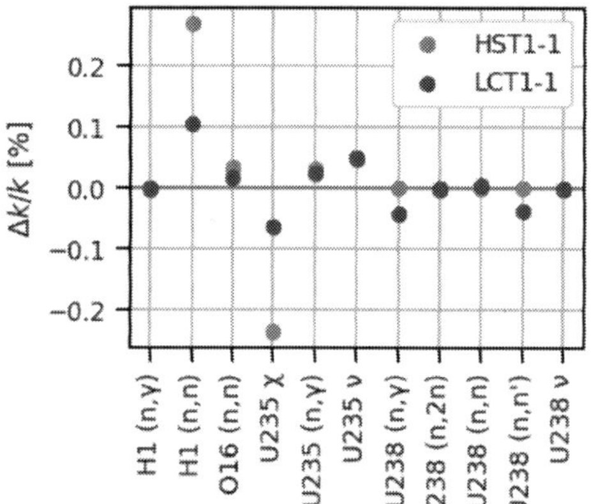

Figure 4. The average amount of criticality adjustment due to each nuclide reaction by 84,120 cross-section sets ($\Delta k/k > 1.0 \times 10^{-3}$ %)

The standard deviations of criticality obtained by substituting the nearest covariance matrix into equation (7) are shown in Table 7. As shown in Table 7, the standard deviation arising from U235 χ is larger than 0.09 % (the amount of change in criticality by the cross-section adjustment method). Therefore, the negative adjustment by U235 χ is consistent with the covariance data. Furthermore, it is also consistent with the covariance data that criticality is adjusted by H1 (n,n), which is the second largest when criticality is not adjusted by U235 χ.

Table 7. The standard deviation of criticality arising from each reaction obtained from the nearest covariance matrix

Nuclide	Reaction or parameter	Standard deviation of criticality [%]
U235	χ	0.47
H1	Elastic scattering	0.44
U235	ν	0.26
U235	(n,γ)	0.19
O16	Elastic scattering	0.18

In the cross-section adjustment method, the best-estimated cross-section set is obtained, where criticality is adjusted mainly by U235 χ due to the largest standard deviation. However, the results obtained by using the random sampling method indicate that the cross-section sets that adjust criticality mainly in H1(n,n) is also significant.

4 Conclusion

In this study, the significance of various cross-section sets that accurately evaluate nuclear characteristics was assessed. The random sampling method was used to obtain various cross-section sets. The random cross-section sets, which accurately evaluate nuclear characteristics, were compared to the cross-section set obtained by the cross-section adjustment method. The comparison indicated that there are other significant candidates that are consistent with the covariance data. By using the random sampling method, significant candidates can be obtained without changing the benchmark problems used for the cross-section adjustment method. Therefore, it may be useful when the result of the cross-section adjustment method is unreliable.

References

1. J. B. Dragt, J. W. M. Dekker, H. Guppelaar, et al., "Methods of adjustment and error evaluation of neutron capture cross sections; application to fission product nuclides", Nucl. Sci. Eng., **62**, 117-129 (1977).
2. T. Takeda, A. Yoshimura, T. Kamei, "Prediction uncertainty evaluation methods of core performance parameters in large liquid-metal fast breeder reactors", Nucl. Sci. Eng., **103**, 157-165 (1989).
3. E. Bauge, D. A. Rochman, "Cross-observables and cross-isotopes correlations in nuclear data from integral constraints", EPJ Nucl Sci Technol., **4**, 35 (2018).
4. G. Palmiotti, M. Salvatores, K. Yokoyama, et al., "Methods and approaches to provide feedback from nuclear and covariance data adjustment for improvement of nuclear data files", EPJ Web Conf., **239**, 13001 (2020).
5. M. Salvatores, G. Palmiotti, G. Aliberti, et al., "Methods and Issues for the Combined Use of Integral Experiments and Covariance Data: Results of a NEA International Collaborative Study", Nuclear Data Sheets, 118, 38-71 (2014).
6. T. Watanabe, T. Endo, A. Yamamoto, et al., "Cross section adjustment method based on random sampling technique", J Nucl Sci Technol., **51**, 590-599 (2014).
7. T. Endo, A. Yamamoto, T. Watanabe, "Bias factor method using random sampling technique", J Nucl Sci Technol., **53**, 1494-1501 (2016).
8. L. Plevnik, G. Žerovnik, "Computer code ENDSAM for random sampling and validation of the resonance parameters covariance matrices of some major nuclear data libraries", Annals of Nuclear Energy, **94**, 510-517 (2016).
9. M. Herman, P. Oblozinsky, C.M. Mattoon, et al., "AFCI-2.0 Neutron Cross Section Covariance Library", Brookhaven National Laboratory, BNL-94830-2011 (2011).
10. R. Vanhanen, "Computing Positive Semidefinite Multigroup Nuclear Data Covariances", Nucl. Sci. Eng., **179**, 411-422 (2015).
11. N. Higham, "Computing a Nearest Symmetric Positive Semidefinite Matrix", Linear Algebra Appl., **103**, 103-118 (1988).
12. OECD/NEA, "International Handbook of Evaluated Criticality Safety Benchmark Experiments", OECD/NEA, NEA/NSC/DOC(95)03 (2016).
13. C. J. Werner (editor), "MCNP Users Manual - Code Version 6.2", Los Alamos, LA-UR-17-29981 (2017).
14. B. C. Kiedrowski, F. B. Brown, J. L. Conlin, et al., "Whisper: Sensitivity/Uncertainty-Based Computational Methods and Software for Determining Baseline Upper Subcritical Limits", Nucl. Sci. Eng., **181**, 14-47 (2015).
15. M. B. Chadwick, M. Herman, P. Obložinský, et al., "ENDF/B-VII.1: Nuclear Data for Science and Technology: Cross Sections, Covariances, Fission Product Yields and Decay Data", Nucl. Data Sheets, **112**, 2887-2996 (2011).
16. B. L. Broadhead, I. A. Kodeli, "ZZ SCALE6.0/COVA-44G, 44-group cross section covariance matrix library extracted from SCALE6.0", NEA Data Bank, USCD1236/03 (2011).

Covariance of resonance parameters ascribed to systematic uncertainties in experiments

Shunsuke Endo[1,*], *Atsushi* Kimura[1], *Shoji* Nakamura[1], *Osamu* Iwamoto[1], *Nobuyuki* Iwamoto[1], and *Gerard* Rovira[1]

[1]Nuclear Science and Engineering Center, Japan Atomic Energy Agency, Tokai, Ibaraki 319-1195, Japan

Abstract. In resonance analyses, experimental uncertainties affect the accuracy of resonance parameters. The resonance analysis code REFIT can consider the statistical uncertainty of the experimental data when evaluating the resonance parameter uncertainty. However, since the systematic uncertainties are not independent at each measured energy, they must be treated differently from the statistical uncertainty. In the present study, we developed a new method to incorporate the systematic uncertainty coming from sample thickness into the uncertainty of resonance parameters. We applied this method to transmission of natural zinc measured at ANNRI of MLF in J-PARC and derived the systematic uncertainty of resonance parameters. We found that some of resonance parameters have larger systematic uncertainties than the statistical ones.

1 Introduction

The experimental uncertainties of cross-section measurements in neutron time-of-flight (TOF) method consist of statistical and systematic uncertainties. Furthermore, the systematic uncertainty can be separated into neutron-energy dependent and independent terms. The neutron-energy dependent term contains the uncertainties of, for example, incident-neutron beam spectrum and correction for self-shielding effect, etc. On the other hand, as the neutron energy independent term, the uncertainties of the sample thickness and normalization, etc., can be considered. These independent uncertainties can be represented by only one value and give a uniform relative uncertainty over all energy regions.

Although the resonance analysis code, REFIT [1], can treat the statistical uncertainty, it cannot consider the systematic uncertainty in the resonance analysis. Some methods to treat the systematic uncertainties have been proposed in the literature [2–4]. However, many studies of cross-section measurements have not yet included the systematic uncertainty in their resonance analyses.

We propose a new method to evaluate the systematic uncertainty and correlation of resonance parameters using REFIT. In particular, the uncertainty of sample thickness is discussed because it gives the largest uncertainty in many cases of transmission measurements. In the present work, the resonance analysis of the transmission of natural zinc (Zn) is used as an example.

2 Measurement

The transmission measurement was performed in the Accurate Neutron-Nucleus Reaction measurement Instrument (ANNRI) of the Materials and Life Science Experimental Facility (MLF) in the Japan Proton Accelerator Research Complex (J-PARC). The accelerator in J-PARC with a proton beam power of 700 kW injected two proton pulses (so-called double-bunch mode) with an inter-pulse of 0.6 μs into the mercury target to generate neutrons. The moderated neutrons were used for the present TOF measurements. A natural Zn sample with the dimensions of 50 $\times$ 50 $\times$ 6 mm and an areal density of n_t = 3.92 ± 0.05 atoms/barn was used. For the measurements, two different types of Li-glass detectors were employed. A ^{6}Li-enriched glass detector was used to measure transmitted neutrons, whereas a ^{7}Li-enriched glass detector was utilized to estimate the background events due to gamma-rays. The details of transmission measurements are given in Ref [5].

3 Transmission analysis

The transmission analysis was performed in the same manner as described in the past analysis in Ref [5]. Figure 1 shows the pulse height spectrum of the ^{6}Li- and ^{7}Li-glass detectors. The events in the filled color region, where single-hit and double-hit events by ^{6}Li(n,α) reactions were found, were adopted for the present analysis. The dead-time correction was employed using the extended dead-time model [5, 6]. The frame-overlap backgrounds were evaluated by fitting the TOF spectrum between 37 to 40 ms by the following function; $p_1 \exp(-p_2 t) + p_3$. The TOF spectra of two Li-glass detectors after dead-time correction and the estimated frame-overlap spectrum are shown in Fig. 2. To remove gamma-ray backgrounds, the TOF spectrum of the ^{7}Li-glass detector was subtracted from that of the ^{6}Li-glass detector. The TOF spectrum of the ^{7}Li-glass detector was normalized by a factor of 2.2 ± 0.2,

*e-mail: endo.shunsuke@jaea.go.jp

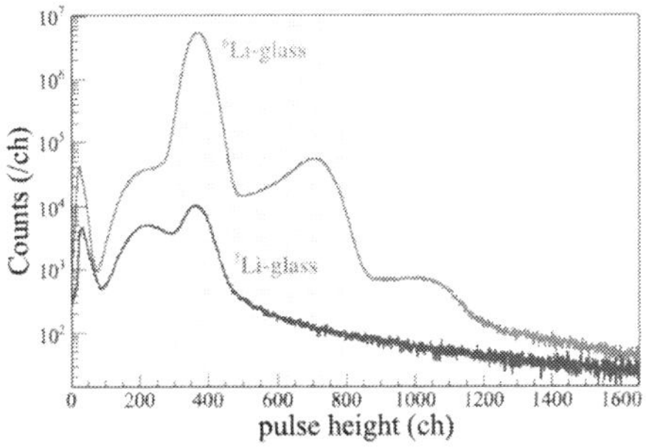

Figure 1. Pulse height spectra of the ^{6}Li- and ^{7}Li-glass detectors. The gated region shown by green is used in the present analysis.

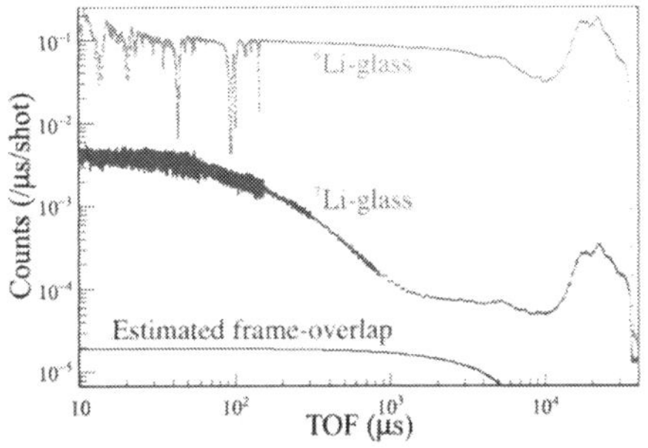

Figure 2. TOF spectra of the ^{6}Li- and ^{7}Li-glass detectors after dead-time correction and estimated frame-overlap backgrounds.

which was derived from the black-resonance in a notch-filter inserted measurement, to correct the difference of the detection efficiencies. The transmission was obtained by dividing the sample-in spectrum from the sample-out spectrum. The obtained transmission is shown in Fig. 3.

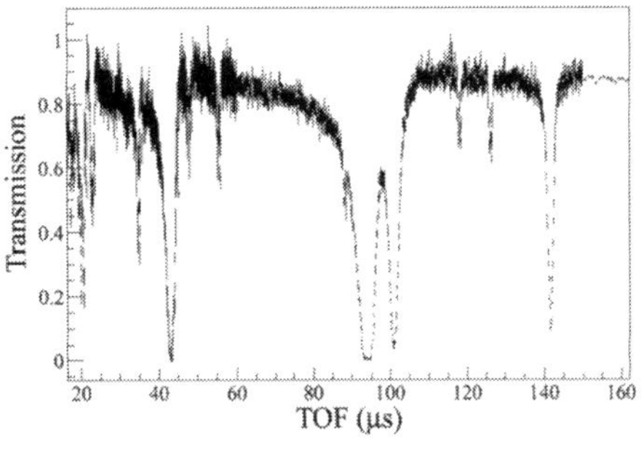

Figure 3. Measured natural Zn transmission.

The reduced total cross section, which includes the resolution function of the MLF and the Doppler broadening, can be calculated by

$$\sigma_{\text{tot}} = -\frac{1}{n_t} \ln T, \tag{1}$$

Table 1. Relative uncertainties of total cross section at two neutron energies.

Neutron Energy [eV]	234	586
TOF [μs]	141	90
Statistical Unc.	3.3%	3.0%
Sample thickness Unc.	1.4%	1.4%
Other systematic Unc.	1.1%	1.2%

where T is transmission. The relative uncertainties of total cross section are listed in Table 1 at two neutron energies. The other systematic uncertainty contains the uncertainties of dead-time correction, beam intensity, and the spectrum normalization factor of the ^{7}Li-glass detector.

4 Resonance analysis and covariance evaluation

The resonance analysis was made using REFIT. As mentioned in Sec. 1, since REFIT does not currently have the ability to evaluate uncertainty of resonance parameters caused by the systematic uncertainty, we derived the sets of resonance parameters with varying sample thickness for the obtained transmission data. The sample thickness of a REFIT input was changed from $n_t - \alpha\Delta n_t$ to $n_t + \alpha\Delta n_t$ dividing into N cases. The systematic uncertainty was calculated as

$$\Delta\Gamma_\eta = \sqrt{\frac{\sum_{i=1}^{N} w_i(\Gamma_{\eta,i} - \overline{\Gamma_\eta})^2}{\sum_{i=1}^{N} w_i}}, \tag{2}$$

where $\Gamma_{\eta,i}$ is the obtained resonance parameter in i-th sample thickness; $\overline{\Gamma_\eta}$ is the obtained resonance parameter for the nominal sample thickness; w_i is the weight calculated by

$$w_i = \frac{1}{\sqrt{2\pi}} \exp(-\beta_i^2), \tag{3}$$

where β_i is calculated by

$$\beta_i = \left(\frac{2(i-1)}{N-1} - 1\right)\alpha, \tag{4}$$

and means that the i-th sample thickness is $n_t + \beta_i\Delta n_t$. The correlation between resonance parameters Γ_η and Γ_ζ was determined as

$$C_{\eta,\zeta} = \frac{\sum_{i=1}^{N} w_i(\Gamma_{\eta,i} - \overline{\Gamma_\eta})(\Gamma_{\zeta,i} - \overline{\Gamma_\zeta})}{\sqrt{\sum_{i=1}^{N} w_i(\Gamma_{\eta,i} - \overline{\Gamma_\eta})^2}\sqrt{\sum_{i=1}^{N} w_i(\Gamma_{\zeta,i} - \overline{\Gamma_\zeta})^2}}. \tag{5}$$

Applying this method, the systematic uncertainty and correlations were estimated. In this estimation, we used $\alpha = 4$ and $N = 9$, i.e. using sample thickness $n_t - 4\Delta n_t$, $n_t - 3\Delta n_t, \cdots, n_t + 4\Delta n_t$. The neutron width and resonance energy were fitted with fixing the gamma-width to the value in JENDL-5 [7]. Figures 4 and 5 show the fitting result and the definition of resonance number. Because of the double-bunch effect in MLF, some resonances make two dips, such as for resonance No. 6. Resonance No. 11 partially overlapped with resonance No. 12. The obtained

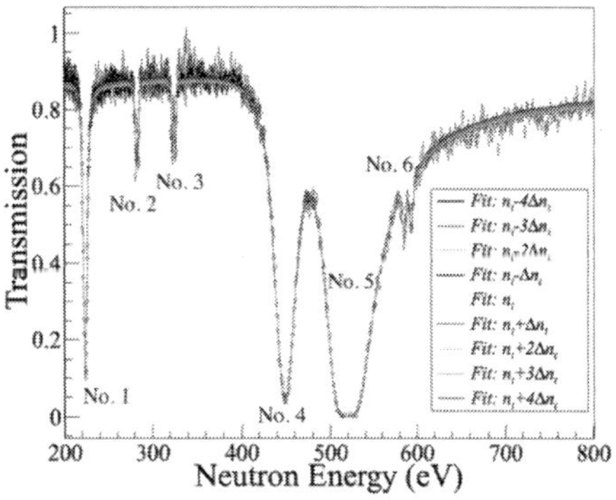

Figure 4. Fitting results for each sample thickness between 200 to 800 eV. The number shows the resonance considered in the present analysis.

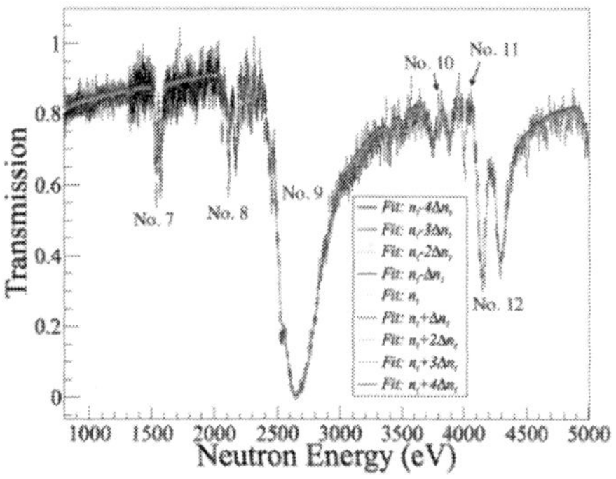

Figure 5. Same as Fig. 4, but in the neutron energy region between 800 to 5000 eV.

correlation of neutron width is shown in Fig. 6. The resulting resonance parameters and uncertainties are listed in Table 2.

5 Discussion

Figure 7 shows the obtained resonance energy of resonance No. 4 for each sample thickness. The error bar represents the fitting uncertainty only considering statistical uncertainty. The resonance energies are consistent regardless of the sample thickness. According to this study, the systematic uncertainty of resonance energy coming from the sample thickness is negligible compared to the statistical uncertainty.

Figure 8 displays the neutron width of resonance No. 4 for each sample thickness. As expected, the neutron width

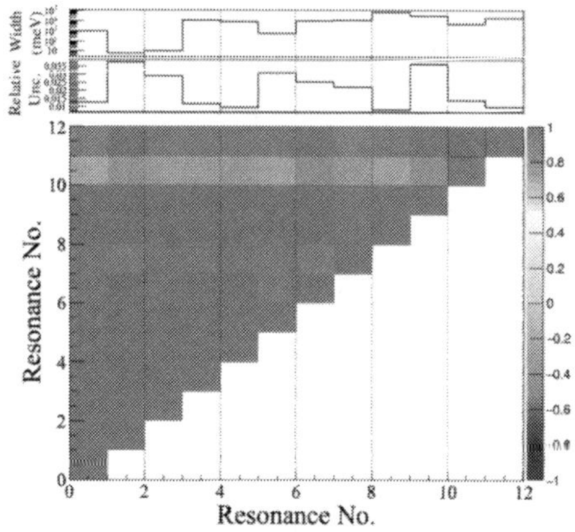

Figure 6. Obtained correlations between neutron widths of all fitted resonances.

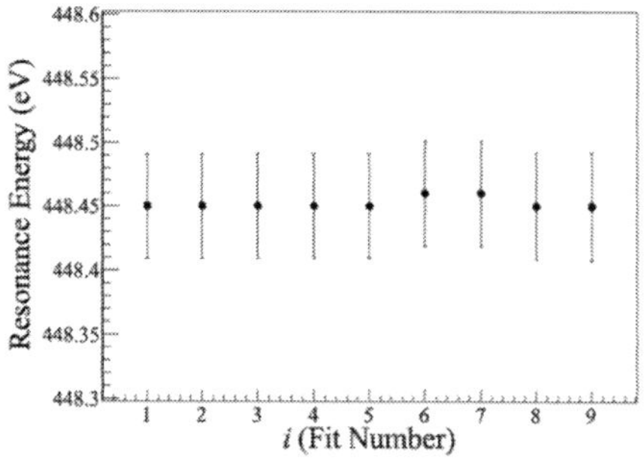

Figure 7. Resonance energy of resonance No. 4 using each sample thickness. The fit number represents which sample thickness was used in fitting such as $n_t - 4\Delta n_t$ for $i = 1$.

decreases as the sample thickness used in the fitting increases. The systematic uncertainty defined by Eq. (2) corresponds to the slope of this plot. Moreover, the neu-

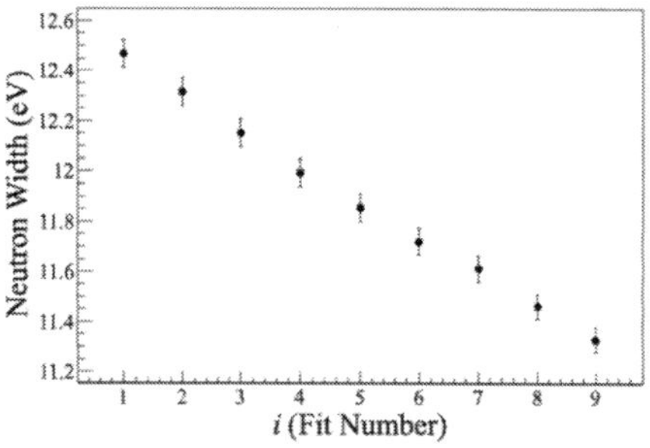

Figure 8. Neutron width of resonance No. 4 for each sample thickness.

tron widths for each sample thickness of resonance No. 11 are shown in Fig. 9. In this case, it is difficult to evaluate the systematic uncertainty and correlations among the res-

Table 2. Obtained resonance parameters. The gamma width was adopted from JENDL-5 [7]. In the neutron width, the first uncertainty represents the fitting uncertainty, and the second uncertainty represents the systematic uncertainty evaluated by Eq. (2). The uncertainty of the resonance energy was deduced from uncertainties of fitting, flight length and initial time delay.

Resonance No.	Reference	Resonance energy (eV)	Neutron width (eV)	Gamma width (eV)
1	Present	223.50 ± 0.02	0.931 ± 0.006 ± 0.012	0.350
	JENDL-5	223	0.771	
2	Present	282.31 ± 0.06	$(6.13 \pm 0.25 \pm 0.23) \times 10^{-3}$	0.294
	JENDL-5	281	0.006	
3	Present	323.84 ± 0.08	$(11.1 \pm 0.7 \pm 0.3) \times 10^{-3}$	0.186
	JENDL-5	323.5	0.011	
4	Present	448.45 ± 0.05	11.86 ± 0.05 ± 0.14	0.650
	JENDL-5	448.2	10.08	
5	Present	517.04 ± 0.07	8.81 ± 0.02 ± 0.08	0.160
	JENDL-5	516.4	8.5	
6	Present	587.95 ± 0.17	0.537 ± 0.034 ± 0.017	0.440
	JENDL-5	587	0.462	
7	Present	1547.7 ± 0.5	8.93 ± 0.33 ± 0.23	0.440
	JENDL-5	1546.2	7.2	
8	Present	2472.2 ± 1.5	12.2 ± 1.7 ± 0.3	0.440
	JENDL-5	2469	10.457	
9	Present	2624.6 ± 0.9	66.2 ± 0.4 ± 0.5	0.500
	JENDL-5	2627	60	
10	Present	3782.5 ± 2.9	31.5 ± 2.6 ± 1.1	0.350
	JENDL-5	3789	31.714	
11	Present	4036.6 ± 1.7	4.60 ± 0.32 ± 0.06	0.198
	JENDL-5	4046	5.2	
12	Present	4165.9 ± 1.3	19.6 ± 0.4 ± 0.2	0.676
	JENDL-5	4170	19.4	

onance parameters since the fitting uncertainty is dominant due to the large statistical fluctuation.

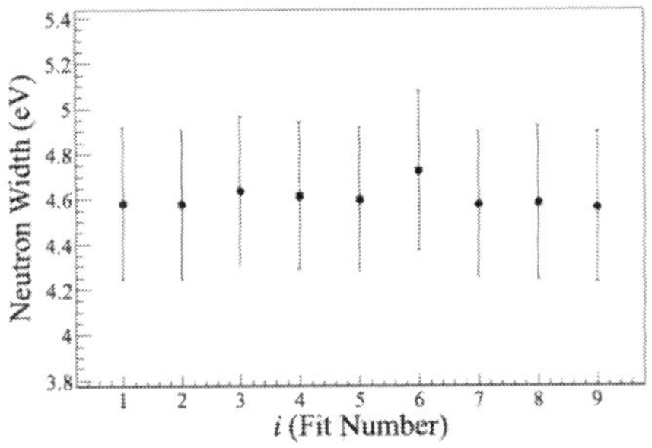

Figure 9. Neutron width of resonance No. 11 for each sample thickness.

If the experimental systematic uncertainty is not considered in the resonance analysis, the total uncertainty of neutron width in some resonances, such as resonances No. 1, 4 and 5, is underestimated. Therefore, it is significant to evaluate the systematic uncertainty in the resonance analyses.

Positive correlations among many resonances were found in Fig. 6. This is an expected result from the following consideration. When the input value of sample thickness in REFIT becomes small, the calculated transmission from a cross-section increases. To reproduce the experimental transmission results, the cross section has to become larger. Therefore, the resonance parameters, especially neutron width, become larger. Such behavior makes the correlation among many resonances positive. On the other hand, a weak negative correlation between resonance No. 11 and the other resonances was found. According to Eq. (5), the correlation with resonance No. 11 should be weak because fitted neutron widths have a flat distribution as seen in Fig. 9. Such "negative" correlation may be incidental.

This technique with reliable resonance parameters is applicable to determining unknown sample thicknesses and sample temperature. The sample thickness can be estimated from the χ^2 distribution by varying an input sample thickness and performing a fit to measured data. Moreover, in the same way as the case of sample thickness, the sample temperature can be deduced by varying an input sample temperature. This way makes use of the resonance broadening due to the Doppler effect as in Kai et al. [8]. The applications to those are underway.

6 Summary

We proposed a new method to derive the systematic uncertainty and correlations among the resonance parameters in the resonance analyses. This is the simple method that the sets of resonance parameters were obtained by changing the input value of the sample thickness in REFIT. The results show that it is essential to consider the systematic uncertainty when deriving the resonance parameters, especially the neutron width, because its contribution to the total uncertainty might be higher than that of the fitting uncertainty.

References

[1] M. C. Moxon, T. C. Ware, C. J. Dean. REFIT-2009 A Least-Square Fitting Program for Resonance Analysis of Neutron Transmission, Capture, Fission and Scattering Data Users' Guide for REFIT-2009-10. UKNSF. 243 (2010).

[2] P. Schillebeeckx et al., Nuclear Data Sheets **113**, 3064-3100 (2012).

[3] C. De Saint Jean et al., Nucl. Sci. Eng. **161**, 363-370 (2009).

[4] B. Habert et al., Nucl. Sci. Eng. **166**, 276-287 (2010).

[5] S. Endo et. al., J. Nucl. Sci. Technol. **59**, 318-333 (2022).

[6] A. Kimura et. al., J. Nucl. Sci. Technol. **56**, 479-492 (2019).

[7] O. Iwamoto et al., J. Nucl. Sci. Technol. **60**, 1-60 (2023).

[8] T. Kai et al., Physics Procedia **88**, 306-313 (2017).

XSUN-2022/SUSD3D n/γ sensitivity-uncertainty code package with recent JEFF-3.3 and ENDF/B-VIII.0 covariance data

Ivan A. Kodeli[1*]

[1]UKAEA, Culham Centre for Fusion Energy, Abingdon, OX14 3DB, United Kingdom

Abstract. A new version of the XSUN-2022 computer code package is under preparation to be released through NEA Data Bank which includes the latest updates of the SUSD3D sensitivity/uncertainty (S/U) code and the corresponding cross section and covariance matrix libraries based on the JEFF-3.3, ENDF/B-VIII.0 and FENDL-3.2 evaluations. The XSUN code package is a user–computer interface environment, written in Xbase++ ® and using Compiler 1.90.331 and Alaska 32-Bit Linker, for user-friendly pre- and post-processing of the input and output data for a complete and self-consistent set of deterministic codes. The flowchart integrates the codes TRANSX, PARTISN, and SUSD3D, all available from the OECD/NEA Data Bank and RSICC. The codes are used for the preparation of multigroup nuclear cross-section, neutron and gamma transport calculations for criticality and shielding calculations and nuclear data sensitivity and uncertainty calculations, respectively. XSUN-2022 is the 3rd released version, after XSUN-2013 and -2017.

1 Introduction

The development of the XSUN (Cross-section Sensitivity and Uncertainty) [1,2] Windows interface environment started in 2012 in order to facilitate the use of a suite of deterministic particle transport and nuclear data sensitivity and uncertainty codes. Deterministic codes are at present less frequently used than Monte Carlo (M/C) due to inherent approximations and complexity of their use. However, due to their fast algorithms, deterministic codes are still attractive and useful in many applications, be it for fast radiation transport providing a complete flux distribution, for the cross-section sensitivity and uncertainty analyses, deep penetration problems and validation of Monte Carlo codes and methods. XSUN package therefore integrates a complete set of deterministic tools for transport and S/U analysis, including nuclear data.

The XSUN-2013 and -2017 [1,2] code systems were released respectively in spring 2014 and in 2017 through OECD Nuclear Energy Agency Data Bank (Package-ID NEA-1882) and RSICC (Code Number CCC-825). The packages can be obtained from the two data centers, by contacting the RSICC or NEA Data Bank Computer Program Services.

The package is a user–computer interface environment for Windows and integrating the codes TRANSX [3], PARTISN [4] and SUSD3D [5-7] (see Fig. 1), all available from the OECD/NEA Data Bank and RSICC. The system was developed to assist the users in the preparation of input cards, rapid modification and execution of this complete chain of codes for S/U analysis. XSUN-2017 package included, among others, the following nuclear data:

- Nuclear cross section data library in the ECCO 33-neutron energy group structure in MATXS format. The library includes data for 109 isotopes at different temperatures (300 K, 550 K, 600 K, 650 K, 700 K, 800 K) and for different self-shielding factors. The data were produced using the NJOY-99 code and are based mostly on the ENDF/B-VII.1 nuclear data evaluation (with a few minor data from the -VII.0 release).
- Multi-group covariance matrix libraries in the ECCO 33 energy groups based on the ENDF/B-VII.1 [8], JENDL-4.0 [9] and SCALE-6.0m [10] evaluations. The matrices were processed using the NJOY-99 [11] and the ANGELO [12] codes.

An updated version of the code system is under final testing and will be released through the NEA data Bank. Modifications and improvements were introduced in the TRANSX-2.15 code used to produce cross sections compatible with many discrete-ordinates (S_N) transport codes (such as PARTISN), in the SUSD3D nuclear data S/U code and in the transport cross sections and covariance matrices used by the above codes. The 33-group nuclear data libraries and the material coverage were found to be insufficient for fusion applications, which therefore required extending the transport cross section and covariance matrix libraries both in terms of fine-group and material coverage. Multigroup cross-section and covariance matrix libraries for S/U analysis in 33- and 175-group structure were prepared based on JEFF-3.3 [13], ENDF/B-VIII.0 [14] and FENDL-3.2a [15] evaluations. Processing of covariance matrices included detailed verification and validation of their mathematical properties (positive definiteness, rank, etc.). The package was applied to shielding benchmark

* Corresponding author: ivan.kodeli@ukaea.uk

analysis to test the new features and cross section evaluations.

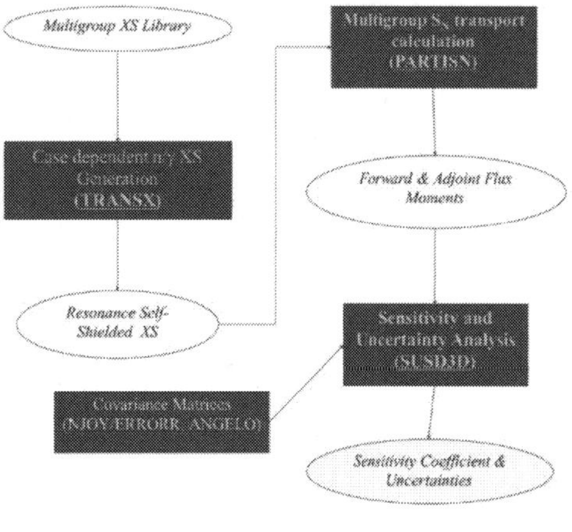

Fig. 1. XSUN computer code environnement.

2 SUSD3D perturbation code

SUSD3D [5-7] is a multi-dimensional nuclear cross-section sensitivity and uncertainty code suitable for complex one-, two- and three-dimensional transport studies. Sensitivity coefficients and standard deviation in the calculated detector responses or design parameters of interest (k_{eff}, β_{eff}, reaction rates) due to input cross section data and their uncertainties are calculated based on the first-order generalised perturbation theory. Several types of uncertainties can be considered, i.e. those due to: (1) neutron multi-group cross sections, (2) energy-dependent response functions, (3) secondary angular distribution (SAD) or secondary energy distribution (SED) uncertainties.

The particle transport calculations are done externally using the existing codes such as DOORS [16], DANTSYS [17], PARTISN [4], and the information on the direct and adjoint fluxes is passed to SUSD3D via the neutron flux moment files. Extensions to ADVANTG/DENOVO and DRAGON codes were also developed in the scope of PhD theses [18, 19]. At present, the XSUN package supports only the interface with the PARTISN code. Interface with the DANTSYS system can be done easily by the user, while coupling the SUSD3D/DOORS codes requires manual preparation of the DOORS and parts of the SUSD3D input cards by the user. The sensitivity profiles are folded with the cross-section covariance matrices to determine the variance in an integral response of interest. The input deck is prepared automatically by the XSUN system taking the geometry description from the PARTISN input and the isotopic cross-section composition as defined in the TRANSX input as well as the self-shielding factors calculated from the latter code. 1-, 2-, and 3-dimensional test cases are included in the package.

The development of SUSD3D started in the early 1990-ies in the scope of the French pressure vessel surveillance programme and the fusion projects of the European Commission. Examples of the use include criticality (k_{eff}, β_{eff}) and shielding (fusion, fission, such as PV surveillance, FNG benchmark pre- and post-analyses). k-ratio derivation sensitivity method [20-21] was proposed and implemented in 2011 to calculate S/U in the effective delayed neutron fraction – β_{eff}. Both the unconstrained and constrained sensitivity methods [22, 23] can be used to propagate the uncertainties in fission spectra. The latter was implemented in 2008 after demonstrating the equivalence between the zero-sum correction and constrained sensitivity.

Several improvements were introduced in the SUSD3D code since the last XSUN-2017 version, such as:

- SUSD3D code and XSUN utility were updated to be able to treat of up to 440 nuclides as included in the new extended cross section libraries. Moreover, the cross section and covariance file treatment in SUSD3D was modified allowing a considerable acceleration of calculations, by up to several order of magnitudes in case of using ~200 fine energy group nuclear data libraries and many nuclides. A parameter "libx" also allows a larger flexibility of the use of nuclear data.
- Extension of the sensitivity analysis to gamma-ray quantities such as gamma flux and heating. Gamma relevant nuclear data (MF=16, MF=23, MF=26) can be processed allowing S/U for coupled n/γ problems.
- Normalisation of the sensitivities and uncertainties is done internally in the code if requested by activating the option "resp=-1".
- Many minor corrections and modifications (e.g. printed output, plot files etc.). Top 10 reactions and isotopes causing the highest sensitivity and uncertainty are listed in the SUSD3D output. SAD S/U analysis can make use of the P_1 (mu-bar, MT251) data processed by NJOY-2016 or higher P_N (N=1 to 5) terms processed by the ERRORR34 [24] code.

SUSD3D code now covers the sensitivity to different cross-section types: MF=3 (neutron gain term), MF=6 (neutron loss term), MF=2 (direct term-response function), MF=4 (SAD), MF=5 (SED), MF16 (gamma production), MF23, MF26 (gamma loss and gain terms).

Likewise, the TRANSX code was modified to account for the increased number of energy groups and cross section matrices.

3 Multi-group nuclear data in XSUN-2022

Nuclear data libraries included in the XSUN-2013 and -2017 packages were mostly based on ENDF/B-VII.1 and including a limited number of nuclear materials. The XSUN-2022 package will include data from recent cross section evaluations in both VITAMIN-J fine and broad ECCO group structures, with larger number of nuclides (up to 440):

- JEFF-3.3 in ECCO 33 and VITAMIN-J 175 neutron group structures
- ENDF/B-VIII.0 (ECCO 33 and VITAMIN-J 175)

- FENDL-3.2 (211 neutron/42 gamma groups); the library is suitable for coupled neutron/gamma S/U analysis.

Older nuclear data, in particular covariances from ENDF/B-VII.1, JENDL-4.0u, COMMARA-2, SCALE-6.0 and IRDFF for a limited number of isotopes will be also included. An extension to the new JENDL-5 evaluation is planned.

Cross section covariance data

An important factor limiting the use of S/U analysis is the availability and the quality of cross-section covariance data. The availability of covariance matrix data in several recent cross section evaluations, such as ENDF/B-VIII.0, JEFF-3.3, JENDL-4/5 etc., has been considerably improved which demonstrates an increased awareness of the importance of these data among the nuclear community.

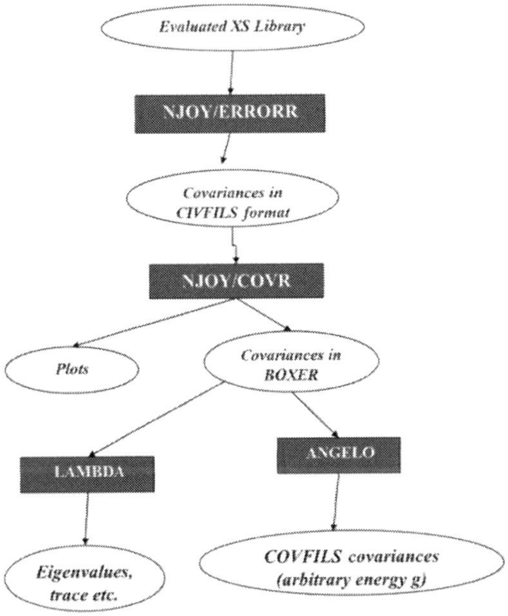

Fig. 2. Covariance matrix processing and verification scheme. Covariance matrices in the Covfils format processed by NJOY and ANGELO can be used in the SUSD3D code.

Although covariance matrices cannot be measured directly, and are governed by probability theory and statistics, with the inherent difficulty of estimating systematic uncertainties, they must still satisfy some basic mathematical properties such as positive semi-definiteness. The correct implementation of covariance matrix methodology in nuclear data evaluation can be therefore judged only indirectly by mathematical verification of the matrices and by the consistency, in a statistical sense, between the observed C/E values and the calculated nuclear data uncertainties.

The complete sets of recent JEFF-3.3, ENDF/B-VIII.0 and FENDL-3 covariance matrix evaluations were processed and partly verified for mathematical properties to be included in the XSUN-2022 package.

Several computer codes were used for the processing, plotting, mathematical verification and identification of ill-defined covariance data. The computational flow-chart is shown in Figure 2 including the following codes:

• NJOY-99, -2016 and -21 codes [11], ERRORR and COVR modules,

• LAMBDA [12] code: verification of mathematical properties of covariance matrices (eigenvalues, trace, correlation coefficients exceeding ±1),

• ANGELO-2 [12] code: interpolation of covariance matrices to user defined energy group structure; only file-MF33 covariance matrices can be treated using flat flux weighting.

The LAMBDA code was used to identify the matrices containing negative eigenvalues and correlation coefficients exceeding the range between -1 and +1. Verification of mathematical properties was done for the standard covariance (MF=33). Large majority of the processed matrices, in particular those relative to the important isotopes, were found to be mathematically correct, with a few exceptions listed below. Processing problems were also encountered for a few nuclides (in brackets cases with small negative eigenvalues, possibly due to numeric):

JEFF-3.3:

- covariance data were processed for 447 available isotopes; Processing difficulties were observed for the following nuclides: ^{232}Pa, ^{238}Np, ^{236}Np, ^{148}Nd, ^{95}Nb, ^{92,94,96,97,98,100}Mo, ^{153}Eu;
- number of nuclides with negative eigenvalues: 22 (using the present BOXER format), 18 (with an increased computational precision in BOXER code†): ^{192}Ir, ^{103}Rh, ^{185}W, ^{88}Y, (^{190,195}Pt, ^{170}Er).

ENDF/B-VIII.0:

- covariance data for 244 available isotopes were processed. Processing difficulties: ^{154,159}Dy, ^{78}Kr, ^{93}Nb, ^{184,186,187,188,189,190,192}Os, ^{70}Zn, ^{192}Pt, ^{54}Fe;
- number of nuclides with negative eigenvalues: 5: ^{192}Ir, (^{190,195}Pt, ^{239}Pu).

FENDL-3.1:

- covariance data for 58 isotopes;
- FENDL-3.1d: number of nuclides with negative eigenvalues: 2: ^{58}Ni, ^{50}Ti.

On the other hand, large correlation coefficients exceeding ±1.0 by a factor of up to 10^8 were observed for threshold (but not only) reactions in large number of files processed using the NJOY code, which could be (partly) due to processing problems. A similar problem was already observed in the past, e.g. in [25] it was concluded on a "...possible problem in ERRORR in case of covariance matrix processing for threshold reactions. ERRORR seems to unphysically reduce the standard deviation in the user energy group containing the threshold when the corresponding lower energy

† In the COVR module the data precision of the BOXER format is controlled by the "nvf" parameter. It can be increased by changing the following two parameters:
- change nvf=7 to nvf=8: format '(11f7.4)' is replaced by '(10f8.5)'
- change nvf=10 to nvf=13: format '(1p8e10.3)' is replaced by '(1p6e13.6)'.
The above modifications were found to resolve some (smaller) negative eigenvalues resulting from numeric problems.

group boundary extends below the threshold energy (averaging problem?)." Further studies are necessary to clarify the origin of the problems.

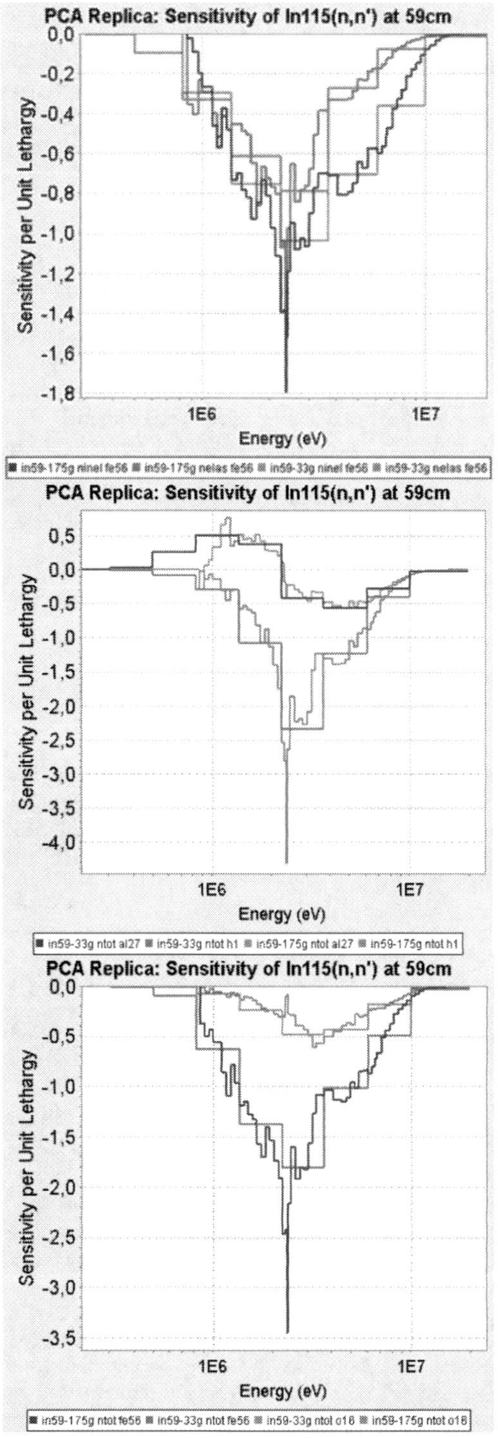

Fig. 3. Comparison of the sensitivity profiles for the reaction rates measured in the ASPIS PCA Replica benchmark. The sensitivities were obtained using the 175- and 33- energy group nuclear data.

Several integral experiments, including critical, kinetics and shielding, were studied using the SUSD3D code to verify if the matrices provide statistically credible, and quantitatively accurate nuclear data uncertainties. Shielding cases include fusion relevant (FNG) and fission reactor (ASPIS) benchmarks. Differences between various covariance matrix evaluations, both recent and older, were evaluated, which allows to judge on the progress made. An example of the analysis of the ASPIS PCA Replica benchmark experiment taken from the SINBAD database [26] - [28] is presented in Fig. 3 and Table 1 comparing the S/U analyses performed using the fine 175- and the broad 33-group nuclear data. For this fission reactor problem, a reasonable agreement is observed for most reactions. However, cases of larger differences were observed in fusion benchmark analyses, in particular for some threshold reactions such as (n,2n), (n.α), where the 33-group structure may be too coarse.

Table 1. Uncertainties in calculated ^{115}In(n,n') detector reaction rates due to Fe, H, O and Al cross-section uncertainties

^{115}In(n,n')	Sensitivity (%/%)		Uncertainty (%)			
			33 group		175 group	
Reaction	33 group	175 group	JEFF3.3	ENDF/B-VIII	JEFF3.3	ENDF/B-VIII
^{56}Fe(n,n')	-1.58	-1.58	4.3	7.3	4.4	6.4
^{56}Fe(n,n)	-1.10	-1.03	6.0	3.1	5.8	3.0
^{1}H(n,n)	-2.74	-2.40	5.0	1.0	4.5	0.9
^{16}O(n,n)	-0.67	-0.58	0.8	0.8	0.7	0.7
^{27}Al(n,n')	-0.48	-0.46	2.1	3.6	1.9	3.4
^{27}Al(n,n)	0.45	0.26	3.5	1.2	3.0	0.8
TOTAL			9.7	9.0	9.1	8.0

In spite of the progress achieved recently, some unresolved problems and gaps persist, such as for example

- Lack of covariances relative to gamma nuclear data,
- availability and processing of covariances for the secondary angular and energy neutron distributions (SAD/SED, MF=34,35) are limited, e.g. only P_1 terms in MF34 can be typically processed,
- limited information is available for some specific applications, e.g. uncertainties and correlations in delayed nu-bars (MF=31), correlations among isotopes and materials, etc.

4 Examples of analyses

The SUSD3D code was used since the early 1990-ies in different applications ranging from the pressure vessel surveillance, fusion and fission benchmark analysis for the nuclear data validation and guiding the design, reactor criticality (k_{eff}) and kinetics (β_{eff}) studies etc. Recent updates allow to extend the sensitivity and uncertainty analysis to gamma quantities such as gamma flux and heating. SUSD3D was recently used to determine the nuclear data uncertainty in material activation of irradiated nuclear steels.

4.1 Gamma-ray sensitivity of KFK γ-ray leakage benchmark

Gamma-ray leakage spectra from a set of three iron spheres of diameters 25, 30 and 35 cm with a ^{252}Cf neutron source in the centre were measured in Karlsruhe in 1977. Absolute gamma-rays in the energy range between 300 keV and 3 MeV were measured with the Si(Li) Compton spectrometer positioned on the outer surface of the iron shells. The measurement uncertainties were around 10-20%.

SINBAD evaluation of KFK-1977 measured gamma spectra from Ø25, 30 and 35 cm Fe spheres using bare ^{252}Cf(s.f.) source was prepared by Stanislav Simakov within the WPEC SG47 activities, including detailed descriptions of facility, methods and final numerical results with uncertainties and MCNP model.

Comparison between the gamma spectra calculated using the PARTISN and MCNP codes and the first results of the gamma flux S/U analysis are presented in Figs. 4 and 5, respectively.

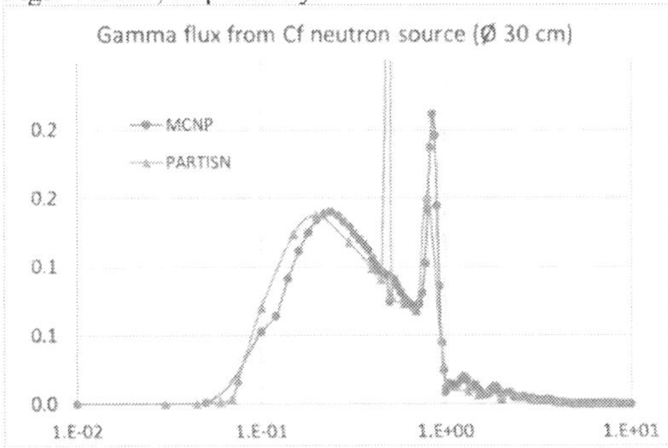

Fig. 4. Comparison between the gamma spectra calculated using the MCNP and PARTISN codes.

4.2 Material activation analysis

Nuclear data uncertainty in material activation of irradiated nuclear steels in structural fusion reactor components were studied to estimate the safety margin to be added to the calculated mean time to reach low-level waste (LLW) disposal limit in order to account for uncertainties.

The S/U analysis on dominant radio-nuclide production rates provides uncertainty of the individual nuclide generation. The safety margin (Δt) can be obtained from:

$$A_{lim} = \sum_i N_{0i}\left(1 + \frac{\Delta R}{R}\right)\lambda_i e^{-\lambda_i(t+\Delta t)}$$

where A_{lim} refers to the specified activity level limit for LLW disposal, and R is the reaction rate for the isotope production.
In the case of a single dominant radioisotope (i=1) the uncertainty in time to reach LLW limit can be obtained analytically from:

$$\Delta t = \frac{ln\left(1 + \frac{\Delta R}{R}\right)}{\lambda}$$

In the general case of several dominant nuclides (i>1) the solution is obtained by iteration.

A study simulating the neutron activation of the reduced activation ferritic/martensitic (RAFM) steels Eurofer and F82H, and two traditional steels - G91 and SS 316L(N) IG in a model of a spherical reactor geometry based on the helium-cooled pebble bed design is presented in [29].

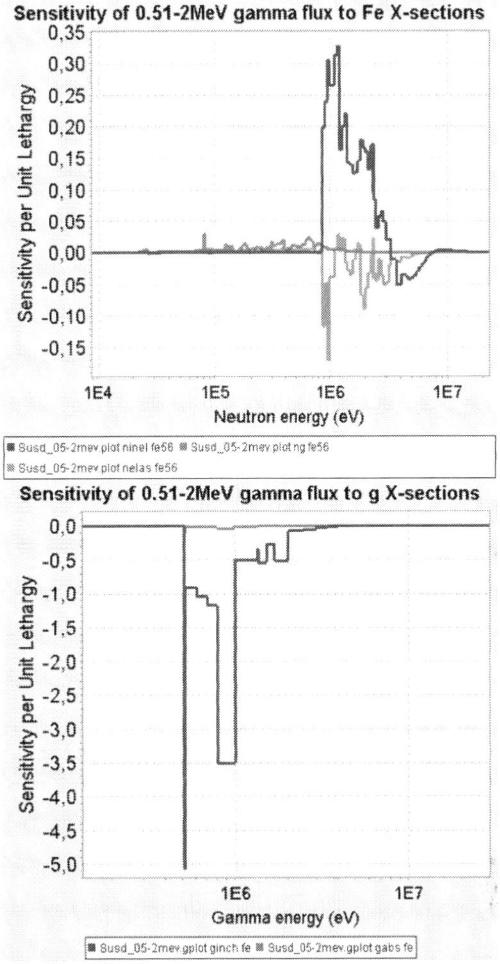

Fig. 5. Sensitivity of gamma flux between 0.51 - 2 MeV at radius 16.3 - 16.8cm to neutron inelastic, elastic and capture (above) and gamma incoherent and absorption (below) cross sections (results for Ø 30 cm sphere)

4.3 XSUN-2022 test cases to be provided

Test cases included in the XSUN-2022 package cover both critical and shielding applications:

- Fast critical assembly (SNEAK-7B): 1D, 2D and 3D models for the transport, sensitivity and uncertainty analysis. Nuclear data from the JEFF-3.3, ENDF/B-VIII.0 in VITAMIN-J 175- and ECCO 33-group structures are used,
- 1D shielding case: S/U of material activation in a typical fusion environment with 14 MeV neutron source involving large number of materials.

5 Conclusions

Deterministic codes are less frequently used today comparing to M/C due to inherent approximations and complexity of their use. However, several advantages such as lower CPU time and convenience of powerful and fast S/U analyses, still make them attractive. XSUN, an integrated environment for running a suite of deterministic codes, was developed to facilitate their use for transport and S/U analysis. The package includes the SUSD3D nuclear data S/U code which was used for the analysis of fission and fusion reactor design and safety parameters since the early 1990-ies.

Several improvements and new features were introduced in the new XSUN-2022 package and the code SUSD3D:

- SUSD3D code was extended to S/U of gamma related quantities (gamma flux, heating, etc.)
- Nuclear data from JEFF-3.3, ENDF/B-VIII.0 in 175- and 33-groups and FENDL-3.2 for 211 neutron/42-gamma groups are included, covering large number of materials (~440).
- Considerably faster treatment of cross section and covariance data files in SUSD3D to allow faster calculations for problems with large number of nuclides and energy groups.
- Many improvements and modifications in SUSD3D algorithm, such as automatic normalization for fixed-source problems, extended SAD S/U analyses, printing top 10 sensitive reactions and nuclides, and several other improvements and modifications.

Mathematical verification of the JEFF-3.3, ENDF/B-VIII.0 and FENDL-3.1 covariance matrices revealed a few negative matrices (< 2-4%), but large number of inconsistent correlation coefficients. Most negative eigenvalues are small in magnitude, and it is unlikely they would have major impact on the uncertainty calculations. However, some larger negative eigenvalues may pose problems for nuclear data adjustment studies and should be corrected before their use in data assimilation studies. For several nuclides negative eigenvalues of the same magnitude were observed in several different nuclear data evaluations, suggesting a probable common origin of these covariance data. This was the case even though the corresponding nuclear cross sections differed.

On the other hand, many covariance matrices processed by the NJOY code were found to contain correlation coefficients exceeding ±1.0. This was in particular the case for the threshold reactions in the vicinity of the threshold, but not only.

New versions of SUSD3D and XSUN-2022 are under preparation and will be available from NEA Data Bank.

Acknowledgements

This work has been partly funded by the RCUK Energy Programme [grant number EP/T012250/1]. To obtain further information on the data and models underlying this paper please contact PublicationsManager@ukaea.uk. Some fusion benchmark analyses were partly funded by the EUROfusion project within EC fusion programme (2022).

References

1. I. Kodeli, S. Slavič, "XSUN-2013, Windows interface environment for transport and sensitivity-uncertainty software TRANSX-2, PARTISN and SUSD3D", OECD NEA code package NEA-1882 (http://www.oecd-nea.org/tools/abstract/detail/nea-1882/) and RSICC code package CCC-825
2. I. Kodeli, S. Slavič, Science and Technology of Nuclear Installations, Volume **2017**, Article ID 1264736, (2017) 16 p. *https://doi.org/10.1155/2017/1264736*
3. R. E. MacFarlane, TRANSX 2.15, Code System to Produce Neutron, Photon, and Particle Transport Tables for Discrete-Ordinates and Diffusion Codes from Cross Sections in MATXS Format, RSICC Peripheral Shielding Routine Collection, PSR-317 (1995)
4. R. E. Alcouffe, R.S. Baker, J.A. Dahl, S.A. Turner, R.C. Ward, PARTISN 5.97, 1-D, 2-D, 3-D Time-Dependent, Multigroup Deterministic Parallel Neutral Particle Transport Code, LA-UR-08-07258, NEA/Data Bank CCC-0760/01 computer code collection, November 2008
5. I. Kodeli, Nucl. Sci. Eng. **138** (2001) 45-66
6. I. Kodeli, *The SUSD3D Code for Cross-Section Sensitivity and Uncertainty Analysis – Recent Development*, Transactions of the American Nuclear Society, 104, Hollywood, Florida, June 26–30, 2011
7. I. Kodeli, SUSD3D, 1-, 2-, 3-Dimensional Cross Section Sensitivity and Uncertainty Code, NEA-1628/03, OECD/NEA DB Computer code collection, 2008
8. M. B. Chadwick et al., Nucl. Data Sheets **112** (2011) 2887
9. K. Shibata et al., J. Nucl. Sci. Technol. **48** (2011)1.
10. ZZ-SCALE6.0/COVA-44G, 44-group cross-section covariance library, USCD1236/03, OECD/NEA DB Computer code collection, 2012.
11. R. E. MacFarlane, D. W. Muir, *The NJOY Nuclear Data Processing System*, Manual LA-12740-M (1999)
12. I. Kodeli, ANGELO-LAMBDA, Covariance matrix interpolation and mathematical verification, NEA-DB Computer Code Collection, NEA-1798/02 (2008)
13. A.J.M. Plompen, et al., European physical journal. A, Hadrons and nuclei. 2020, vol. **56**, art. 181, 108 str. ISSN 1434-601X. DOI: /10.1140/epja/s10050-020-00141-9.
14. D. Brown, et al., Nucl. Data Sheets **148**, 1 (2018)
15. FENDL-3.2, https://www-nds.iaea.org/fendl/
16. W. A. Rhoades and D. B. Simpson, DOORS, 1-, 2-, 3-dimensional discrete-ordinates system for deep-penetration neutron and photon transport, OECD/NEA Computer Code Collection, CCC-0650

17. R.E. Alcouffe et al., DANTSYS3.0, 1-D, 2-D, 3-D MultiGroup Discrete Ordinate Method Transport, OECD/NEA Computer Code Collection, CCC-0547
18. P. Sabouri, Application of Perturbation Theory Methods to Nuclear Data Uncertainty Propagation using the Collision Probability Method, PhD Thesis, Université de Grenoble, France (2013)
19. B. Kos, A. Čufar, I. Kodeli, Nuclear Engineering and Technology **53,** 2151-2161 (2021)
20. I. Kodeli, *Sensitivity and Uncertainty in the Effective Delayed Neutron Fraction β_{eff} (Method & SNEAK-7A Example),* Proc. of the 5th Workshop for the OECD Benchmark for Uncertainty Analysis in Best-Estimate Modelling (UAM) for Design, Operation and Safety Analysis of LWRs (UAM-5), Stockholm (April 13-15, 2011), NEA-1769/04 package, OECD-NEA Data Bank.
21. I. Kodeli, Nuclear Instruments and Methods in Physics Research A **715** (2013)70–78
22. I. Kodeli, A. Trkov, R. Capote, Y. Nagaya, V. Maslov, Nuclear Instruments and Methods in Physics Research A **610** (2009) 540–552
23. Y. Nagaya, I. Kodeli, G. Chiba and M. Ishikawa, Nucl. Inst. & Meth. Phys. Res. A **603** (2009) 485-490
24. I. Kodeli, Computational Tools for Cross-Section Sensitivity and Uncertainty Evaluation. Final Report on Task EFF-BB4-F, EFF-DOC-446 (1996)
25. I. Kodeli, *Experience with EFF-3 Cross Section Processing Using ERRORR Module*, NJOY Workshop and User Group Meeting, 15 May 2001, Aix-en-Provence.
26. I. Kodeli, E. Sartori, Annals of Nuclear Energy **159** (2021) 108254
27. J. Butler, M.D. Carter, I.J. Curl, M.R. March, A.K. McCracken, M.F. Murphy, A. Packwood, The PCA Replica Experiment PART I, Winfrith Measurements and Calculations, AEEW-R 1736 (1984)
28. I. Kodeli, S. van der Marck, *Consistency among the results of the ASPIS Iron88, PCA Replica and PCA benchmark results*, Proc. ICRS14 RPSD22 Conference, Seattle, US, Sept. 25 - 29, 2022
29. S. O. von Tiedemann, D. M. Collins, M. R. Gilbert, I. A. Kodeli, Nuclear Data Uncertainty Propagation and Implications for Radioactive Waste Management of Fusion Steels, (to be published in Fusion Eng. Design)

Fast-neutron capture cross section data measurement of minor actinides for development of nuclear transmutation systems

Tatsuya Katabuchi[1], *Osamu* Iwamoto[2], *Jun-ich* Hori[3], *Atsushi* Kimura[2], *Nobuyuki* Iwamoto[2], *Shoji* Nakamura[2], *Gerard* Rovira[2], *Shunsuke* Endo[2], *Yuji* Shibahara[3], *Kazushi* Terada[3], *Yu* Kodama[1], *Hideto* Nakano[1], *Yaoki* Sato[1], *Shota* Matsuura[1]

[1]Laboratory for Zero-Carbon Energy, Institute of Innovative Research, Tokyo Institute of Technology, Ookayama, Meguro, Tokyo 152-8550, Japan
[2] Nuclear Science and Engineering Center, Japan Atomic Energy Agency, Tokai-mura, Ibaraki 319-1195 Japan
[3] Institute for Integrated Radiation and Nuclear Science, Kyoto University, Kumatori, Osaka 590-0494 Japan

Abstract. The neutron capture cross sections of ^{237}Np, ^{243}Am and ^{241}Am in the keV energy region were measured by the time-of-flight method at the Japan Proton Accelerator Research Complex (J-PARC). Characterization of the minor actinide samples were made to reduce the uncertainties of the cross section. A neutron beam filter system was developed and installed in ANNRI to solve the double-bunch mode issue of J-PARC. The pulse-height weighting technique was employed to determine the neutron capture cross sections. The neutron capture cross sections of ^{237}Np, ^{243}Am and ^{241}Am were determined with higher accuracies than the past experiments.

1 Introduction

Long-lived minor actinides (MAs) in nuclear waste from nuclear power plants are a long-standing issue to continue nuclear energy production. In order to solve the issue, researchers have suggested nuclear transmutation, in which long-lived radionuclides are transmuted into stable or shorter-life nuclides via neutron-induced nuclear reactions [1]. Development of nuclear transmutation systems such as an accelerator-driven system requires accurate neutron nuclear reaction data [2,3]. The present research project entitled *"Study on accuracy improvement of fast-neutron capture reaction data of long-lived MAs for development of nuclear transmutation systems"* has been conducted as a joint collaboration including Tokyo Tech, Japan Atomic Energy Agency and Kyoto University [4]. This project focuses on the neutron capture reaction of MAs, especially ^{237}Np, ^{241}Am and ^{243}Am, in the fast neutron energy region. The final goal of this project is to improve the accuracy of the neutron capture cross sections of ^{237}Np, ^{241}Am and ^{243}Am by employing a high-intensity neutron beam from a spallation source of the Japan Proton Accelerator Research Complex (J-PARC) [5]. To achieve the goal, a neutron beam filter system in J-PARC, sample characteristic assay, and theoretical reaction model were developed. A part of the results of the project are reviewed in this paper.

2 Experiments

The neutron beam filter system was designed to solve the double-pulse beam issue in J-PARC. Neutrons are produced in the spallation target with a cycle of 25 Hz in the Materials and Life Science Experimental Facility (MLF) of J-PARC. The proton accelerator is operated in a special operational mode such that two 100-ns wide proton beam pulses separated by 600 ns are injected into the spallation target every 40 ms as shown in Fig. 1. The purpose of the double-pulse operation is to increase the thermal neutron flux. The double-pulse beam time structure disappears in the neutron slowing-down process in moderators and thereby does not affect a neutron time-of-flight (TOF) spectrum in the thermal energy region. However, in the higher energy region (> 10 eV), the double-pulse beam is crucially problematic [6]. Resonance peaks on a TOF spectrum become doublet due to the double pulse beam time structure. This makes resonance analysis more complicated, requiring the implementation of a resolution function of the double-pulse beam into resonance analysis.

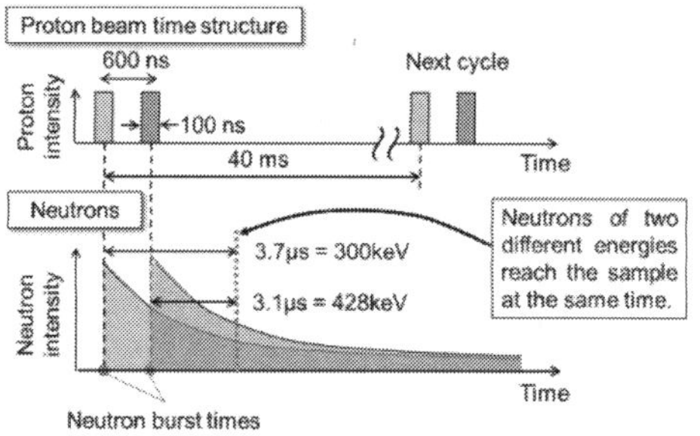

Fig. 1. Time structure of double-pulse mode operation of J-PARC/MLF and neutron generation.

Situation is even worse in the fast neutron energy region that is above the resolved resonance region. The cross section is a smooth function of neutron energy.

Neutrons that have two different energies associated with the double proton beam pulses reach the sample at the same time. It is difficult to deconvolute the measured neutron capture yield into single-pulse components.

To solve the double-pulse beam issue in J-PARC/MLF, a neutron beam filter system has been developed and installed into the Accurate Neutron Nucleus Reaction Measurement Instrument (ANNRI) [7]. The neutron beam filter technique has been often used in nuclear reactors, to make a mono-energetic neutron beam [8,9]. A total cross section minimum caused by an s-wave resonance interference is used as an energy window of neutron energy filtering. Combining optimal filter materials, a neutron energy spectrum can be tailored to be mono-energetic.

In the present project, the neutron beam filter technique was employed to separate two different neutron energy components at the same TOF channels. The sample position of ANNRI was 27.9 m from the neutron source. The neutron beam filter system was placed at a flight length of 18 m as shown in Fig. 2. The selected filter materials were Fe, Bi, Al, Si, Cr and Sc. They have resonances which are suitable for filtering neutrons in the keV energy region. The available resonance energies of the filter materials are summarized in Table 1. The neutron filter materials were fabricated and tested in neutron beam experiments at J-PARC. In the test experiments, transmission neutron spectra were measured by detecting prompt γ-rays from the $^{10}B(n,\alpha\gamma)^{7}Li$ reaction placing a boron sample at the sample position. The 478-keV γ-rays from the $^{10}B(n,\alpha\gamma)^{7}Li$ reaction were detected with a NaI(Tl) detector. Figures 3 and 4 show neutron TOF spectra for 20-cm thick Fe filter and 20-cm thick Si filter, respectively. The neutron spectra show well-filtered peaks at 24, 54 and 146 keV.

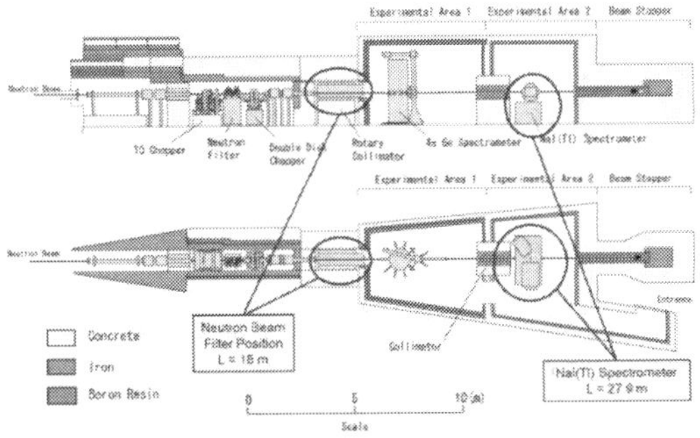

Fig. 2. Accurate Neutron Nucleus Reaction Measurement Instrument (ANNRI), and the positions of neutron beam filter system and NaI(Tl) spectrometer.

This project used NaI(Tl) detectors for measurements of neutron capture cross sections. The ANNRI-NaI(Tl) detectors consist of two differently-sized NaI(Tl) detectors [10]. The larger detector (crystal size: 330-mm diameter, 203-mm long) was placed at a scattering angle of 90° as respect to the beam axis. The smaller detector (203-mm diameter 203-mm long) was at an angle of 125°. The detectors were shielded from background γ-rays and neutrons with various shielding materials such as lead, borated polyethylene and paraffin, cadmium and ^{6}Li-enriched lithium compounds. A sample was placed in a carbon-fiber beam duct and set at a flight length of 27.9 m. The beam duct was filled with helium gas to reduce scattering neutron background by air.

Table 1. Available resonance energies of the filter materials

Material	Resonance energies (keV)
Fe	24
Bi	2.2, 5.1, 11.5, 15.3, 32.9
Al	27, 125, 265
Si	54, 146
Cr	56, 82
Sc	7.5, 10.5

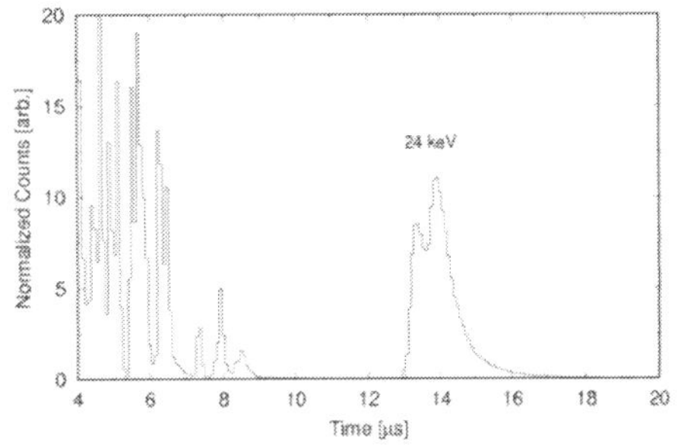

Fig. 3. Neutron TOF spectrum of neutron beam filtered with the 20-cm thick Fe filter.

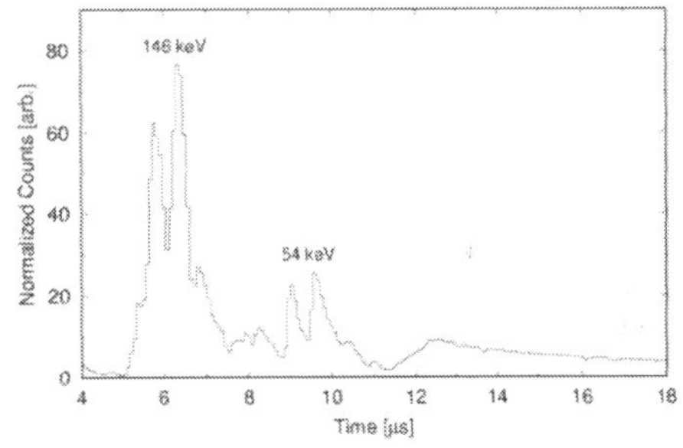

Fig. 4. Neutron TOF spectrum of neutron beam filtered with the 20-cm thick Si filter.

One of the main sources of systematic uncertainties of cross section is sample characteristics. The sample characteristic assay, especially, precise mass spectrometry of isotopes and impurities lowers sample systematic uncertainties. Isotope compositions are sometimes off from the rated values. Radioactive samples often have impurities not listed in a certificate sheet.

In this project, a thermal ionization mass spectrometer (TIMS) of the Institute for Integrated Radiation and Nuclear Science of Kyoto University was used to analyse the isotope compositions and impurities of samples. All MA samples for measurement at J-PARC were sealed in containers. For analysis, unsealed MA solution taken from the same batch as the sealed MA samples were prepared. The results of the impurity analysis are summarized in Table 2. The impurity ratios were determined very accurately, reducing the uncertainties of the final cross sections.

The radioactivities of ^{237}Np, ^{243}Am and ^{241}Am were 5.18, 278.2 and 957.4 MBq, respectively. The radioactivities were determined from heat measurement by calorimeter and γ-ray measurement by a Ge detector [11].

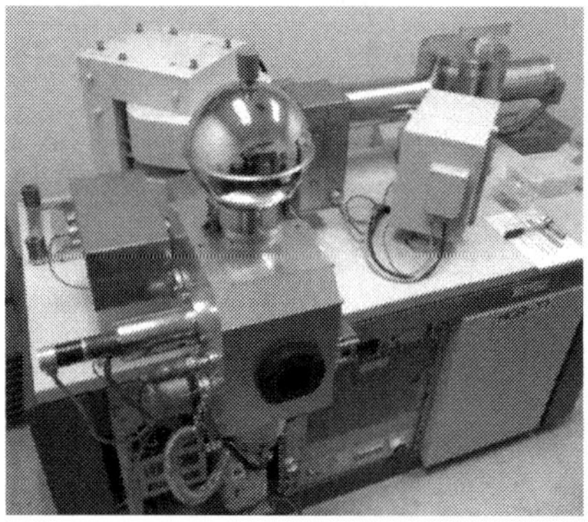

Fig. 5. Thermal ionization mass spectrometer (TIMS).

Table 2. Impurity ratios of the ^{237}Np, ^{241}Am and ^{243}Am samples measured with TIMS.

	^{237}Np sample	^{241}Am sample	^{243}Am sample
^{237}Np	1	-	-
^{241}Am	-	1	$(2.33\pm0.01)\times10^{-2}$
^{242m}Am	-	-	$(0.48\pm0.05)\times10^{-3}$
^{243}Am	-	$(1.4\pm0.7)\times10^{-6}$	1
^{238}Pu	$(6.1\pm0.6)\times10^{-7}$	-	$(1.4\pm0.2)\times10^{-5}$
^{239}Pu	$(5.1\pm0.3)\times10^{-4}$	$(9.2\pm2.2)\times10^{-3}$	$(4.0\pm0.1)\times10^{-3}$
^{240}Pu	$(3.9\pm0.4)\times10^{-5}$	$(3.1\pm0.8)\times10^{-5}$	$(8.6\pm0.2)\times10^{-3}$
^{241}Pu	$(2.0\pm1.4)\times10^{-6}$	-	-
^{242}Pu	$(2.4\pm1.6)\times10^{-6}$	$(5.2\pm1.3)\times10^{-5}$	$(9.9\pm2.7)\times10^{-6}$

3 Results

The neutron capture cross sections of ^{237}Np, ^{243}Am and ^{241}Am were measured using the neutron beam filter system of ANNRI at J-PARC. The experimental results are shown in Figs. 6-8. Past experimental data [12-17] and evaluated cross sections are also plotted for comparison. Part of the experimental results have been reported in Refs. [18-20]. The details of the experiments and data analysis can be found therein.

Three neutron filters, Fe, Si and Cr, were used for measurement of ^{237}Np. For ^{243}Am and ^{241}Am, two neutron filters, Fe and Si, were used. The pulse-height weighting technique (PHWT) [21, 22] was employed to derive the neutron capture yield. Measurements of gold samples were also made and the relative ratios of the capture yields of the sample to the gold measurements were obtained. The absolute values of the capture cross sections of the samples were determined from the JENDL-4.0 evaluated cross section of the ^{197}Au(n,γ)^{198}Au reaction.

The errors of the present cross sections plotted in Figs. 6-8 include statistical and systematic uncertainties. Uncertainties of total sample mass, impurity abundances, neutron transport coefficients, standard cross section and weighting function of PHTW were considered as the systematic uncertainty. As seen in Figs. 6-8, the capture cross sections obtained in the present experiments have less uncertainties than the past measurements. In particular, the results of ^{243}Am and ^{241}Am achieved much higher accuracies than the past measurements.

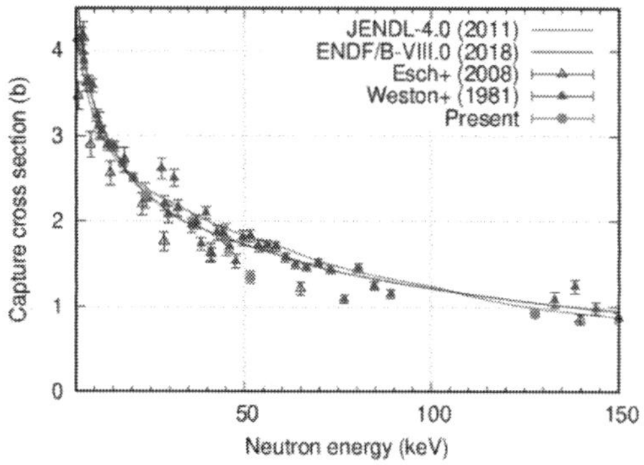

Fig. 6. Neutron capture cross section of ^{237}Np. The present experimental results are compared with past experimental data and evaluated cross sections.

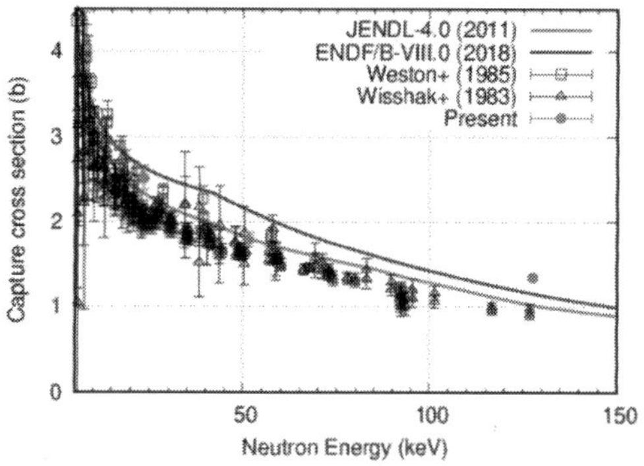

Fig. 7. Neutron capture cross section of ^{243}Am. The present experimental results are compared with past experimental data and evaluated cross sections.

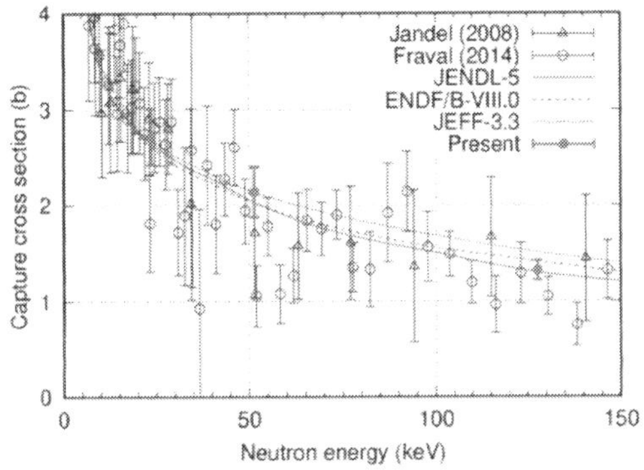

Fig. 8. Neutron capture cross section of ^{241}Am. The present experimental results are compared with past experimental data and evaluated cross sections.

4 Conclusions

The neutron capture cross sections of ^{237}Np, ^{243}Am and ^{241}Am in the keV energy region were measured by the TOF method at J-PARC/ANNRI. Characterization of the MA samples were made to reduce the uncertainties of the cross section measurement. To solve the double-bunch mode issue of J-PARC, a neutron beam filter system was developed and installed in ANNRI. The accuracies of the neutron capture cross sections of ^{237}Np, ^{243}Am and ^{241}Am were improved.

The neutron experiments at the MLF of the J-PARC were performed under the user programs (Proposal No. 2018A0213, 2018B0195, 2019A0188, 2019B0143, 2020A0273 and 2021B0313). This work was supported by the Innovative Nuclear Research and Development Program from the Ministry of Education, Culture, Sports, Science and Technology of Japan [JPMXD0217942969].

References

1. C. D. Bowman, Annu. Rev. Nucl Sci. **48**, 505 (1998)
2. M. Salvatore, R. Jacqmin, *Uncertainty and target accuracy assessment for innovative systems using recent covariance data evaluations*, WPEC-26, OECD NEA, San Diego, USA (2008)
3. H. Iwamoto, K. Nishihara, T. Sugawara, et al., Nucl Sci Technol. **50**, 856 (2013)
4. T. Katabuchi et al., EPJ web conf. **239**, 01044 (2020)
5. F. Maekawa, et al., Nucl. Instrum. Meth. Phys. Res. A **620**, 159 (2010)
6. K. Kino, M. Furusaka, F. Hiraga, et al., Nucl. Instrum. Methods. Phys. Res. A **736**, 66 (2014)
7. G. Rovira, A. Kimura, S. Nakamura et al., Nucl. Inst. Meth. Phys. Res. Sect. A **1003**, 165318 (2021)
8. F. Y. Tsang and R. M. Brugger, Nucl. Instrum. Meth. **134**, 441 (1976)
9. R. C. Greenwood and R. E. Chrien, Nucl. Instrum. Meth. 138, 125 (1976)
10. M. Igashira, Y. Kiyanagi, M. Oshima, et al. Nucl. Instrum. Meth. Phys. Res. A **600**, 332 (2009)
11. Terada K, Nakao T, Kimura A, et al. Proceedings of the International Conference on Nuclear Data For Science and Technology (ND2016), 11–16 Sept 2016, Bruges, Belgium. EPJ Web of Conferences **146**, 03019 (2017)
12. L. W. Weston and J. H. Todd, "Neutron Capture Cross Section of 237Np", Nuclear Science and Engineering, **79**, 184-196 (1981).
13. E.-I. Esch et al., "Measurement of the 237Np(n,γ) cross section from 20 meV to 500 keV with a high efficiency", highly segmented 4π BaF2 detector, Physical Review C, **77**, 034309 (2008).
14. K. Wisshak, F. Kappeler, Nucl. Sci. Eng. 85, 251 (1983)
15. L. W. Weston LW, J. H. Todd. Nucl. Sci. Eng. **91**, 444 (1985)
16. K. Fraval et al., Phys. Rev. C **89**, 044609 (2014)
17. M. Jandel et al., Phys. Rev. C **78**, 034609 (2008)
18. Y. Kodama et al., J. Nucl. Sci. Technol., **58**, 1159 (2021)
19. G. Rovira et al., J. Nucl. Sci. Technlo., **59**, 110 (2022).
20. G. Rovira et al., J. Nucl. Sci. Technol. (to be published)
21. R. L. Macklin, J. H. Gibbons, Phys Rev. **159**, 1007 (1967)
22. Mizuno S, Igashira M, Masuda K. J Nucl. Sci. Technol. **36**, 439 (1999)

Generation and analysis of independent fission yield covariances based on GEF model code

Zerun Lu, *Tiejun* Zu*, *Liangzhi* Cao, and *Hongchun* Wu

School of Nuclear Science and Technology, Xi'an Jiaotong University, Xi'an, Shaanxi 710049, China

Abstract. The fission yield data provided by the evaluated nuclear data files do not contain covariance information, which is not conducive to uncertainty analysis. To generate covariance information, the model parameters of the code GEF which describes the fission process are sampled and the independent fission yield samples are calculated. The covariances of independent fission yields of ^{235}U, ^{239}Pu, and ^{241}Pu thermal neutron-induced fissioning systems are generated individually based on the above samples. This method is verified by comparing the uncertainties of burnup-related responses based on fission yield samples calculated by GEF and based on fission yield samples generated with the covariances. The influence of correlations among fissioning systems is also quantified and the joint covariances among different fissioning systems calculated with GEF are demonstrated correct. In addition, the Bayesian Monte Carlo method is adopted to adjust the model parameters of GEF, and the numerical results prove the effectiveness of the adjustment.

1 Introduction

Fission yields are the basic nuclear data in nuclear fuel cycle calculation. As input data of reactor burnup calculation, fission yields play an important role in the prediction of inventories of fission product nuclides. The inventories of fission products determine the calculation accuracy of macro parameters of reactor physics, such as eigenvalue. The fission yield data are mainly obtained through theoretical model evaluation [1]. During the evaluation process, imperfect model parameters propagate uncertainties to fission yields. Therefore, it is necessary to further analyze the uncertainties of fission yields propagated to macro parameters. However, the most commonly used fission yield data in ENDF-6 format [2] only contain the mean value and uncertainty of each fission yield itself, lacking the correlations among different fission yields of fission products or fissioning systems. This affects the accuracy of the uncertainties of the macro parameters.

To provide the correlations among fission yield data and quantify the uncertainties more accurately, many studies have been carried out worldwide. Subgroup 37 (SG37) of the Working Party on International Nuclear Data Evaluation Co-operation (WPEC) was established with the goal of developing "Improved Fission Product Yield evaluation methodologies" [3]. Katakura proposed a Bayesian/general least-squares method to construct covariances among fission yields of fission products with the same mass number [4]. Fiorito et al. considered the different physical constraints of fission yields and generated the covariances of independent fission yields through an iterative generalized least-square method [5]. At the Oak Ridge National Laboratory (ORNL), Wahl's systematics and its random parameters were used to generate random fission yields, and a Bayesian approach based on constraints from the mass yields, the conservation of the mass and charge number was proposed to generate the covariances among different fission yields [6]. Recently, the covariance matrices of independent fission yields were determined based on the generalized least-square method by Kohsuke Tsubakihara, et al [7].

GEF (**GE**neral description of **F**ission observables) code [8, 9] is a semi-empirical model designed to give a complete description of the fission process [10]. GEF code determines the corresponding fission products of different fission events through model parameters and determines the yields of different fission products through the Monte-Carlo process. Some important model parameters of GEF exist in the form of normal distribution, so the parameter samples are obtained by perturbation based on the probability distribution, and then fission yield samples are obtained by running GEF with the parameter samples. In this paper, the independent fission yield covariances of ^{235}U, ^{239}Pu, and ^{241}Pu thermal neutron-induced fissioning systems are generated respectively using the fission yield samples. These covariances are verified based on uncertainties of burnup-related responses. At the same time, the influence of the correlations among fissioning systems on the uncertainties of the burnup-related responses is analyzed. In addition, the model parameters of GEF are preliminarily adjusted through the Bayesian Monte Carlo method [10, 11]. All uncertainty calculations in this paper are based on random sampling.

This paper is organized as follows. Section 2 expounds on the detailed process of generating and

* Corresponding author: tiejun@mail.xjtu.edu.cn

verifying independent fission yield covariances for individual fissioning systems. Section 3 shows the influence of correlations among fissioning systems. Section 4 introduces the preliminary adjustment of GEF model parameters. Conclusions and perspectives are summarized in Section 5.

2 Fission yield covariance generation for individual fissioning systems

2.1 Generation of fission yield covariances

The process of generating independent fission yield covariances mainly includes three steps, which are summarized below.

(1) Sample the GEF model parameters.
(2) Calculate independent fission yield samples with the above parameter samples.
(3) Calculate covariances among different fission products with the above fission yield samples.

The GEF code version used in this paper is 2021.1.1 [12]. The GEF2021.1.1 determines the normal distribution of the 23 most important parameters for calculating fission yields, and these parameters are independent of each other. In the present work, we sample these parameters. The specific information on these model parameters is shown in Table 1.

Table 1. Specific information on sampled model parameters in GEF2021.1.1 [12].

Model parameter	GEF name	Mean	Standard deviation
Position of the shell for the S_1 channel	P_DZ_Mean_S1	-0.3	0.06
Position of the shell for the S_2 channel	P_DZ_Mean_S2	-0.4	0.06
Position of the shell for the S_3 channel	P_DZ_Mean_S3	0.2	0.06
Position of the shell at $Z \approx 42$	P_DZ_Mean_S4	0.0	0.06
Shell effect for the S_1 channel	P_Shell_S1	-3.0	0.06
Shell effect for the S_2 channel	P_Shell_S2	-4.4	0.06
Shell effect for the S_3 channel	P_Shell_S3	-7.2	0.12
Shell effect at $Z \approx 42$	P_Shell_S4	-1.1	0.03
Rectangular contribution to the width of S_2 channel	P_A_Width_S2	11.5	3%
Shell effect at mass symmetry	Delta_S0	0.0	0.06
Curvature of shell for the S_1 channel	P_Z_Curv_S1	0.38	3%
Curvature of shell for the S_2 channel	P_Z_Curv_S2	0.098	3%
Curvature of shell for the S_3 channel	P_Z_Curv_S3	0.09	3%
Curvature of shell at $Z \approx 42$	P_Z_Curv_S4	0.05	3%
Weakening of the S_1 shell	T_low_SL	0.31	0.006
$(\hbar\omega)_{eff}$ for tunneling of the S_1 channel	T_low_S1	0.31	0.006
$(\hbar\omega)_{eff}$ for tunneling of the S_1 channel	T_low_S2	0.31	0.006
$(\hbar\omega)_{eff}$ for tunneling of the S_1 channel	T_low_S3	0.31	0.006
$(\hbar\omega)_{eff}$ for tunneling at $Z \approx 42$	T_low_S4	0.31	0.006
Width of the fragment distribution in N/Z	HOMPOL	1.3	6%
Charge polarization	POLARadd	0.35	0.06
Asymmetry in diffuseness of S_2 mass peak	S2leftmod	0.75	3%
General scaling of fragment angular moment	Jscaling	1.0	6%

1000 sets of model parameter samples are generated and based on these samples, independent fission yield samples of ^{235}U, ^{239}Pu, and ^{241}Pu thermal neutron-induced fissioning systems are calculated individually. The number of fission events of GEF is 2×10^5. Based on these samples, the respective fission yield covariances of the three fissioning systems are calculated.

2.2 Verification of fission yield covariances

To verify the fission yield covariances, 1000 samples of fission yield are generated respectively for the three fissioning systems by resampling the above-obtained covariances. Besides, as mentioned above, there are

1000 samples of fission yield by sampling the GEF model parameters. Based on the two types of samples, neutronics-burnup coupling calculation is conducted for the TMI-1 pin cell problem in UAM project [13]. The fuel material of this problem is UO_2, and the ^{235}U enrichment is 4.85%. The calculation is based on the hot full power condition and the power density is 33.58 MW/tU. The burnup of the problem reaches 60 GW·d/tU. Figure 1 shows the geometry of the TMI-1 pin cell. The relative uncertainties of the responses, the infinite multiplication factor (k_{inf}), and the number densities of important fission product nuclides, are calculated based on the fission yield samples resampled based on the above-obtained covariances. The same uncertainties are also calculated based on the fission yield samples by sampling the GEF model parameters, which are taken as the reference results. The neutronics-burnup coupling calculation tool is the high-fidelity reactor physics code NECP-X [14]. Other burnup data and cross section data are extracted from ENDF/B-VII.1 evaluated nuclear data library [15].

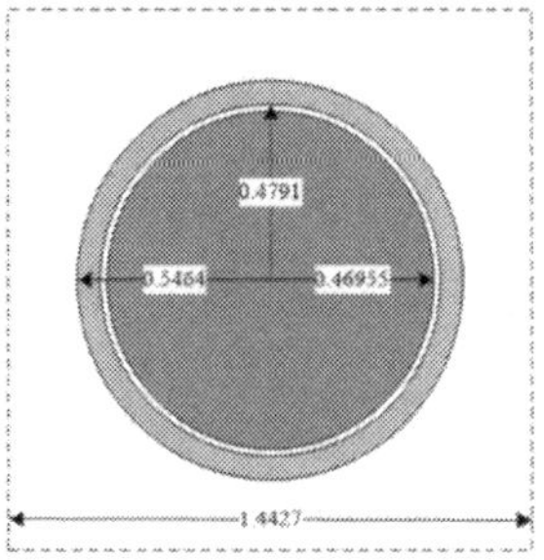

Fig.1. Geometry of the TMI-1 pin cell (unit: cm).

Figure 2 to Figure 4 show the results of the uncertainty comparison. The results show that for the three fissioning systems, the uncertainties of resampling based on the fission yield covariances are all consistent with the reference results. This demonstrates that the method of generating independent fission yield covariances for individual fissioning systems is correct.

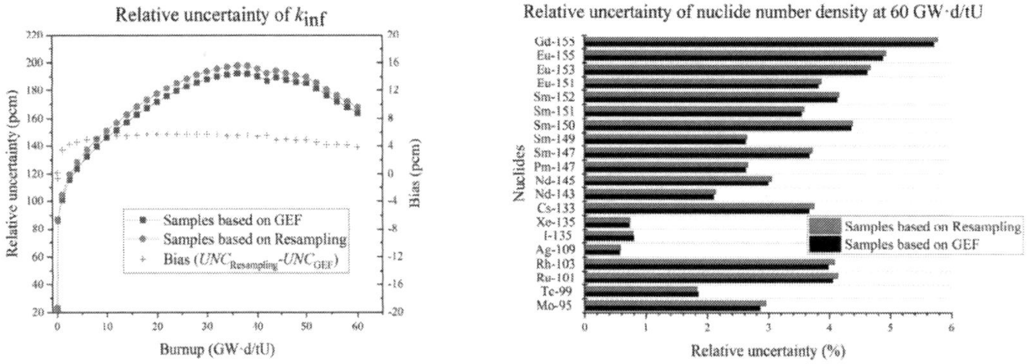

Fig.2. Comparison of uncertainties of the TMI-1 problem based on fission yield samples of $^{235}U+n_{th}$.

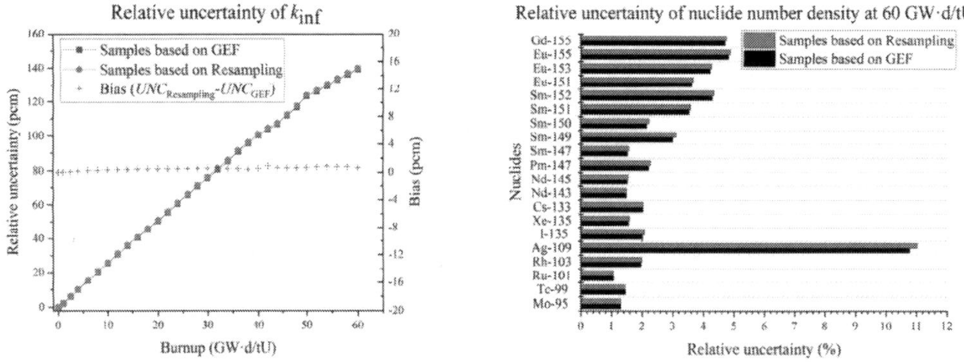

Fig.3. Comparison of uncertainties of the TMI-1 problem based on fission yield samples of $^{239}Pu+n_{th}$.

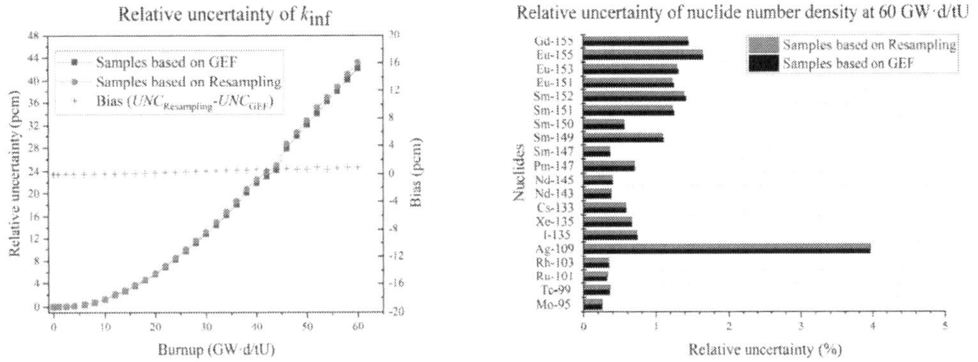

Fig.4. Comparison of uncertainties of the TMI-1 problem based on fission yield samples of $^{241}Pu+n_{th}$.

3 Influence of correlations among fissioning systems

Generally, there are correlations among multiple fissioning systems. According to the analysis in Section 2, since the same set of model parameter sample of GEF can calculate one set of fission yield sample for each of the three fissioning systems ($^{235}U+n_{th}$, $^{239}Pu+n_{th}$, and $^{241}Pu+n_{th}$) simultaneously, the fission yield samples calculated by GEF actually include the correlations among fissioning systems. To analyze the influence of these correlations, two cases are calculated. Case 1 uses fission yield samples of three fissioning systems calculated by GEF in one random run of the TMI-1 problem. Case 2 uses fission yield samples of one fissioning system calculated by GEF in one random run of the TMI-1 problem and calculates the sum of the uncertainties of three fissioning systems with equation (1) :

$$\Delta_{sum} = \sqrt{\left(\Delta_{U235}\right)^2 + \left(\Delta_{Pu239}\right)^2 + \left(\Delta_{Pu241}\right)^2} \quad (1)$$

where Δ is relative uncertainties of burnup-related responses (k_{inf} or nuclide number densities).

In other words, Case 1 considers correlations among three fissioning systems and Case 2 assumes that the three fissioning systems are independent. Figure 5 shows the uncertainty results for the two cases. It can be seen that the uncertainties of Case 1 considering correlations are significantly higher than that of Case 2 assuming independence. The bias of relative uncertainty of k_{inf} reaches about 100 pcm at the end of irradiation. For uncertainties of nuclide number densities, the two cases also differ greatly. Therefore, the influence of the correlations among fissioning systems on the quantification of uncertainty cannot be ignored.

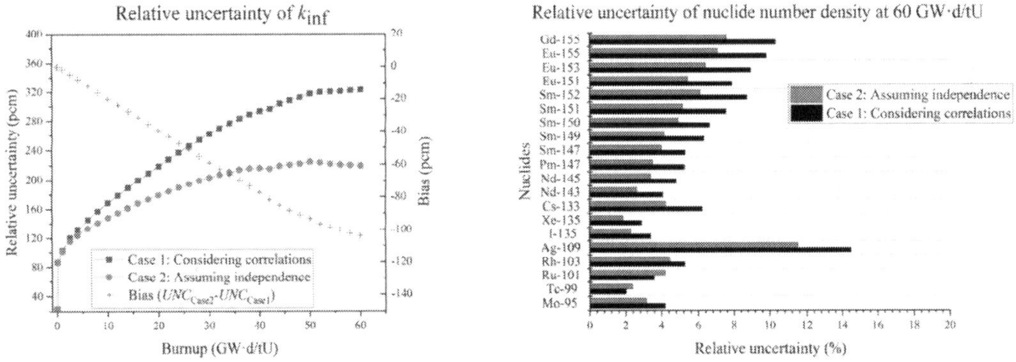

Fig.5. Uncertainty results of the TMI-1 problem for influence of correlations among fissioning systems.

This section also adopts the method like Section 2 to calculate the joint covariances among independent fission yields of the three fissioning systems based on the fission yield samples calculated with GEF model parameter samples. The method to verify the joint covariances is also like Section 2. The fission yield samples of three fissioning systems are obtained based on the resampling with joint covariances, and the samples of three fissioning systems are used simultaneously in one random run of the TMI-1 problem to calculate the uncertainties. The reference uncertainty results are also based on fission yield samples calculated with GEF. The comparison of uncertainty results is shown in Figure 6. The uncertainty results of resampling show good consistency with the reference results, which proves that the GEF calculation process for the correlations among fissioning systems is also correct. Thus, when nuclear data users need the covariances of fission yield data, the GEF code can provide covariance data for individual fissioning systems and multiple fissioning systems at the same time.

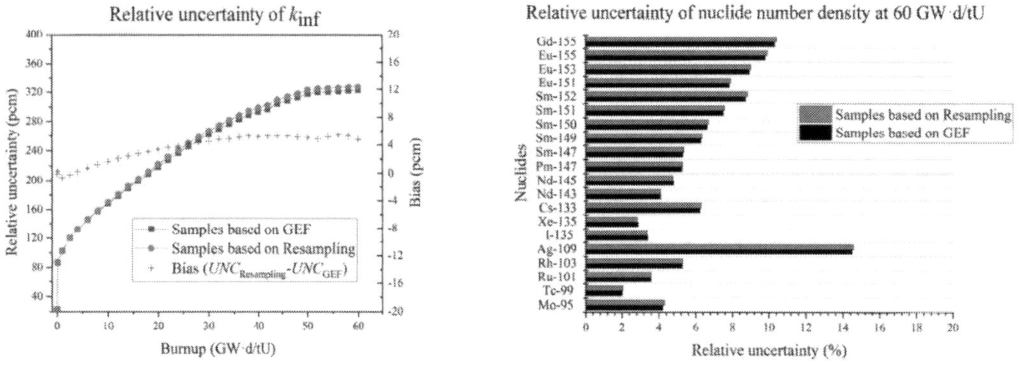

Fig.6. Comparison of uncertainties of the TMI-1 problem based on fission yield samples from joint covariances.

4 Preliminary adjustment of GEF model parameters

This section adopts the Bayesian Monte Carlo method to adjust 23 parameters with normal distribution in GEF2021.1.1 which are shown in Table 1 and presents some preliminary numerical results.

4.1 Bayesian Monte Carlo method

The Bayesian Monte Carlo method combines Bayesian inference theory with Monte Carlo sampling and adjusts the prior nuclear data by calculating the weight of each sample.

Based on the work in this section, Bayesian inference theory is shown in equation (2):

$$p(\boldsymbol{\sigma}' | \mathbf{E}) \propto \mathrm{L}(\mathbf{E} | \boldsymbol{\sigma}) p(\boldsymbol{\sigma}) \tag{2}$$

where $\boldsymbol{\sigma}$ is the prior GEF model parameter vector and $\boldsymbol{\sigma}'$ is the posterior parameter vector; $\mathbf{E}$ is the experimental data vector; $p(\boldsymbol{\sigma})$ is the prior probability density function of parameters; $\mathrm{L}(\mathbf{E}|\boldsymbol{\sigma})$ is the likelihood function to use experimental information to adjust prior parameters; $p(\boldsymbol{\sigma}'|\mathbf{E})$ is the posterior probability density function of parameters. According to equation (2), Bayesian inference theory adjusts the probability distribution of prior parameters by introducing experimental data. In our work, the experimental data are the inventories of fission product nuclides.

The Bayesian Monte Carlo method first calculates the fission yield samples based on the model parameter samples of GEF and calculates the fission product nuclide inventory samples based on the fission yield samples. Next, the likelihood function in the form of weight for each model parameter sample based on comparing the calculated and experimental values of fission product nuclide inventories is calculated. Finally, the model parameters are adjusted based on Bayesian inference theory. To calculate the weight of each sample, first calculate the square of Mahalanobis distance χ_i^2 for each sample, as shown in equation (3):

$$\chi_i^2 = (\mathbf{E} - \mathbf{C}_i)^{\mathrm{T}} \mathbf{V}_{\mathrm{E}}^{-1} (\mathbf{E} - \mathbf{C}_i) \tag{3}$$

where $\mathbf{C}_i$ is the calculated fission product inventory sample vector based on prior parameter sample i; $\mathbf{V}_{\mathrm{E}}$ is the covariance matrix of experimental data. Then, ω_i, the weight of sample i is calculated with equation (4):

$$\omega_i = \exp\left(-\frac{\chi_i^2}{\chi_{\min}^2}\right) \tag{4}$$

where $\chi_{\min}^2$ is the minimum value of all the χ_i^2 values.

We can obtain the posterior model parameters and the posterior parameter covariances through equation (5) and equation (6):

$$\boldsymbol{\sigma}' = \frac{\sum_{i=1}^{N} \omega_i \boldsymbol{\sigma}^i}{\sum_{i=1}^{N} \omega_i} \tag{5}$$

$$\mathrm{cov}(\sigma_l', \sigma_m') = \frac{\sum_{i=1}^{N} \omega_i (\sigma_l^i - \sigma_l')(\sigma_m^i - \sigma_m')}{\sum_{i=1}^{N} \omega_i}, l, m = 1, \ldots\ldots, N_\sigma \tag{6}$$

where N is the number of samples; N_σ is the number of model parameters; σ_l^i is the i-th prior sample of model parameter l.

The general steps for adjusting GEF model parameters are:

(1) Sample GEF model parameters.
(2) Calculate fission yield samples with the above parameter samples.
(3) Calculate fission product inventory samples $\mathbf{C}$ corresponding to experimental data $\mathbf{E}$.
(4) Calculate the weight of each sample with equation (3) and equation (4).
(5) Adjust the model parameters with equation (5) and equation (6).

4.2 Generation of pseudo-experimental data

The Bayesian Monte Carlo method has high requirements for experimental data, which requires that experimental data and corresponding calculated data have high consistency [10]. In addition, equation (3) shows that the covariances of experimental data are needed. Nevertheless, common integral experimental data, such as nuclide inventory data in SFCOMPO-2.0 spent fuel experimental database [16], usually have some outliers, and these experimental data do not contain covariances, which will seriously affect the effect of nuclear data adjustment.

Therefore, this section generates pseudo-experimental data based on the fission yield data in the evaluated nuclear data files (ENDF). The accuracy of pseudo-experimental data is high, and covariance data are obtained. The specific steps of generation are:

(1) Generate fission yield samples based on mean values and uncertainties of fission yields in ENDF.
(2) Calculate nuclide number density samples of a burnup problem based on the above samples.
(3) Calculate the sample mean values and covariances based on the above nuclide number density samples.

After this process, the sample mean values are the values of the pseudo-experimental data, and the sample covariances are the covariances of the pseudo-experimental data.

In the process of generating pseudo-experimental data for this work, fission yield data of ^{235}U thermal neutron-induced fission in JEFF-3.3 evaluated nuclear data library [17] are selected for sampling, and the sample size is 10000. The burnup problem is the TMI-1 pin cell and the calculation tool is NECP-X code. The nuclide number densities of 13 fission product nuclides at 60 GW·d/tU are taken as the final pseudo-experimental data. These nuclides are: ^{95}Mo, ^{99}Tc, ^{101}Ru, ^{103}Rh, ^{133}Cs, $^{143,145}Nd$, $^{151,152}Sm$, $^{151,153,155}Eu$, and ^{155}Gd. These nuclides are important in the uncertainty analysis of the burnup credit system [18], so they are selected. The correlation matrix of the pseudo-experimental data of this work is shown in Figure 7. Correlations are shown for the number densities of different nuclides.

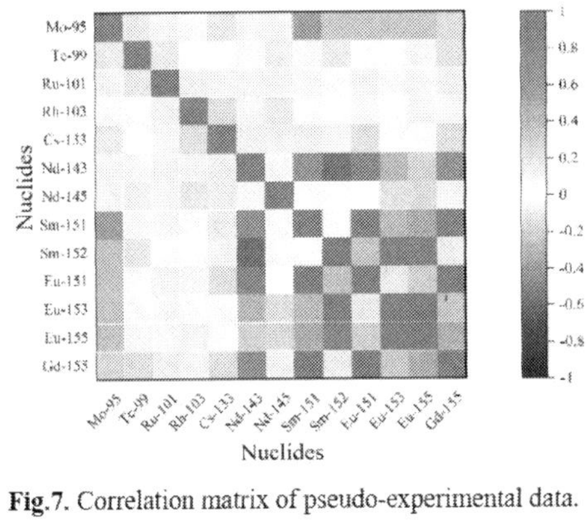

Fig.7. Correlation matrix of pseudo-experimental data.

4.3 Numerical results

^{235}U thermal neutron-induced fission is selected as the fissioning system for adjustment of GEF model parameters. The number of model parameter samples is 10000 and each run of GEF uses 2×10^5 fission events. To maintain consistency with the pseudo-experimental data, the calculation of GEF-based prior fission yield samples also uses the burnup data of JEFF-3.3.

The relative adjustment of model parameters is shown in Figure 8. The results show that the adjustment range of all parameters is reasonable and within the prior uncertainties. The uncertainty results of prior and posterior parameters are shown in Figure 9. For most parameters, the posterior uncertainties are reduced, and the effect is more obvious for parameters with larger prior uncertainties.

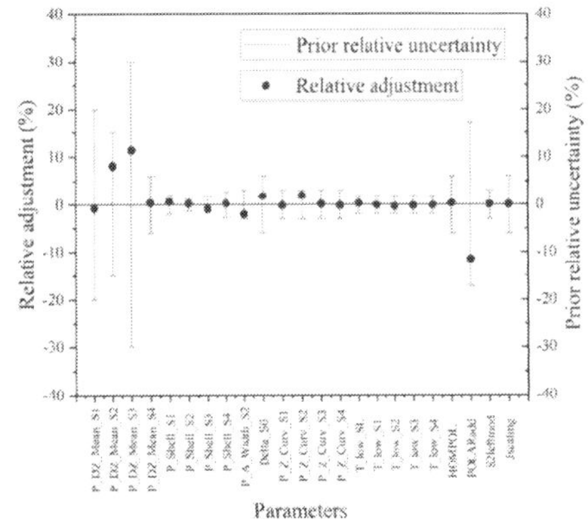

Fig.8. Relative adjustment of model parameters.

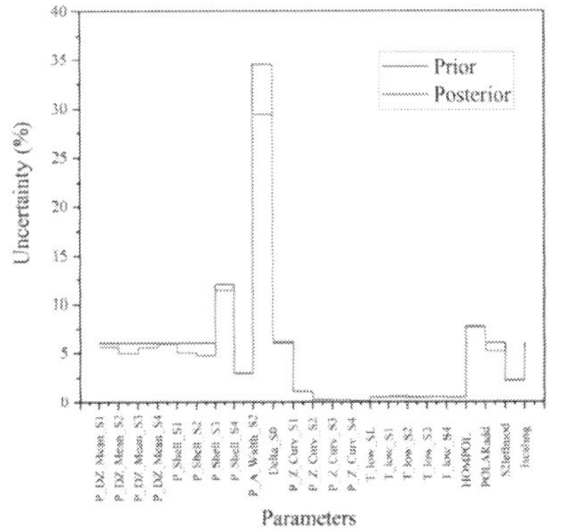

Fig.9. Comparison of uncertainties of prior and posterior parameters.

The fission yield covariances of ^{235}U thermal neutron-induced fission calculated by GEF and the uncertainties of burnup-related responses calculated based on prior and posterior parameters are compared. The prior and posterior lower triangular correlation matrices of fission yields are shown in Figure 10. There are 67 fission product nuclides compared: $^{87,88}Br$, $^{88\text{-}92}Kr$, $^{91\text{-}94}Rb$, $^{92\text{-}97}Sr$, $^{95,96m,97,97m,98,98m,99}Y$, $^{97\text{-}102}Zr$, $^{101,103}Nb$, $^{103,104}Mo$, $^{131\text{-}133}Sb$, $^{132,133,133m,134\text{-}136}Te$, $^{135,136,136m,137,138}I$, $^{136\text{-}141}Xe$, $^{139\text{-}143}Cs$, $^{141\text{-}145}Ba$, $^{144,145}La$, and ^{148}Ce. The fission yields of these nuclides are generally greater than 1%, so the calculation accuracy of GEF is higher. Figure 10 shows that the posterior correlations have decreased for most nuclides. Figure 11 shows the prior and posterior uncertainty results of the TMI-1 problem based on fission yield samples of ^{235}U thermal neutron-induced fission calculated by GEF. The basic setup for the calculation is the same as in Figure 2. The results show that the posterior uncertainties of k_{inf} and nuclide number densities are significantly reduced. The reduction of uncertainty of k_{inf} up to about 80 pcm throughout the process of irradiation. The above numerical results prove that the preliminary adjustment of GEF model parameters is effective.

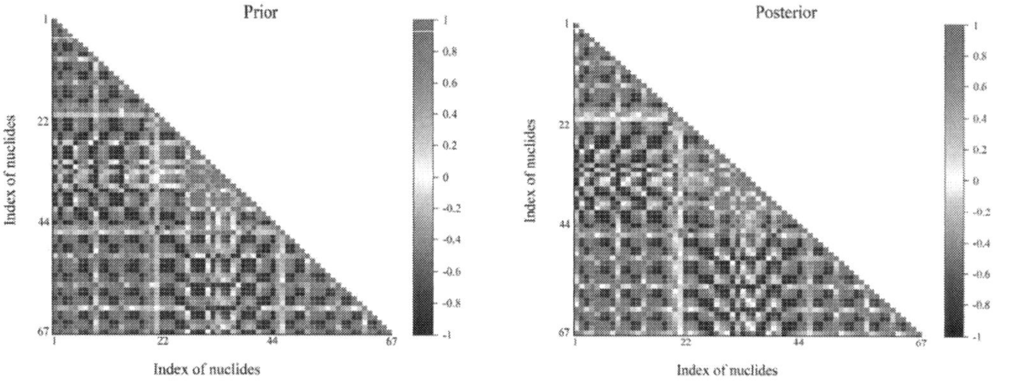

Fig.10. Comparison of prior and posterior fission yields correlation matrices.

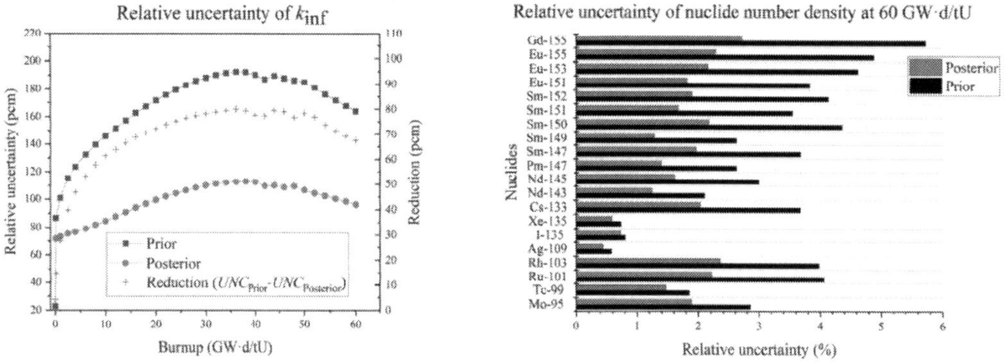

Fig.11. Comparison of prior and posterior uncertainties of the TMI-1 problem.

5 Conclusions and perspectives

The independent fission yield covariances are generated based on the GEF model code in this paper. The samples of fission yield are calculated by sampling model parameters, and the covariances are calculated based on the samples of fission yield. The above process is carried out for ^{235}U, ^{239}Pu, and ^{241}Pu thermal neutron-induced fissioning systems, individually. To verify the correctness of the process, the fission yield samples calculated by GEF and the fission yield samples generated based on the covariances of this work are respectively applied to the random run of the TMI-1 pin cell burnup problem. The uncertainties of the infinite multiplication factor (k_{inf}) and nuclide number densities propagated by two types of fission yield samples are in good consistency for three fissioning systems separately. The correctness of the method of generating fission yield covariance is verified.

The influence of correlations among fissioning systems is also analyzed and quantified. The numerical results show that the influence cannot be ignored. The joint independent fission yield covariances of multiple fissioning systems are calculated with GEF and verified. Finally, the parameters of the GEF model code are preliminarily adjusted with the Bayesian Monte Carlo method based on the pseudo-experimental data. The results show that the adjustment was effective.

The following conclusions are summarized:

(1) Independent fission yield covariances are obtained based on the sampling of GEF model parameters. The influence of correlation among fissioning systems cannot be ignored.

(2) When nuclear data users need the covariances of fission yield data, the GEF code can provide correct covariance data for individual fissioning systems and multiple fissioning systems simultaneously.

(3) Preliminary adjustment of GEF model parameters is performed with the Bayesian Monte Carlo method and the effect is good.

In future work, we will do more numerical tests on the current nuclear data adjustment method and strive to improve the applicability of the method. We will try to apply the adjustment method to other nuclear data, not limited to fission yield-related data.

This research is supported by National Natural Science Foundation of China (No. 11790323 & 12075183).

References

1. B. Voirin et al., EPJ Nuclear Sci. Technol. **4**, 26 (2018)
2. A. Trkov, M. Herman, D. A. Brown, Report BNL-90365-2009 Rev.2 (2012)
3. R. Mills, WPEC subgroup 37 description, WPEC – Subgroup 37 (2013)
4. J. Katakura, J. Nucl. Sci. Technol. **50**, 799 (2013)
5. L. Fiorito et al., Ann. Nucl. Energy **88**, 12 (2016)
6. M. T. Pigni et al., NEA, Issy-les-Moulineaux, OECD/NEA WPEC Subgroup-37 meeting (2013) http://www.oecdnea.org/download/wpec/sg37/Meetings/2013_May/32_Pigni_Cov.pdf
7. K. Tsubakihara et al., J. Nucl. Sci. Technol. **58**, 151 (2021)
8. K. H. Schmidt et al., Nucl. Data Sheets **131**, 107 (2016)
9. K. H. Schmidt, B. Jurado, C. Amouroux, Nuclear Energy Agency Data Bank NEA/DB/DOC(2014)1 (2014)
10. D. Siefman et al., EPJ Nuclear Sci. Technol. **6**, 52 (2020)
11. E. Bauge, S. Hilaire, P. Dossantos-Uzarralde, Proceedings of ND2007 (2007)
12. K.H. Schmidt, GEF2021.1.1: General description of fission observables. Released: June 2, 2021 http://www.khschmidts-nuclear-web.eu/GEF-2021-1-1.html
13. K. Ivanov et al., Nuclear Energy Agency Nuclear Science NEA/NSC/DOC(2013)7 (2013)
14. J. Chen et al., Ann. Nucl. Energy **116**, 417 (2018)
15. M. Chadwick et al., Nucl. Data Sheets **112**, 2887 (2011)
16. F. Michel-Sendis et al., Ann. Nucl. Energy **110**, 779 (2017)
17. A. J. M. Plompen et al., Eur. Phys. J. A **56**, 181 (2020)
18. I. Gauld, U. Mertyurek, NUREG/CR-7251, ORNL/TM-2018/782 (2018)

Uncertainty quantification calculations of nuclide number densities of Gd-bearing fuel rods in light water reactors

Go Chiba[1,*]

[1]Hokkaido University, Faculty of Engineering, Kita 13 Nishi 8, Sapporo, Hokkaido 060-8628, Japan

Abstract. The Post-Irradiation Examination (PIE) data are quite useful to validate the evaluated nuclear data. Recently, a new experimental program, REGAL (the Rod-Extremity and Gadolinia AnaLysis), has been proposed and launched. During this program, the PIE data are acquired for the Gd-bearing fuel rods during (or after) the initial Gd burnout. These PIE data would be different from those for the normal UOX or MOX fuel rods from a viewpoint of nuclide transmutation process. In the present work, we attempt to quantify this difference through calculating sensitivity coefficients of nuclide number densities of the Gd-bearing rods using the depletion perturbation theory and performing uncertainty quantification calculations with the sensitivity coefficients and the nuclear data covariance data. Through the numerical analyses, we found that the impact of the uncertainties of the nuclear data of Gd isotopes on the number densities of actinoids is negligible. The nuclear data-induced uncertainties of Gd isotopes number densities are relatively large during the Gd depletion, and those become small after the Gd burnout. After the Gd burnout, the relative standard deviations of Gd-155 and -157 number densities are approximately 6% and 13%, respectively, and the nuclear data of the parent nuclides of these Gd isotopes, Gd-154 and -156, are also important as well as those of Gd-155 and -157.

1 Introduction

The Post-Irradiation Examination (PIE) data are quite useful to validate the evaluated nuclear data, and the PIE data which have been accumulated in the world so far have been utilized in benchmarking the newly-developed evaluated nuclear data files. Recently, a new experimental program, REGAL (the Rod-Extremity and Gadolinia AnaLysis), has been proposed and launched[1]. In this program, the PIE data are acquired for the Gd-bearing fuel rods during (or after) the initial Gd burnout. These PIE data would be different from those for the normal UOX or MOX fuel rods from a viewpoint of nuclide transmutation process since the neutron flux energy spectra in these Gd-bearing rods are somewhat different from those in the normal fuel rods. In the present work, we attempt to quantify this difference through calculating sensitivity coefficients of nuclide number densities of the Gd-bearing rods and the normal fuel rods using the depletion perturbation theory (DPT) and performing uncertainty quantification calculations with the sensitivity coefficients and the covariance data of the nuclear data.

2 Brief description of numerical methods and tools

2.1 Sensitivity calculations and uncertainty quantification

The relative sensitivity of the nuclide number density N to nuclear data i, σ_i, is defined as

$$s_i = \frac{dN}{d\sigma_i} \cdot \frac{\sigma_i}{N}. \quad (1)$$

Since quite a large number of nuclear data, such as neutron-induced reaction cross sections, fission yields, decay constants and decay branching ratios, are used in nuclear fuel depletion calculations, numerical differentiation to obtain these sensitivities to all of the relevant nuclear data is unrealistic. DPT is a powerful method to efficiently calculate sensitivities of N with respect to nuclear data during nuclear fuel depletion. The fundamental theory of DPT has been established in many years ago[2, 3], and it has been recently adopted to the fuel depletion problems of the light water fuel assembly containing burnable absorbers[4]. The details of DPT are omitted here, and interested readers are recommended to refer the literatures listed at the end of this paper.

Nuclear data-induced uncertainties of N can be easily quantified by the well-known sandwich rule as

$$\Delta N/N = \sqrt{\sum_i \sum_{i'} s_i s_{i'} \mathrm{cov}(\sigma_i, \sigma_{i'})}, \quad (2)$$

*e-mail: go_chiba@eng.hokudai.ac.jp

where $\text{cov}(\sigma_i, \sigma_{i'})$ is a relative covariance between σ_i and $\sigma_{i'}$. When we define a sensitivity vector $\mathbf{s}$ as $\mathbf{s} = (s_1\ s_2\ \cdots\ s_I)^T$, where T is for vector transposition, $\Delta N/N$ can be rewritten using $\mathbf{s}$ and $\mathbf{V}_\sigma$, which is a covariance matrix covering all the nuclear data considered, as

$$\Delta N/N = \sqrt{\mathbf{s}^T \mathbf{V}_\sigma \mathbf{s}}. \quad (3)$$

2.2 Reactor physics code system CBZ

Numerical calculations of the nuclide number density sensitivities with DPT are generally based on numerical methods for nuclear fuel depletion calculations, which consist of neutron flux distribution calculations by solving the neutron transport equation and nuclides transmutation calculations. In the present study, all these calculations were carried out with a reactor physics code system CBZ[5] which has been developed at Hokkaido University.

At each fuel burnup point, medium-wise 107-group cross section data were generated, and neutron transport calculations with the 107-group cross sections including generalized adjoint neutron flux calculations for DPT were performed with a neutron transport calculation module of CBZ based on the method of characteristics, and reaction rates were calculated for all the nuclear fuel media. Energy structure of the 107-group is that of the SRAC code developed for thermal reactor analyses. Note that angular integration of bilinear functions of neutron flux and generalized adjoint neutron flux in DPT should be carefully carried out in a system containing strong neutron absorbers[4]. Scattering anisotropy was taken into account by the P0 transport approximation.

Nuclides transmutation calculations were carried out with the simplified nuclide chain model consisting of 21 actinoids and 138 fission products (FP). Nuclides transmutation equations including adjoint problems in DPT were solved by the matrix exponential method with the minimax polynomial approximation method[6]. To reduce the time-discretization error, the predictor-corrector method was employed. Only for the normal (or forward) nuclides transmutation calculations, the advanced optimally-weighted predictor-corrector method[7] was used with the coarse burnup mesh of 2.0 GWD/t.

All these calculations were carried out with the JENDL library: JENDL-4.0[8] for neutron-induced reaction cross sections, JENDL/FPY-2011 and JENDL/FPD-2011[9] for fission yield data and decay data of FP. Calculated sensitivities were then used to quantify nuclear data-induced uncertainties of nuclide number densities. The present work focuses on the number densities of actinoids and Gd isotopes, uncertainties of reaction cross sections of actinoids and Gd isotopes were taken into account. In JENDL-4.0, covariance data are evaluated for almost all the important actinides, and those were taken into account in the present study. However, JENDL-4.0 provides no covariance data for reaction cross sections of gadolinium isotopes, and thus we used the covariance data for these nuclear data taken from ENDF/B-VIII.0[10].

3 Problem specification

Sensitivities of nuclide number densities with respect to nuclear data during nuclear fuel depletion were calculated for a 3×3 multicell model which had been used in our previous study[11]. In this model, a Gd-bearing fuel rod whose Gd enrichment is 5.0 wt% is located at the center position and it is surrounded by normal uranium fuel rods whose uranium-235 enrichment is 3.9 wt%. Geometrical and material specifications are taken from a BWR fuel assembly model developed through the OECD/NEA burnup credit benchmark phase-IIIC[12]. Reflective boundary conditions were adopted. As for coolant condition, void fraction of 0% was assumed. The geometrical specification of the 3×3 multicell model is shown in **Fig. 1**.

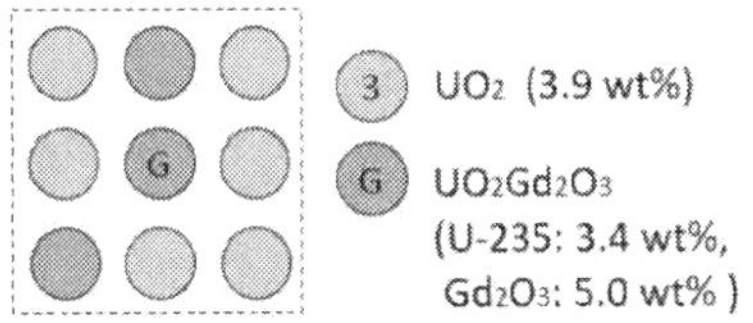

Figure 1. Geometric specification of a 3×3 multicell problem

The infinite neutron multiplication factor during fuel depletion of this 3×3 multicell model is shown in **Fig. 2**. The reactivity peak is observed at around 13 GWD/t. Volume-integrated number densities of U-235, Pu-239, Gd-155, and Gd-157 of the Gd-bearing rod are shown in **Fig. 3**. The Pu-239 build up and the burnout of Gd-155 and -157 can be found. The fuel burnup values corresponding to the burnout of Gd-155 and -157 are approximately 18 and 12 GWD/t, respectively.

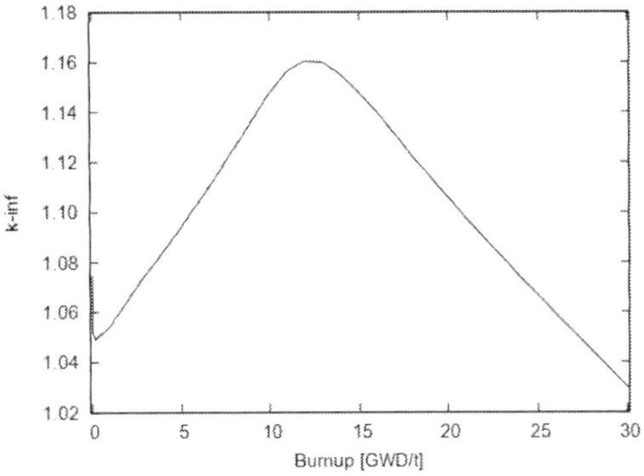

Figure 2. Infinite neutron multiplication factor during fuel depletion

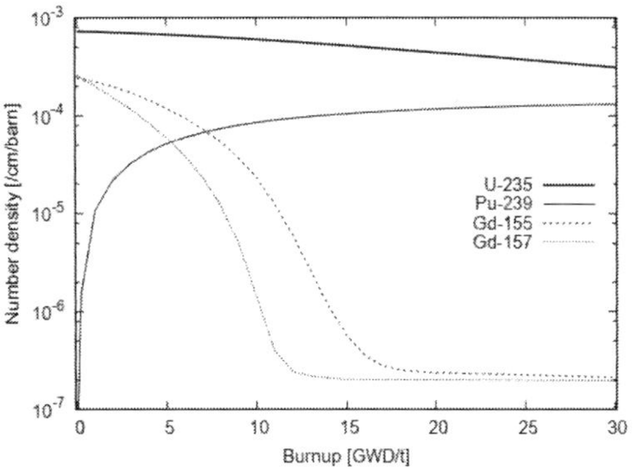

Figure 3. Volume-averaged nuclide number densities in the Gd-bearing rod

4 Numerical results

4.1 Verification of depletion perturbation theory against Gd-bearing fuel rod

The sensitivity calculation capability of CBZ based on DPT has been carefully verified in our previous works[4, 13]. In these works, however, sensitivities of the number densities of nuclides contained in the Gd-bearing fuel rods have never been addressed. In the present work, sensitivities of the nuclide number densities in the Gd-bearing rod at the fuel burnup of 10 GWD/t were calculated with DPT, and those were compared with the reference sensitivities calculated by the direct numerical differencing; the small perturbation of 1% was given to the infinite dilution cross section of specific nuclide, reaction, and energy group, and the change in the nuclide number density was observed. In the calculations with DPT, three different fuel burnup mesh schemes were adopted. **Figure 4** shows sensitivities of U-235 number density to Gd-157 (n,γ) cross section. Sensitivities calculated with the coarse burnup mesh differ from the references, but the good agreement is obtained when the fine burnup mesh of 0.5 GWD/t was adopted. Next, **Figure 5** shows sensitivities of Pu-239 number density to Gd-155 (n,γ) cross section. Better agreements are obtained when finer burnup mesh is adopted similar as Figure 4, but the non-negligible differences are still observed in the resonance energy range. This comes from the implicit effect which is not taken into account in CBZ. The same comparisons were made under the infinite dilution condition where no self-shielding effect was considered during the depletion. **Figure 6** shows sensitivities of Pu-239 number density to Gd-155 (n,γ) cross section under the infinite dilution condition, and the difference observed in the resonance range disappears. From the above comparisons, it is found that the DPT capability of CBZ works properly for the number densities of nuclides in Gd-bearing rods, and that the implicit effect is observed in those to Gd-155 (n,γ) cross sections. Since the contribution of the resonance range to the total capture reaction rate is limited in the Gd-155 (n,γ) reaction, conclusion of our work shown in the following would be unchanged if the implicit effect is properly taken into account.

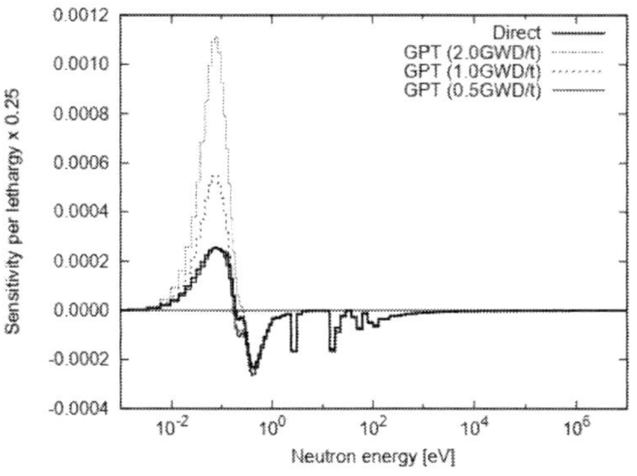

Figure 4. Sensitivities of U-235 number density to Gd-157 (n,γ) cross section

4.2 Uncertainty quantification calculation

Using the sensitivities calculated with the DPT capability of CBZ and the covariance data of the nuclear data, uncertainties of the nuclide number densities can be quantified. **Figures 7** and **8** show nuclear data-induced uncertainties of U-235 and Pu-239 number densities during the depletion. In addition to the uncertainties coming from the all nuclear data considered, those from the Gd-155 and -157 covariance data are presented also in these figures. Relative uncertainties of the nuclide number densities due to the Gd-155 and -157 nuclear data uncertainties are small, and their contributions are negligible in the total uncertainties. The same trends were observed in the uncertainties of number densities of other actinoids, Pu, Am, and Cm isotopes, and the maximum relative uncertainties of 1% were observed in some Am and Cm isotopes around the beginning of the fuel depletion. Correlations in the number densities between the Gd-bearing rod and the non Gd-bearing

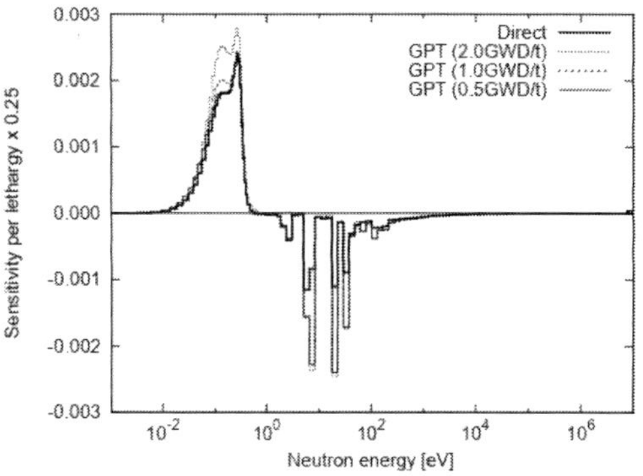

Figure 5. Sensitivities of Pu-239 number density to Gd-155 (n,γ) cross section

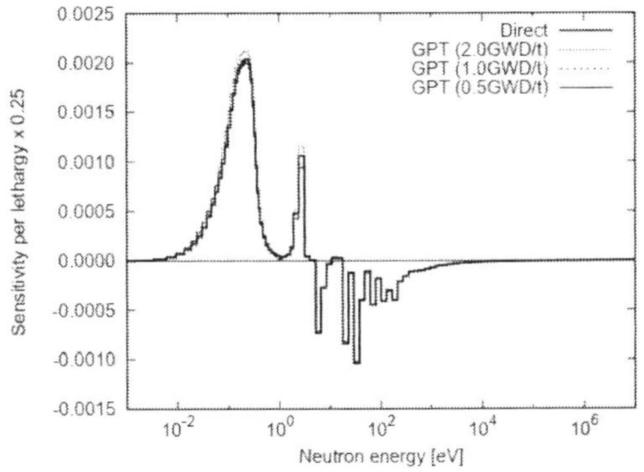

Figure 6. Sensitivities of Pu-239 number density to Gd-155 (n,γ) cross section (infinite dilution condition)

rod, which is neighboring to the central Gd-bearing rod, were also calculated. **Figure 9** shows a correlation matrix of Am-241 number density, and strong positive correlations can be found. The same trends were observed in correlation matrices of number densities of the other nuclides. The above results suggest that the nuclide transmutation process in Gd-bearing fuel rods before and after the Gd burnout are very similar to that in normal fuel rods when the JENDL-5 and ENDF/B-VIII.0 covariance data are used.

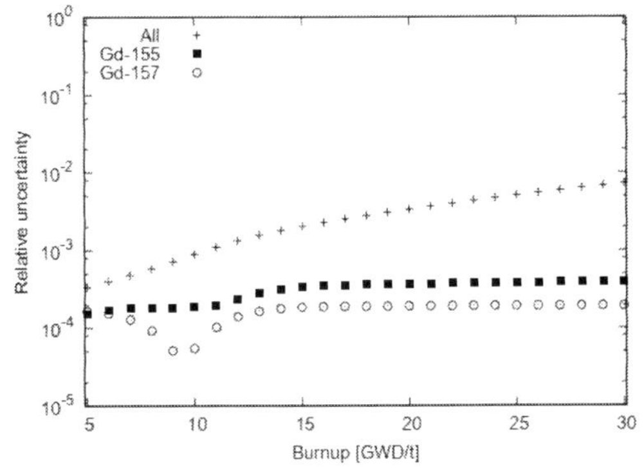

Figure 7. Nuclear data-induced uncertainty in U-235 number density during burnup

Figures 10 and **11** show nuclear data-induced uncertainties of Gd-155 and -157 number densities during the depletion. In these figures, contributions of the nuclear data of the parent nuclide, Gd-154 for Gd-155 and Gd-156 for Gd-157, are presented. Before the Gd burnout, the principal contributor is the concerned Gd isotope itself, but after the burnout, the contribution of the parent nuclide becomes also significant. The nuclear data-induced uncertainties of the number densities of Gd-155 and -157 after the burnout are 6% and 13%, respectively. The PIE data of Gd-bearing rods before and after the Gd burnout would be useful to validate the nuclear data of the Gd isotopes.

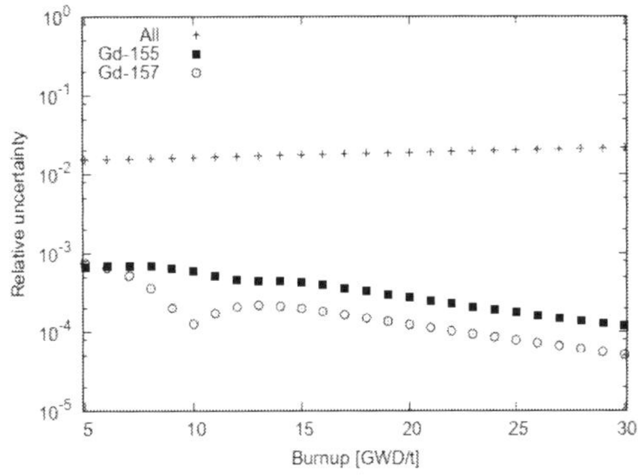

Figure 8. Nuclear data-induced uncertainty in Pu-239 number density during burnup

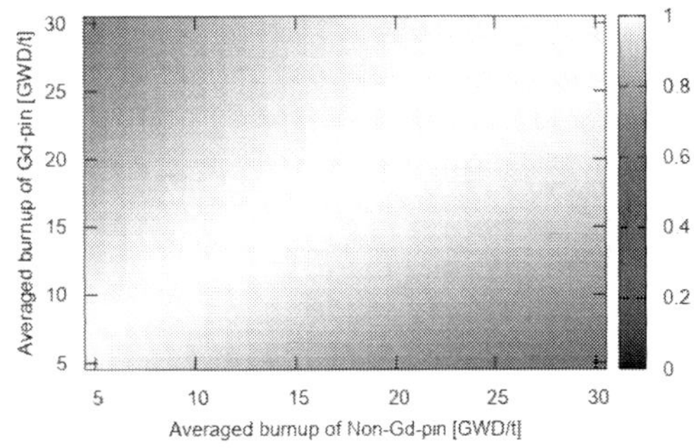

Figure 9. Correlation matrix of Am-241 number density uncertainty between the Gd-bearing and non Gd-bearing rods

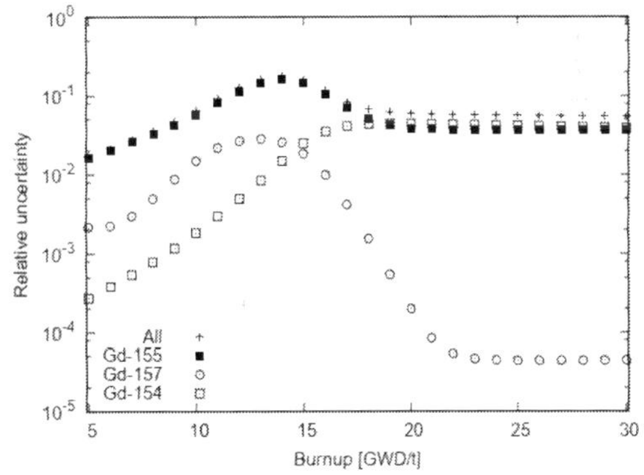

Figure 10. Nuclear data-induced uncertainty in Gd-155 number density during burnup

 https://doi.org/10.1051/epjconf/202328100016

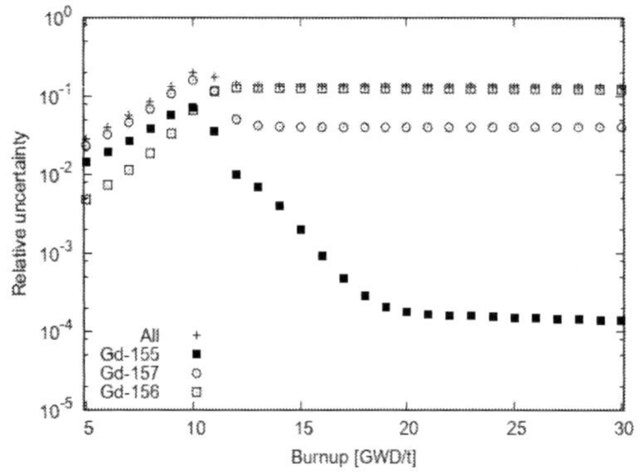

Figure 11. Nuclear data-induced uncertainty in Gd-157 number density during burnup

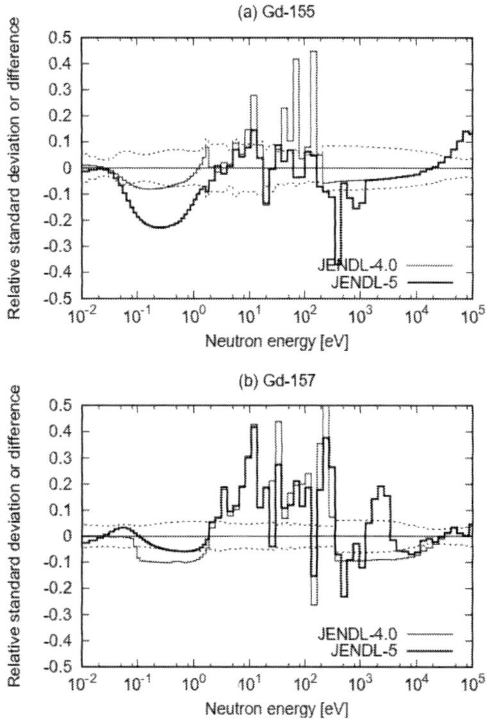

Figure 12. Relative difference in group-wise (n,γ) cross section of Gd-155 and -157 from the ENDF/B-VIII.0 evaluation

It would be useful to observe the covariance data of the Gd-155 and -157 (n,γ) cross sections and to compare them with the differences of these nuclear data among the different evaluations. **Figure 12** shows relative difference in group-wise (n,γ) cross sections of Gd-155 and -157 from the ENDF/B-VIII.0 evaluation. In these figures, dotted lines represent relative standard deviations given in the ENDF/B-VIII.0 evaluation. It is interesting to point out that the difference in the Gd-155 (n,γ) cross section between ENDF/B-VIII.0 and JENDL-5 is significantly large in comparison with the covariance data in ENDF/B-VIII.0.

5 Conclusion

In order to observe and quantify the difference in the nuclide transmutation process between Gd-bearing and non Gd-bearing fuel rods, sensitivities of nuclide number densities of Gd-bearing and non Gd-bearing rods were calculated with DPT, and by using them nuclear data-induced uncertainties of these number densities were quantified. Numerical results suggested that the nuclide transmutation process in Gd-bearing fuel rods before and after the Gd burnout are very similar to that in normal fuel rods. After the Gd burnout, the relative standard deviations of Gd-155 and -157 number densities were estimated at approximately 6% and 13%, respectively, and it was found that the nuclear data of the parent nuclides of these Gd isotopes, Gd-154 and -156, are also important as well as those of Gd-155 and -157.

References

[1] J. Eysermans, et al., Ann. Nucl. Energy, **172**, 109057 (2022)
[2] A. Gandini, Nucl. Sci. Eng., **38**, 1 (1969)
[3] M.L. Williams, Nucl. Sci. Eng., **70**, 20 (1979)
[4] G. Chiba, J. Nucl. Sci. Technol., **55**, 290 (2018)
[5] G. Chiba, Proc. Int. Conf. on Physics of Reactors, PHYSOR 2020, (2020)
[6] Y. Kawamoto, et al., Ann. Nucl. Energy, **80**, 219 (2015)
[7] J. Sasuga, G. Chiba, et al., J. Nucl. Sci. Technol., **58**, 1134 (2021)
[8] K. Shibata, et al., J. Nucl. Sci. Technol., **48**, 1 (2011)
[9] J. Katakura, *JENDL FP decay data file 2011 and fission yields data file 2011*, JAEA-Data/Code 2011-025, Japan Atomic Energy Agency (2011).
[10] D.A. Brown, et al., Nucl. Data Sheets, **148**, 1 (2018)
[11] G. Chiba, Ann. Nucl. Energy, **151**, 107949 (2021)
[12] Nuclear Energy Agency (NEA), *Burn-up credit criticality safety benchmark phase III-C*. NEA/NSC/R/(2015)6, OECD/NEA, Paris (2015).
[13] G. Chiba, Ann. Nucl. Energy, **96**, 313 (2016)

https://doi.org/10.1051/epjconf/202328100017

Updates to the AMUR code for R-matrix analyses on heavy nuclei

Satoshi Kunieda[1,*], *Shunsuke* Endo[1,**], and *Atsushi* Kimura[1,***]

[1]Japan Atomic Energy Agency, Shirakata 2-4, Tokai-mura, Ibaraki 319-1195, Japan

Abstract. The AMUR code, that had been initially designed for R-matrix analysis on light nuclei, was updated to perform resonance analyses on heavy nuclei by including approximated theories with experimental corrections such as resolution function. As a feasibility study, we tested these updates by performing a covariance analysis on J-PARC/ANNRI measured data for n+^{93}Nb reactions. In this work we present preliminary results obtained from the fitting procedure including both R-matrix resonance parameters and parameters related to the experimental corrections. We also discuss about physical constraint from the theory for forthcoming in-depth analysis to the measurements.

1 Introduction

The AMUR code [1, 2], which is based on the R-matrix nuclear reaction theory [3] and the Kalman filtering method [4] for the data-assimilation, is under development in JAEA for the cross-section evaluation with the covariance data in the resolved resonance energy region. The code was initially designed for the analysis of the light-nuclei to narrow large differences of cross-sections among evaluations in the world (e.g., on n+^{16}O [1]). Also, it was applied to the development of JENDL-5 [5] for a number of light-mass isotopes such as ^{12,13}C, ^{15}N, ^{16}O and ^{19}F. Recently, the authors extended capability of the code toward the analysis of heavier nuclei by introducing the Reich-Moore approximation [6] to calculate the radiative-capture and fission cross-sections. The code is also able to calculate the Doppler broadening with the free-gas approximation. In addition to those capabilities, we understand that in-depth resonance analysis requires a number of corrections which depend on the experimental conditions.

In this work, we challenge a resonance analysis of neutron cross-section data measured in J-PARC/ANNRI [7] with AMUR. The self-shielding and multiple-scattering effects in the target sample should be considered in the resonance analysis, however, the implementation of such correction into the code is under investigation. Therefore, the experimental data to be analyzed in this study are those corrected by the Mote Carlo simulation. Under this precondition, fundamental challenge on the analysis may be consideration of the energy resolution for incident neutrons with the double-bunching problem in the time-of-flight (TOF) measurements at our accelerator facility. In this study, we incorporate the resolution functions estimated by Kino et al. [8] into the code to take account such experimental conditions in the resonance analysis. We devote our much efforts to a feasibility study of the covariance/uncertainty analysis on the J-PARC/ANNRI measurements, where a test analysis was demonstrated on the ^{93}Nb isotope. Preliminary results of fitting parameters and cross-sections are also presented. Finally, let us discuss about physical constraints from the R-matrix theory (constraints from behavior of the phase-shifts and the interference effects) for forthcoming in-depth/sophisticated analysis to the measurements.

*e-mail: kunieda.satoshi@jaea.go.jp
**e-mail: endo.shunsuke@jaea.go.jp
***e-mail: kimura.atsushi@jaea.go.jp

2 Extension on AMUR toward analysis of heavier nuclei

2.1 Basic capabilities on theoretical calculation

AMUR was initially developed for the analysis of the light-nuclei to narrow large differences of cross-sections among evaluations in the world, e.g., on the ^{16}O(n,α) cross-section, by making used of the theoretical constraints [1] from the R-matrix theory. The code was also designed to analyze covariance data of the resonance parameters and cross-sections, where both the experimental and theoretical knowledge were reflected to the nuclear data evaluation as much as possible [2]. The analysis on the heavier nuclei is different from that for the light nuclei in general as follows.

- A great number of levels should be analyzed, although the number of channel spin is limited.
- The fission reaction should be included for the actinide nuclei.
- It is much important to take account of the experimental conditions.

We thought that AMUR could be applied to analysis on the heavier nuclei if the code is updated for those differences, learning a number of experiences on the other codes such as SAMMY [9] and REFIT [10].

Recently, the authors extended capabilities of the code toward the analysis of medium- and heavy-mass nuclei to fully cover range of isotopes required for nuclear science and engineering. In the theoretical part, AMUR is now able to read resonance parameter set stored in the ENDF-6 format (MF=2) files, then performs reconstruction with the Breit-Wigner (LRF=1,2) or the Reich-Moore type formulae (LRF=3,7), which means the code is able to calculate radiative capture and fission cross-sections with approximated/reduced R-matrix formulae. The code is also able to perform the Doppler broadening with the free-gas approximation. Once the areal density of sample material is given, the transmission is calculated from the total cross-sections. Figure 1 illustrates neutron transmission on ^{233}U measured by Guber *et al.* [11] and calculated results of AMUR with resonance parameter set (LRF=3) in ENDF/B-VIII.0 [12]. We have not yet made thorough comparisons with other codes, this figure visually suggests AMUR is getting equivalent with the other resonance analysis codes in terms of basic capabilities for calculating/reconstructing the cross-sections.

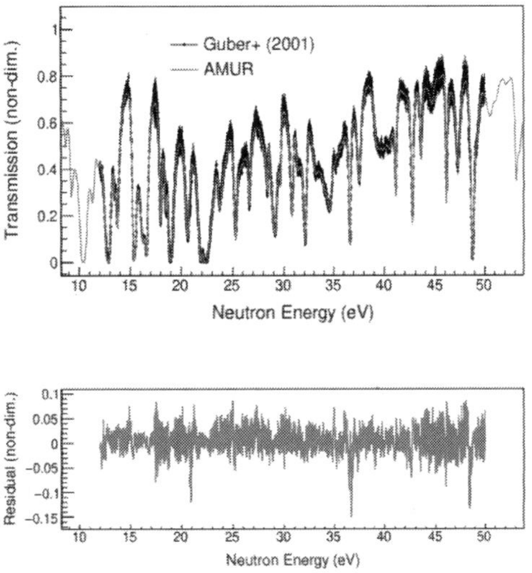

Figure 1. Neutron transmission on ^{233}U measured by Guber *et al.* [11] and calculated results of AMUR with resonance parameter set (LRF=3) in ENDF/B-VIII.0 [12] where the residuals are plotted in the bottom panel.

2.2 Resolution function

In general, experimental data are different from true values of the cross-sections, because the data are measured under intrinsic experimental conditions such as on the accelerator facility, detection system, composition of the sample, and so forth. Therefore, it is necessary to simulate such experimental conditions as much as possible in the resonance analysis or the nuclear data evaluation. In this regard, time/energy structure of the incident neutrons is one of the most important factors in the cross-sections measurements.

The cross-section measurements are underway in the J-PARC/MLF facility with the TOF method and the ANNRI detection system to measure neutron cross-sections for minor actinides and structural material in the resolved resonance energy region. The facility shows its own time structure of neutron beam, where the operation of the proton accelerator is usually in the double-bunch mode, which eventually results in degeneration of resonant structure on the cross-sections.

The resolution function of J-PARC/MLF had been estimated by Kino *et al.* [8]. The study had been based on the Mote Carlo simulation that was validated by a number of experimental campaigns at the same facility. The function had already been applied to the resonance analysis code REFIT [10, 13], which enabled us to have a number of experiences on the resonance analysis (The most recent work is found in Ref. [14]). Recently, those works were followed by one of the authors and the function was incorporated into AMUR. Figure 2 illustrates 2D plots of the calculated resolution functions in the single- and double-bunch modes. With the AMUR code, energy distribution of the incident neutrons are calculated at an arbitrary flight time. The figure shows that we successfully followed the work of Kino *et al.* [8] where significance of the double-bunching effect is seen above 10 eV of the incident neutrons.

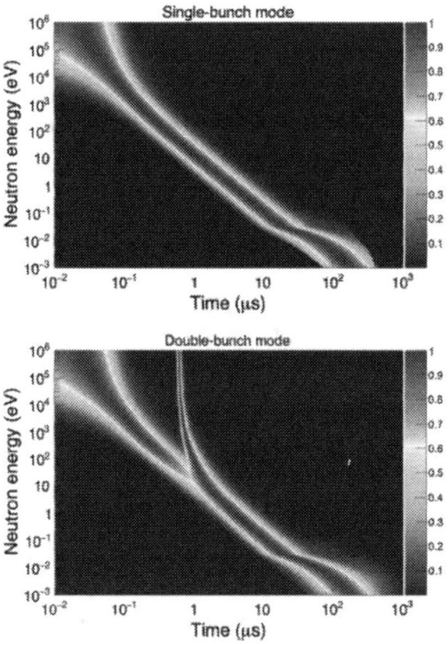

Figure 2. Resolution functions of the J-PARC/MLF facility in the single- and double-bunch modes calculated by the AMUR code base on the work of Kino *et al.* [8]

In this work, the code was eventually updated to analyze TOF measurements directly. For example, values of the time-dependent neutron transmission which are to be compared with experimental data are calculated by the equations as follows,

$$T_{rans}(t) = \int_0^{\infty} R(L, t + t_0, E) T_{rans}(E) dE, \quad (1)$$

where the symbol $R(L, t + t_0, E)$ denotes the resolution function that also depends on the flight path L and initial delay t_0 in the measurements. The energy dependent transmission $T_{rans}(E)$ is calculated with target thickness n, composition of i-th isotope in the sample c_i and the total cross-section $\sigma_i(E, T)$ reconstructed at temperature T :

$$T_{rans}(E) = \exp\left(-n \sum_i c_i \times \sigma_i(E, T) \right). \quad (2)$$

3 Preliminary analysis of J-PARC/ANNRI measurements

3.1 Experimental data to be analyzed

Quite recently, some of the authors measured transmission and (n,γ) cross-sections on ^{93}Nb [14] at J-PARC/MLF. This measurement is one of the good practices for the updated version of AMUR because of reasons as follows.

- ^{93}Nb is a mono isotope in which possible effects from the other isotopes are expected to be minimized.
- In the experimental campaign, both the transmission and (n,γ) cross-sections are measured, which allows us to obtain values of the resonance parameters both for neutron and γ-ray reaction channels, consistently.
- The experiments had been performed in the double-bunch mode of accelerator. That is a good practice for AMUR with new option of the resolution function.

In this study, we analyze those experimental data up to 400 eV, in which both the theoretical and experimental parameters are searched for, simultaneously.

3.2 Prior parameters and covariances

Among a number of the approximated resonance formulae which are stemmed from the R-matrix theory, we adopted the Reich-Moore type (ENDF-6/MF=2/LRF=3) to calculate/reconstruct the cross-sections, as a first step of AMUR being applied to heavier nuclei. Initial of the resonance parameter set was taken from the preceding work by Endo *et al.* [14] in which parameter values had been obtained from the same measurements with the modified REFIT code that accommodates to the double-bunching issue.

In the analysis, based on our experiences, prior uncertainty values were assumed 1 eV for the resonance energies E_r and 10% for neutron and radiation widths, Γ_n and Γ_γ. With those prior parameters, the calculated total cross-section and its correlations matrix are shown in Fig. 3, where energy grid in both plots are obtained so as to retain accuracy of 0.1% for the interpolation. The matrix is very complicated, showing both the short- and long-range correlation due to the interference between/among the resonances and the non-resonant processes. Furthermore, value of correlation is suddenly minimized at the peak positions because phase-shifts of the elastic scattering process should be maximized toward $\pi/2$ at the resonant energies. Those correlations are visually suggesting that physical constraints certainly exist in analysis of experimental data. Thus, it is interesting to see how such theoretical constraints are showing up in the resonance analysis, which will be discussed in the next subsection/section.

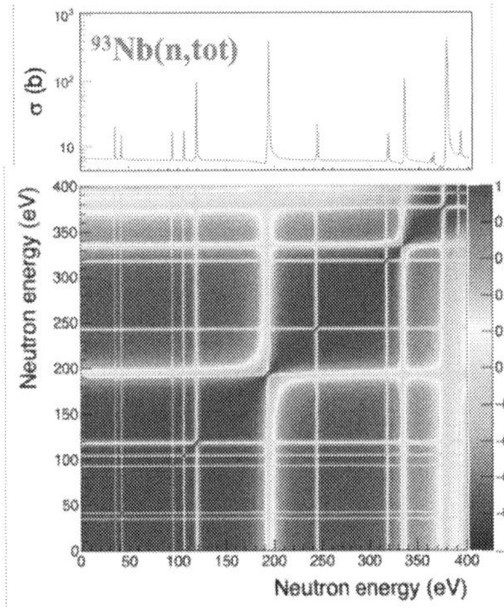

Figure 3. Calculate total cross-section and its correlation matrix on ^{93}Nb with prior parameters and uncertainties described in the texts.

In the present analysis, learning a number of experiences with the SAMMY and REFIT codes, we also introduced the experimental parameters to be searched for. Those parameters preliminary assumed were the flight-path and initial delay of the TOF measurements. The sample thickness/normalization and amount of contaminant isotope (only ^{181}Ta was assumed in this test case) were also treated as free parameters. Although their values could be quantified by measurements, the accuracy is sometimes questionable, which may results in one of the large causes of uncertainty on the evaluated cross-sections. We expected that such experimental parameters could be obtained through resonance analysis because the theory may impose constraints not only on the values of resonance parameters but also on behaviors of the experimental parameters.

3.3 Fitting results

All the resonance parameters and a limited number of the experimental parameters listed in Table 1 were simultaneously searched for, by fitting the experimental data of the ^{93}Nb isotope measured at J-PARC with the updated version of AMUR described in Sec. 2. Figure 4 illustrates fitting results to TOF spectra on the transmission and (n,γ) cross-section measurements. The fitting itself was nearly perfect as we successfully reproduced experimental data, in which the double-bunching and a contaminant (^{181}Ta) effect was simulated well.

Transmission

Parameter	Prior	Posterior
Flight path (m)	28.68 (1%)	28.674 (0.001%)
Initial delay (us)	3.50 (1%)	3.528 (0.069%)
Thickness (atoms/b)	0.01673 (3%)	0.01408 (0.559%)

(n,γ)

Parameter	Prior	Posterior
Flight path (m)	21.52 (1%)	21.490 (0.003%)
Initial delay (us)	4.1 (1%)	4.230 (0.072%)
Normalization to calc.	1.00 (10%)	0.916 (1.01%)
Ta-181/Nb-93	0.0006 (3%)	0.00064 (1.10%)

Table 1. The experimental parameter values and their uncertainties obtained in the present analysis.

The experimental parameter values obtained are listed in Table 1 together with their uncertainties, in which the prior and posterior values are compared for a reference purpose. The flight-path and initial delay were determined within uncertainties less than 0.1% and they were very close to the prior values obtained by the measurement. It is noticeable that the resulting values of the sample thickness/normalization were estimated with uncertainties of $\lesssim 1\%$, but the nominal values are different from the priors by $\sim 10\%$. This time, let us avoid in-depth discussion on such values and the differences because we understand that there could be the other sources in the experimental condition. However, our preliminary result suggests that the theoretical constraint is so large that the resonance analysis allows us to quantify experimental parameters which are difficult to know only by the measurement. For instance, it is virtually impossible to eliminate the systematic uncertainty from the experimental data in general, where the present approach may gives us an useful guide to tackle this problem.

4 Resulting covariances

The covariances of the parameter values were obtained through the fitting procedure. Once we calculate sensitivities of the parameters to the cross-sections, covariance matrices of cross-sections are estimated by the propagation law. The resulting correlation matrices are illustrated in Fig. 5 for total and (n,γ) cross-sections. In respect to analysis of the J-PARC/ANNRI measurements, we are now presenting such correlations of cross-sections for the first time. It is an advantageous feature of AMUR to immediately calculate and visualize correlations with a parallel processing technique.

We firstly see overall features of those plots are quite complicated because of the interference effects on the resonant nuclear reaction. Nevertheless, it is interesting to see the correlation values are drastically minimized/maximized at resonance positions. That is a consequence of quantum mechanical physics in the resonant reaction where the phase-shifts are maximized at the peak position as touched on Sec. 3.2. A strong long-range correlation is seen in the total cross-sections. Since such a trend is not observed in the (n,γ) reaction, such a long-range correlation comes from the elastic scattering. It is true that the shape elastic scattering process, which is calculated by the hard-sphere model in R-matrix, always occurs over the energy range. Medium-range correlations observed in the (n,γ) reaction are due to a number of reasons. For instance, a strong correlation that appears below 100 eV is mainly due to negative levels which exist below neutron separation energy of the compound nuclei. The tails of large resonances produce the other strong correlations observed in higher energy region. Furthermore, the experimental parameters (such as normalization parameter) should be sources of the long-range correlation, as we did see correlations between the resonance and experimental parameters. We believe that it is valuable to see correlation matrices of cross-sections in this way, because they show us physical features of the resonant reactions which could be one of the bases of an elaborate analysis to the experimental data.

Figure 6 illustrates resulting ^{93}Nb(n,γ) cross-sections and their uncertainty values. The uncertainty we obtained was $\sim 5\%$ on average but highly depend on neutron energy or structure of resonances. The fact may be because of the R-matrix theory in which the physical constraint is maximized at peak positions of the resonances. We are not yet fully understood physical background of resulting uncertainty. Nevertheless, that may give a reasonable uncertain of cross-sections since the present analysis is based on knowledges both on the experiment and theory.

5 Summary

The AMUR code was updated for analysis of cross-sections on heavier nuclei in terms of both the theoretical calculation and the experimental correction. In the theoretical part, resonance formulae used in the ENDF-6 format (MF=2/LRF=1,2,3 and 7) were incorpolated to the code together with a capability of the Doppler broadening. In the experimental part, we devoted our much efforts to simulation of the resolution function in the J-PARC/ANNRI measurements. Those options were applied to the resonance analysis of ^{93}Nb, and we found the fitting itself was quite successful in which both resonance and experimental parameters obtained were expected to be reasonable.

In this work, we derived correlation matrices of cross-sections with AMUR for the same isotope. It was first time to perform covariance analysis of cross-sections based on the J-PARC/ANNRI measurement. The resulting correlation matrices were complicated to be fully understood but we certainly saw traces of physical constraints from the R-matrix theory. The same discussion could be possible for the uncertainty of cross-sections obtained in this analysis. This time, let us conclude that we successfully obtained theoretical and experimental parameters with covariance/correlation matrices with AMUR despite the double-bunching problem of the facility. In-depth analysis

https://doi.org/10.1051/epjconf/202328100017

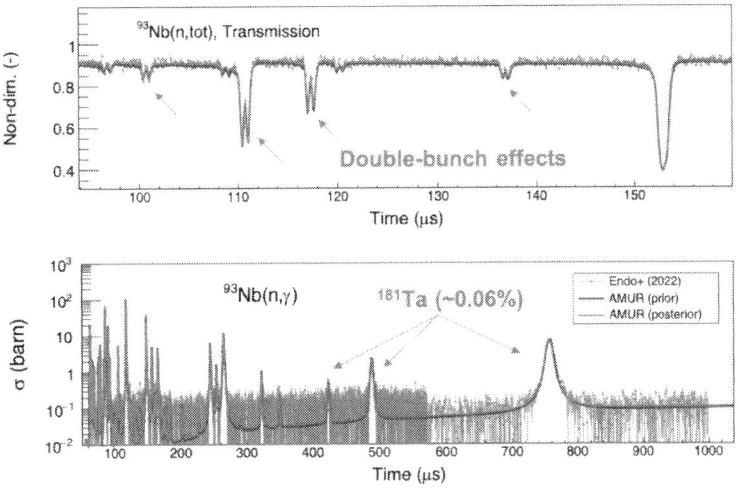

Figure 4. Fitting results to the experimental TOF spectra on transmission and (n,γ) cross-sections measured at J-PARC/ANNRI.

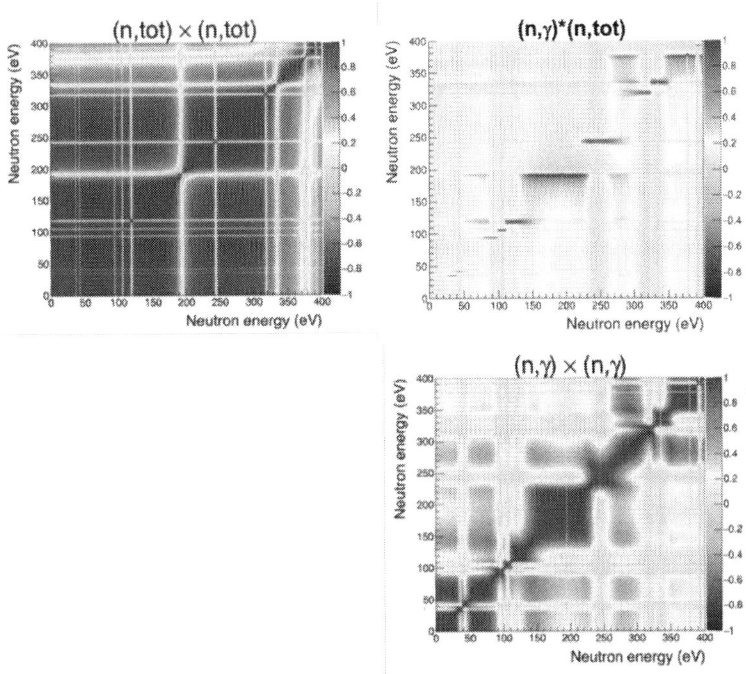

Figure 5. Correlation matrices for n+^{93}Nb total and (n,γ) cross sections.

is to be continued both from the experimental and theoretical points of views.

References

[1] S. Kunieda, T. Kawano, M. Paris, et al., Nucl. Data Sheets **118**, 250 (2014)

[2] S. Kunieda, T. Kawano, M. Paris, et al., Nucl. Data Sheets **123**, 159 (2015)

[3] A.M. Lane, R.G. Thomas, Rev. Mod. Phys. **30**, 257 (1958)

[4] T. Kawano, H. Matsunobu, T. Murata, et al., J. Nucl. Sci. Technol. (Tokyo) **37**, 327 (2000)

[5] O. Iwamoto, N. Iwamoto, S. Kunieda, et al, J. Nucl. Sci. Technol. **60(1)**, 1 (2023)

[6] C.W. Reich, M.S. Moore, Phys. Rev. **111**, 929 (1958)

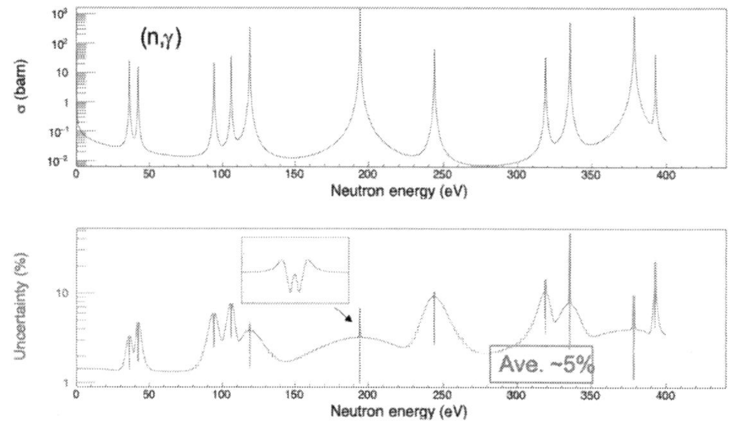

Figure 6. ^{93}Nb(n,γ) cross section (top) and corresponding relative uncertainty in percent (bottom).

[7] M. Igashira, Y. Kiyanagi, M. Oshima, Nuclear Instruments and Methods in Physics Research Section A: Accelerators Spectrometers. Detectors and Associated Equipment **600**, 332 (2009)

[8] K. Kino, M. Furusaka, F. Hiraga, et al., Nuclear Instruments and Methods in Physics Research Section A: Accelerators, Spectrometers, Detectors and Associated Equipment **736**, 66 (2014)

[9] M.N. Larson, Updated user's guide for SAMMY: mul-tilevel R-matrix fits to neutron data using Bayes' equations, ORNL/TM-9179/R4. Oak Ridge National Laboratory; 1998.

[10] M.C. Moxon, T.C. Ware, C.J. Dean, Users' guide for REFIT-2009-10, UKNSF; 2010 April. pp. 243. UK Nuclear Science Forum

[11] K.H. Guber, R.R. Spencer, L.C. Leal, P.E. Koehler, J.A. Harvey, R.O. Sayer, H. Derrien, T.E. Valentine, D.E. Pierce, V.M. Cauley et al., Nuclear Science and Engineering **139**, 111 (2001)

[12] D.A. Brown, M.B. Chadwick, R. Capote, A.C. Kahler, A. Trkov, M.W. Herman, A.A. Sonzogni, Y. Danon, A.D. Carlson, M. Dunn et al., Nuclear Data Sheets **148**, 1 (2018), special Issue on Nuclear Reaction Data

[13] H. Hasemi, M. Harada, T. Kai, et al., Nuclear Instruments and Methods in Physics Research Section A: Accelerators, Spectrometers, Detectors and Associated Equipment **773**, 137 (2015)

[14] S. Endo, A. Kimura, S. Nakamura, et al., Journal of Nuclear Science and Technology **59**, 318 (2022)

Calculated covariance matrices for fission product yields using BeoH

A. E. Lovell[1], * *T.* Kawano[1], and *P.* Talou[1]

[1]Los Alamos National Laboratory, Los Alamos, NM 87545, USA

Abstract. Fission product yields (FPY) are important for a variety of applications (reactor neutronics, spent fuel, dosimetry, radiochemistry, etc.) and are currently included in many of the evaluated libraries around the world. The FPYs in the current US evaluation, ENDF/B-VIII.0, are mainly based on the 1994 evaluation of England and Rider and have only had slight updates—such as the inclusion of a 2 MeV point for ^{239}Pu—since their development. Additionally, only mean values and uncertainties are included in the evaluation, not full correlations. Los Alamos National Laboratory, in collaboration with several other institutes, has been working on an updated evaluation for the FPYs of ^{239}Pu(n,f), ^{235}U(n,f), ^{238}U(n,f), and ^{252}Cf(sf) using the deterministic, Hauser-Feshbach, fission fragment decay code, BeoH. BeoH calculates the FPYs consistently with many other prompt and delayed fission observables, explicitly taking into account multi-chance fission and ensuring consistency between observables. In addition to providing updated means and uncertainties for the FPYs on a pointwise energy grid from thermal to 20 MeV, we calculate correlations between all FPYs at each incident energy and across incident energies. Here, we discuss the development of these covariance matrices, differences in the correlations between FPYs based on the parameters that are included in the model optimization, and correlations across incident energies for neutron-induced fission.

1 Introduction

A number of applications rely on fission product yields (FPYs), such as reactor neutronics, spent fuel, dosimetry, and radiochemistry, and many of the major evaluated libraries [1–3] contain these data. However, only mean values and uncertainties are currently available–covariance matrices have not been developed for any of the libraries, and therefore, a format to store them has not been developed. Additionally, there has been no major update to at least the US evaluation, ENDF/B, since the 1994 evaluation of England and Rider [4], save the addition of a point at 2 MeV for neutron-induced fission of ^{239}Pu [5].

Recently, there has been a revitalization in the development of FPY modeling and experimental work. Modeling efforts have improved to the point where prompt and delayed fission observables–including FPYs–can be calculated consistently in a single framework [6, 7], and large numbers of experimental efforts are underway to measure the energy-dependence of independent and cumulative FPYs [8–10] (and many more recent measurements that are still under analysis), including measurements of photon-induced fission [11, 12]. Stochastic estimations of covariances for the independent and cumulative FPYs in the ENDF/B-VIII.0 and JEFF-3.3 evaluations have been developed [13]. Additionally, there has been a great deal of interest in using machine learning methods in the determination of FPYs [14–18], in particular where data are scarce. A recent International Workshop on Fission Product Yields was held in Santa Fe in 2019, and the International Atomic Energy Agency (IAEA) currently has a Coordinated Research Project (CRP) focusing on modeling FPYs.

Here, we present first results of the cumulative fission product yield covariances calculated using BeoH and a Kalman filter optimization. In Sec. 2, we give a brief overview of BeoH, focusing particularly on the model parameters that are fitted, along with details of the Kalman filter optimization. Results from the covariances of ^{252}Cf spontaneous fission are presented in Sec. 3, and we briefly discuss covariances for fission product yields from neutron-induced reactions. Finally, we conclude in Sec. 4.

2 Theory

BeoH is a deterministic Hauser-Feshbach fission fragment decay model developed at Los Alamos National Laboratory [6, 7]. It consistently calculates the emission of prompt neutrons and γ rays from excited fission fragments using the Hauser-Feshbach statistical theory, then performs a time-independent calculation to determine delayed observables. More details about these calculations can be found in [6, 7], but here, we specifically want to discuss the parametrization of fission fragment initial conditions as these parameters go into the FPY optimization and impact the resulting covariance matrices.

The fission fragment initial conditions are distributions in mass, charge, total kinetic energy, spin, and parity, $Y(A, Z, \mathrm{TKE}, J, \pi)$ and functions of the incident neutron energy, E_{inc}. The mass distribution, $Y(A)$, is a sum of three

*e-mail: lovell@lanl.gov

Gaussians, $Y(A) = G_0(A|E_{\rm inc}) + G_1(A|E_{\rm inc}) + G_2(A|E_{\rm inc})$. $G_0(A|E_{\rm inc})$ is a symmetric Gaussian with

$$G_0(A|E_{\rm inc}) = \frac{W_0(E_{\rm inc})}{\sqrt{2\pi\sigma_0(E_{\rm inc})^2}} \exp\left[-\frac{(A - A_c/2)^2}{2\sigma_0(E_{\rm inc})^2}\right], \quad (1)$$

and $G_{1,2}(A|E_{\rm inc})$ are the asymmetric Gaussians,

$$\begin{aligned} G_i(A|E_{\rm inc}) &= \frac{W_i(E_{\rm inc})}{\sqrt{2\pi}\sigma_i(E_{\rm inc})} \\ &\times \left\{\exp\left[\frac{-(A-\mu_i(E_{\rm inc}))^2}{2\sigma_i(E_{\rm inc})^2}\right]\right. \\ &+ \left.\exp\left[\frac{-(A-(A_c-\mu_i(E_{\rm inc})))^2}{2\sigma_i(E_{\rm inc})^2}\right]\right\}. \end{aligned} \quad (2)$$

In Eqs. (1) and (2), A_c is the mass of the compound nucleus and W_i, μ_i, and σ_i are the energy-dependent weights, means, and standard deviations of the Gaussians. The means and standard deviations are linear in their energy dependence, and $W_{1,2}$ are defined as

$$W_i(E_{\rm inc}) = \frac{1}{1 + \exp[(E_{\rm inc} - w_i^a)/w_i^b]} \quad (3)$$

for $i = 1, 2$, while W_0 is defined from normalization as $2 = W_0 + 2W_1 + 2W_2$. The charge distribution, $Y(Z|A)$, is computed from the Wahl systematics [19], and we allow for scaling factors, f_Z and f_N to adjust the odd-even staggering.

The total kinetic energy, TKE, is a linear function of the incident neutron energy, and allows for a slope change at an incident energy of E_0 as indicated by some experimental data, e.g. [20],

$$\langle{\rm TKE}\rangle(E_{\rm inc}|m) = \begin{cases} a + bE_{\rm inc}, & \text{if } E_{\rm inc} \le E_0 \\ c + dE_{\rm inc}, & \text{if } E_{\rm inc} \ge E_0 \end{cases}, \quad (4)$$

where a, b, d, and E_0 are fit to experimental data and c is determined from the continuity at E_0. The TKE for each mass is a Gaussian with width and mean fitted to experimental data but scaled to reproduce the average TKE for the given incident neutron energy. The TKE is shared between the two fission fragments based on conservation of energy and momentum. The total excitation energy, TXE, of the fragment split is calculated from the Q value, TXE = Q − TKE, and is shared between the two fragments based on a ratio of temperatures, $R_T^2 = (U_l a_h(U_h))/(U_h a_l(U_l))$. The level density parameter, $a(U)$, is energy dependent and calculated by the Gilbert-Cameron level density model [21]. This ratio can be taken to be dependent on the fission fragment mass, $R_T(A)$, to better reproduce the average neutron multiplicity as a function of mass, $\overline{\nu}_p(A)$.

The spin distribution is proportional to the available states in the level density formula, with an adjustable scaling factor, f, on the spin cutoff parameter, $\sigma^2_{l,h}$,

$$R_{l,h}(J) = \frac{J + 1/2}{f^2\sigma^2_{l,h}(U)} \exp\left\{-\frac{(J + 1/2)^2}{2f^2\sigma^2_{l,h}(U)}\right\}, \quad (5)$$

where U is the excitation energy corrected by the pairing energy. Positive and negative parities are assumed to be equally probable. The initial population distribution is constructed for each fragment, and then the Hauser-Feshbach statistical decay is performed, de-exciting the fission fragment through neutron and γ-ray emission. The resulting population of each residual nucleus is collected.

The prompt observables are calculated as the weighted sums of the results of the Hauser-Feshbach calculations for each fission fragment. At this point in the calculation, we keep track of the isomeric states, M, and the independent yields have been calculated. The cumulative yields are calculated as

$$Y_C(A_i, Z_i, M_i) = Y_I(A_i, Z_i, M_i) + \sum_j^N \sum_l^{L_j} Y_C(A_j, Z_j, M_j) b_{jl} \delta_{jl,i}, \quad (6)$$

where b_{jl} are the branching ratios with L_j total decay modes, N is the total number of nuclei that produce the $i^{\rm th}$ nucleus, and $\delta_{jl,i}$ connect $Y_C(A_j, Z_j, M_j)$ with $Y_C(A_i, Z_i, M_i)$. Evaluated decay data is used to determine the branching ratios, keeping consistency within the evaluated libraries. Delayed neutron emission is calculated by summing over all nuclei where the branching ratio includes a neutron emission.

For the calculations of neutron-induced fission, further information about the compound nucleus is needed that will be weighted by the multi-chance fission probabilities. These calculations include information about any pre-fission neutrons emitted, the excitation energy causing fission, and the total kinetic energy at the equivalent incident neutron energy. Again, the fission fragment initial conditions are determined for all fission fragment pairs that can be formed from each compound nucleus that will undergo fission. Each nucleus is then de-excited in the manner described above. Prompt and delayed observables are calculated consistently across incident neutron energies, providing additional constraints on the model, beyond what is available from a single observable alone.

2.1 Kalman filter

To perform the parameter optimization and calculate the fission product yield covariances, we use a Kalman filter routine [22], a type of Bayesian optimization where the likelihood and prior are taken to be multivariate Gaussian distributions. Often, a linear approximation is made, where changes in the model outputs, $\delta\mathbf{f} = f(\mathbf{x}_1) - f(\mathbf{x}_0)$, are assumed to be linearly related to changes in the model parameters, $\delta\mathbf{x} = \mathbf{x}_1 - \mathbf{x}_0$,

$$f(\mathbf{x}_0) = f(\mathbf{x}_1) + \mathbb{C}\delta\mathbf{x}, \quad (7)$$

The parameter and output changes are connected through the sensitivity matrix, $\mathbb{C}$, which is calculated for observables from `BeoH` by

$$\mathbb{C}_{ij} = \left.\frac{\partial f(\mathbf{x})}{\partial x_j}\right|_{\mathbf{x}=\mathbf{x}_0}, \quad (8)$$

where the model difference is calculated for each cumulative fission product yield and the average prompt and delayed neutron multiplicities, $\bar{\nu}_p$ and $\bar{\nu}_d$ at each parameter x_j.

The parameter changes in Eq. (7) are determined through matrix algebra,

$$\delta\mathbf{x} = \mathbb{P}\mathbb{C}^T\mathbb{V}^{-1}(\phi - f(\mathbf{x}_0)), \tag{9}$$

where ϕ is the vector of experimental data to be fitted and the resulting parameter correlation matrix, $\mathbb{P}$, is calculated

$$\mathbb{P} = (\mathbb{X}^{-1} + \mathbb{C}^T\mathbb{V}^{-1}\mathbb{C})^{-1}. \tag{10}$$

Additionally, covariance matrices on the experimental data, $\mathbb{V}$, and prior parameter values, $\mathbb{X}$, are needed. In this work, we take the experimental covariance matrix to be diagonal, with the reported experimental uncertainties on the diagonal, however, a template of expected experimental uncertainties has been developed for fission product yields [23, 24] which should be included as a future improvement. The parameter covariance matrix is also taken to be diagonal with a 1% uncertainty. Although this value is small compared to the uncertainty that we would expect on the model parameters, as discussed in the Sec. 2.2, the optimization is run in several subsequent steps, so the parameters can vary more than 1%.

Once the change in parameter values and updated parameter covariance matrix are calculated, the updated models values can be calculated through Eq. (7) and compared to the exact model solution. A covariance matrix across the observables can be calculated,

$$\mathbb{F} = \mathbb{C}\mathbb{P}\mathbb{C}^T. \tag{11}$$

When the sensitivities are calculated for multiple observables, Eq. (11) provides a way to determine correlations between pairs of observables. In principle, the sensitivity matrices, $\mathbb{C}$, in Eqs. (10) and (11) can be calculated for different observables but are done so in the same way, using Eq. (8).

2.2 Fitting details

We optimize BeoH using the cumulative FPYs, $\bar{\nu}_p$, and $\bar{\nu}_d$. The prompt neutron multiplicities constrain the magnitude of the total kinetic energy, spin cutoff factor, and Wahl scaling factors. The parameters in the pre-neutron emission mass distribution and $R_T(A)$ are additionally constrained. We fix the widths and means of the TKE distribution as a function of mass, and take all information about the compound nucleus (e.g. multi-chance fission probabilities, pre-fission neutron emission, etc.) as given by CoH_3 [25]. The multi-chance fission probabilities can have a large impact on the shape of energy-dependent quantities when crossing the openings of successive fission changes (for example see [7]); however, for the work discussed here, we take the default calculations from CoH_3. Data for the cumulative FPYs, $\bar{\nu}_p$, and $\bar{\nu}_d$ are taken from EXFOR.

3 Results

We first discuss result for the optimization of the fission product yields for the spontaneous fission of ^{252}Cf. Slightly fewer parameters are needed to be included in the fit than for the neutron-induced fission calculations because there is no incident energy dependence to take into account. Examples of the parameter sensitivities for the average prompt neutron multiplicity, $\bar{\nu}_p$ and one representative FPY, ^{147}Nd, are shown in Fig. 1 panels (a) and (b) respectively. We see strong sensitivity to the total kinetic energy, TKE, and spin cutoff parameter, f, in $\bar{\nu}_p$, with some sensitivity to the parameters of the mass distribution. Each FPY is also dependent on the parameters in the mass distribution, as expected; additionally, strong sensitivity to several of the R_T parameters around the mass of the given fission fragment (here, $A \sim 147$) is seen.

These sensitivities are included in the Kalman filter optimization routine, along with the diagonal experimental covariance matrix, the initial parameter values, and a diagonal parameter covariance matrix. We perform two optimizations, one using experimental data taken from the EXFOR database [26] and one using the current ENDF/B-VIII.0 database [1]. For both fits, an iterative procedure is followed, first including $\bar{\nu}_p$, $\bar{\nu}_d$, and FPYs with yields of at least 5%, then including yields down to 1%, then 0.5%. For the fit to experimental data, a further step of including yields down to 0.2% was able to be performed due to the limited coverage of the experimental FPYs. Here, we will discuss both optimizations and the slight differences between them.

In Fig. 2, we show the parameter correlation matrices for (a) the fit to experimental data from EXFOR and (b) the fit to ENDF/B-VIII.0 data. Overall, the parameter correlations are very similar between the two optimizations: there are several strong correlations among the $Y(A)$ parameters, along with TKE and f (through $\bar{\nu}_p$). There are less strong correlations along the diagonal of the correlation matrix, between the various R_T values, that appear to be slightly different between the two optimizations. The differences between the two correlation matrices–$\mathbb{P}$ from the ENDF/B-VIII.0 fit subtracted from $\mathbb{P}$ determined by the fit to experimental data–are emphasized in Fig. 3. These differences are no more than 0.3 and mainly close to the diagonal in the correlations–between the R_T values. This result is expected given the increased number of nuclei that are included in the fit to the ENDF/B-VIII.0 evaluation.

We then use Eq. (11) to calculate the observable covariance matrix between the various fission product yields for ^{252}Cf(sf). Figure 4 shows the FPY correlation matrix from the fit to experimental data. We note first that because the parameter correlation matrices for the two optimizations are very similar and they are sandwiched between the same parameter sensitivity matrix, the FPY correlation matrix shown here is very similar to the correlation matrix for the fit to ENDF/B-VIII.0 (some slight differences will be discussed later in this section). The FPYs along the x and y axes are ordered approximately by increasing ZAID (where ZAID $= 1000 \times Z + A$), which correspond to the indices on the axes (ZAID is not given on the

 https://doi.org/10.1051/epjconf/202328100018

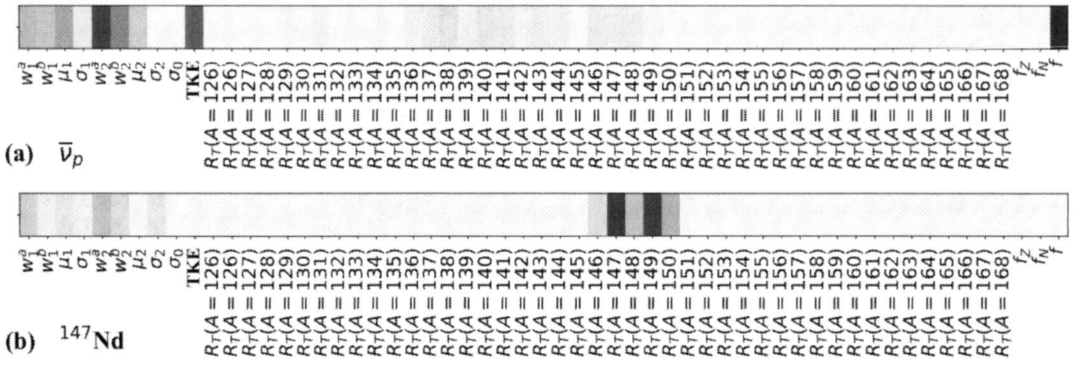

Figure 1. Sensitivities for (a) $\bar{\nu}_p$ and (b) the cumulative fission product yield for ^{147}Nd to the fitted parameters in BeoH. Dark green (purple) indicates regions of high (anti-)correlation; white regions indicate no correlation.

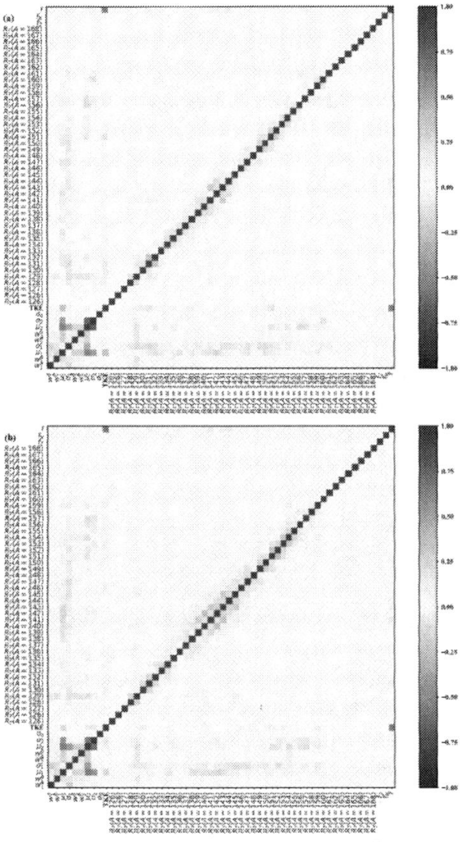

Figure 2. Parameter correlation matrix for (a) the fit to experimental data from EXFOR and (b) the fit to evaluated data from ENDF/B-VIII.0 for ^{252}Cf(sf).

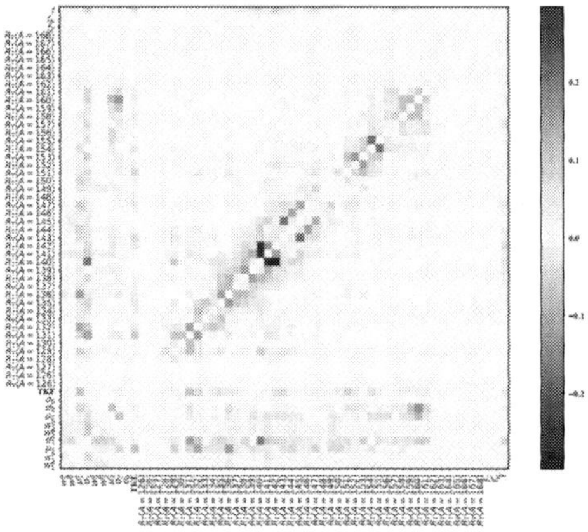

Figure 3. Difference between the parameter correlation matrix using the EXFOR optimization and the ENDF optimization (Fig. 2(b) minus 2(a)).

axes directly). We first notice that the FPYs are largely uncorrelated or have a very weak correlation between pairs, outside of some noticeable structure that will be discuss shortly. Most of the lack of strong correlations comes from the inclusion of $R_T(A)$ in the optimization procedure; when a fixed R_T value is used, the FPY correlations are much stronger, although we do not show that difference in this contribution.

We can investigate the pieces of the matrix in Fig. 4 that do show significant correlations by taking a look at correlations across a mass chain and an isotopic chain. In Fig. 5(a), we show the correlations across the $A = 147$ chain of cumulative fission product yields. All of these fission products are highly correlated, a trend seen in almost all of the other mass chains, although these are not shown in this contribution. Figure 5(b) shows the correlations along an isotopic chain, $Z = 44$. For the isotopic chain, we see more structure among the Ruthenium FPY correlations calculated with BeoH compared to a full pos-

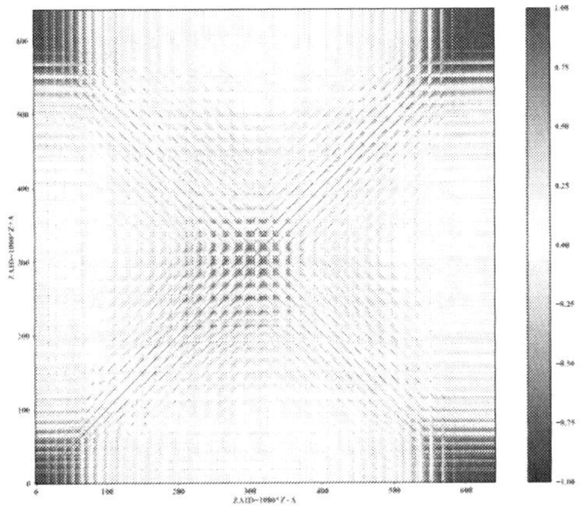

Figure 4. Correlation matrix between the cumulative fission product yields for ^{252}Cf(sf) using the fit to experimental data from EXFOR. Indices are ordered by increasing ZAID = $1000 \times Z + A$ but do not correspond one-to-one to the numbers on the axes.

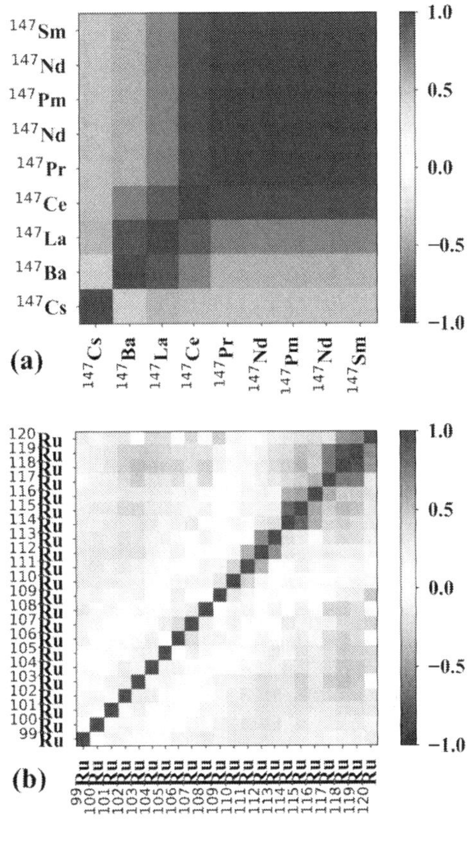

Figure 5. Correlations between isotopes of the (a) A = 147 mass chain and (b) Z = 44 isotopic chain for ^{252}Cf(sf) using the fit to experimental data from EXFOR.

itive correlation. The heavier-mass isotopes are strongly correlated, much more than the lighter-mass isotopes, but there are almost no correlations between the heavier- and lighter-mass isotopes. Between the two optimization procedures, the A correlations vary less than 10% and the Z correlation vary less than 20%. The correlations between the cumulative fission product yields are therefore likely coming mainly from the model and not from the amount of data included in the optimization.

We are also developing covariances for fission product yields resulting from neutron-induced fission reactions, namely on ^{239}Pu, ^{235}U, and ^{238}U. For a given discrete incident neutron energy, the correlations are similar to those shown in Fig. 4, with some structure seen between isotopes but large regions of weak to no correlation. In addition to correlations between the cumulative fission product yields, we can also construct the covariance matrices between incident energies. The fission product yields are highly correlated across incident neutron energies, at least below the opening of second-chance fission, the covariances of which are still under development. However, we expect strong correlations in these higher energy regions, exactly the opposite of what is currently assumed with the lack of correlations in the current evaluated libraries.

4 Conclusions

In this contribution, we show initial results for the calculation of covariance matrices of cumulative fission product yields (FPYs) using the Hauser-Feshbach fission fragment code, BeoH. These covariances are being developed for inclusion in a future release of the evaluated ENDF library. Beginning with the spontaneous fission of ^{252}Cf, we find little differences between the parameter correlation matrices and FPY correlation matrices when using either experimental data from EXFOR or evaluated data from ENDF/B-VIII.0 to constrain the model. The correlations between the FPY of various nuclei are overall weak, but we see more significant structure when considering the correlations between isotopes of the same mass or same charge.

Additionally, energy-dependent covariances are being developed for neutron-induced fission of ^{239}Pu, ^{235}U, and ^{238}U. Along with providing covariances between FPYs for each incident energy considered in the evaluation, covariances can also be constructed between incident energy points, providing cross-energy correlations. Because of the current optimization procedure of fitting parameters separately for each multi-chance fission channel, work must be done to insure that the covariances are reasonable as the multi-chance fission thresholds are crossed. Likely, sensitivities to all parameters will have to be calculated for the full incident energy range (thermal to 20 MeV), even if the fits are only performed within limited energy ranges. The construction of the energy-dependent covariances is still being investigated, along with a reasonable format for inclusion in a future ENDF release.

This work was performed under the auspice of the U.S. Department of Energy by Los Alamos National Laboratory under Contract 89233218CNA000001 and was supported by the Office of Defense Nuclear Nonproliferation Research & Development (DNN R&D), with partial support from the Nuclear Criticality Safety Program, both funded and managed by the National Nuclear Security Administration for the U.S. Department of Energy.

References

[1] D. Brown, M. Chadwick, R. Capote, A. Kahler, A. Trkov, M. Herman, A. Sonzogni, Y. Danon, A. Carlson, M. Dunn et al., Nuclear Data Sheets **148**, 1 (2018), special Issue on Nuclear Reaction Data

[2] A.J.M. Plompen, O. Cabellos, C. De Saint Jean, M. Fleming, A. Algora, M. Angelone, P. Archier, E. Bauge, O. Bersillon, A. Blokhin et al., The European Physical Journal A **56**, 181 (2020)

[3] Iwamoto, Osamu, Iwamoto, Nobuyuki, Shibata, Keiichi, Ichihara, Akira, Kunieda, Satoshi, Minato, Futoshi, Nakayama, Shinsuke, EPJ Web Conf. **239**, 09002 (2020)

[4] England, T. R. and Rider, B. F., *Evaluation and compilation of fission product yields 1993* (1994), lA-UR-94-3106, `94-3106`

[5] M. Chadwick, T. Kawano, D. Barr, M. Mac Innes, A. Kahler, T. Graves, H. Selby, C. Burns, W. Inkret, A. Keksis et al., Nuclear Data Sheets **111**, 2923 (2010), nuclear Reaction Data

[6] S. Okumura, T. Kawano, P. Talou, P. Jaffke, S. Chiba, J. Nucl. Sci. Tech. **55**, 1009 (2018)

[7] A.E. Lovell, T. Kawano, S. Okumura, I. Stetcu, M.R. Mumpower, P. Talou, Phys. Rev. C **103**, 014615 (2021)

[8] C. Bhatia, B.F. Fallin, M.E. Gooden, C.R. Howell, J.H. Kelley, W. Tornow, C.W. Arnold, E. Bond, T.A. Bredeweg, M.M. Fowler et al., Phys. Rev. C **91**, 064604 (2015)

[9] M. Gooden, C. Arnold, J. Becker, C. Bhatia, M. Bhike, E. Bond, T. Bredeweg, B. Fallin, M. Fowler, C. Howell et al., Nuclear Data Sheets **131**, 319 (2016), special Issue on Nuclear Reaction Data

[10] Krishichayan, M. Bhike, C.R. Howell, A.P. Tonchev, W. Tornow, Phys. Rev. C **100**, 014608 (2019)

[11] M. Bhike, W. Tornow, Krishichayan, A.P. Tonchev, Phys. Rev. C **95**, 024608 (2017)

[12] Krishichayan, S.W. Finch, C.R. Howell, A.P. Tonchev, W. Tornow, Phys. Rev. C **98**, 014608 (2018)

[13] E.F. Matthews, L.A. Bernstein, W. Younes, Atomic Data and Nuclear Data Tables **140**, 101441 (2021)

[14] Z.A. Wang, J. Pei, Y. Liu, Y. Qiang, Phys. Rev. Lett. **123**, 122501 (2019)

[15] C.Y. Qiao, J.C. Pei, Z.A. Wang, Y. Qiang, Y.J. Chen, N.C. Shu, Z.G. Ge, Phys. Rev. C **103**, 034621 (2021)

[16] Z.A. Wang, J. Pei, Phys. Rev. C **104**, 064608 (2021)

[17] L. Tong, R. He, S. Yan, Phys. Rev. C **104**, 064617 (2021)

[18] Z.A. Wang, J.C. Pei, Y.J. Chen, C.Y. Qiao, F.R. Xu, Z.G. Ge, N.C. Shu, Phys. Rev. C **106**, L021304 (2022)

[19] A.C. Wahl, Tech. rep., Los Alamos Nat. Lab. LA-13928 (2002)

[20] D.L. Duke, Ph.D. thesis, Colorado School of Mines, The address of the publisher (2015), an optional note

[21] A. Gilbert, A.G.W. Cameron, Canadian Journal of Physics **43**, 1446 (1965), `https://doi.org/10.1139/p65-139`

[22] R.E. Kalman, Transactions of the ASME–Journal of Basic Engineering **82**, 35 (1960)

[23] D. Neudecker, A. Lewis, E. Matthews, *et al.*, *Templates of expected measurement uncertainties* (2019), lA-UR-19-31156, `19-31156`

[24] D. Neudecker, D. Smith, F. Tovesson, R. Capote, M. White, N. Bowden, L. Snyder, A. Carlson, R. Casperson, V. Pronyaev et al., Nuclear Data Sheets **163**, 228 (2020)

[25] T. Kawano, arXiv:1901.05641v1 (2019)

[26] N. Otuka, E. Dupont, V. Semkova, B. Pritychenko, A. Blokhin, M. Aikawa, S. Babykina, M. Bossant, G. Chen, S. Dunaeva et al., Nuclear Data Sheets **120**, 272 (2014)

https://doi.org/10.1051/epjconf/202328100019

Practicalities of Bayesian network modeling for nuclear data evaluation with the nucdataBaynet package

Georg Schnabel[1,*]

[1]NAPC-Nuclear Data Section, International Atomic Energy Agency, A-1040 Vienna, Austria

Abstract. Bayesian networks are a helpful abstraction in the modelization of the relationships between different variables for the purpose of uncertainty quantification. They are therefore especially well suited for the application to nuclear data evaluation to accurately model the relationships of experimental and nuclear models. Constraints, such as sum rules and the non-negativity of cross sections, can be rigorously taken into account in Bayesian inference within Bayesian networks. This contribution elaborates on the practical aspects of the construction of Bayesian networks with the *nucdataBaynet* package for the purpose of nuclear data evaluation.

1 Introduction

In nuclear data evaluation, experimental data and nuclear models are combined by means of a statistical procedure to obtain estimates and uncertainties for quantities of interest, such as cross sections.

The Generalized Least Squares (GLS) method, e.g., [1], implemented in many evaluation codes, such as GANDR [2] and GMA [3], provides a sound framework for the evaluation of nuclear data as long as relationships between variables can be in good approximation linearized. Many of such evaluation codes that are employed above the resonance region do not take into account experimental features, such as resolution broadening or uncertainties in the incident energy, because their influence on evaluation results is usually negligible.

In contrast, the statistical modules employed by R-Matrix codes, such as SAMMY [4], must consider these experimental features in order to properly forward propagate theoretical cross sections to quantities comparable with high-resolution experiments. Relationships in the resonance region are highly non-linear. Consequently the GLS method is not applicable and must be replaced by iterative procedures, which typically use the GLS method in each iteration to guide the optimization process.

The division between the resonance region and the fast energy range is first and foremost due to the physics model, i.e., R-matrix or optical model. From the statistical point of view, these two regions can be treated in the same framework if non-linear relationships are properly taken into account. Furthermore, even within the fast energy range, evaluators may employ different sets of assumptions, such as the inclusion of model defects on cross sections [5] or model parameters [6]. Therefore evaluation work can benefit from the availability of a flexible modeling and inference framework that can be applied over the entire energy range.

Bayesian networks, e.g., [7], are a perspective on Bayesian inference that emphasizes the relationship between variables and can be regarded as such a flexible inference framework. Variables are depicted as nodes and arrows indicate the relationships between those nodes. An extensive paper [8] was recently published that elaborates on the mathematics of Bayesian networks tailored to nuclear data evaluation and also includes evaluation examples.

In this contribution, we provide practical details of how Bayesian networks can be created and inference performed within them with the *nucdataBaynet* R package [9]. This package is hosted on GitHub and publicly accessible under the permissive MIT license. The information in this contribution refers to the version of the package at the time of writing, which is January 2022. The corresponding commit id on GitHub is *e9df5fd*. As an important remark, the aim of this contribution is to give the reader some basic understanding of how the package can be used rather than a comprehensive description of its features.

First, we explain how information about different variables is stored and how these variables or sets of variables are associated with nodes. Second, we elaborate on the creation of links between different nodes. Third, we discuss how inference in Bayesian networks can be performed. Fourth, we give a schematic evaluation example putting everything together. Finally, a summary section is provided.

2 Creating nodes

An essential step in the construction of a Bayesian network is the definition of the nodes. To this end, a data table [10] is defined with all the variables appearing in the Bayesian network. An example of such a data table is given in Ta-

*e-mail: g.schnabel@iaea.org

https://doi.org/10.1051/epjconf/202328100019

IDX	NODE	PRIOR	OBS	REAC	ENERGY	EXPID
1	truexs	1000	NA	(N,TOT)	1.0	NA
2	truexs	1000	NA	(N,TOT)	1.5	NA
3	truexs	1000	NA	(N,TOT)	2.0	NA
4	normerr	0	NA	(N,TOT)	NA	20733
5	exp	0	200	(N,TOT)	1.7	20733

Table 1. Structure of a data table with the definition of the variables of the Bayesian network

ble 1. Many convenience functions exist in the R language to create data tables. Python users can think of these functions as equivalents to the many functions the Python *pandas* package [11] provides for creating data frames.

The columns that must always be present are *IDX*, *NODE*, *PRIOR* and *OBS*. The indices in the column *IDX* establish an order of the variables and are important for the Bayesian inference in the Bayesian network where vectors and covariance matrices need to be constructed. The column *NODE* determines the assignment of variables to nodes. As can be seen in the table, several variables can be associated with the same node. In this example, the variables assigned to the node *truexs* denote cross section values on a computational mesh. The column *OBS* indicates whether the value of a variable was observed or not. Unobserved variables have the special value NA (=Not Answered) whereas observed variables have the observed value in this column.

The meaning of the number in the column *PRIOR* depends on whether a variable has any ingoing links or not. The definition of links will be discussed in the next section, but we need to already introduce some aspects to clarify the meaning of the *PRIOR* column. To this end, let us consider the Bayesian network depicted in Figure 1. The nodes $\vec{x}_1$ and $\vec{x}_2$ do not contain any ingoing links. For instance, $\vec{x}_1$ could be *truexs* in Table 1. In contrast, the nodes $\vec{y}_1$ and $\vec{y}_2$ contain ingoing links. The *nucdataBaynet* package introduces the convention that every node with ingoing links, hence is a function of the values of some other nodes, must also be associated with a node $\vec{\varepsilon}_i$. Those extra nodes do not have any ingoing links and are only connected to exactly one node, such as $\vec{\varepsilon}_1$ to $\vec{y}_1$. Furthermore, the values in those nodes $\vec{\varepsilon}_i$ are an additive contribution for the values in $\vec{y}_i$. The reason for the introduction of those extra nodes by default and the stiff constraints on how they must be connected to other nodes is that it simplifies the implementation of Bayesian inference in the Bayesian network. These nodes are usually used to include statistical errors associated with experimental data.

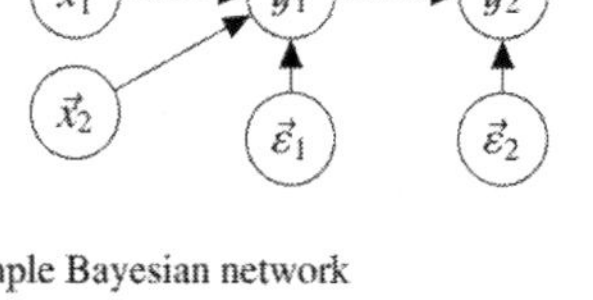

Figure 1. Simple Bayesian network

For nodes without any ingoing links—we may call them source nodes—such as $\vec{x}_1$ and $\vec{x}_2$, the number in *PRIOR* represents the prior best estimate. For all other nodes with ingoing links, such as $\vec{y}_1$ and $\vec{y}_2$, this number denotes the prior best estimate of the extra node, such as $\vec{\varepsilon}_1$ and $\vec{\varepsilon}_2$, respectively. In this construction, the values of the variables associated with nodes with ingoing links, e.g., $\vec{y}_1$ and $\vec{y}_2$ are deterministic functions of the values of all the nodes they depend on, hence also including the values of their $\vec{\varepsilon}_i$ node.

Regarding the node definition, only *IDX*, *NODE*, *PRIOR* and *OBS* are strictly required and all other columns dispensable. However, as will be discussed in the next section, the information in the extra columns may be essential for defining the links between nodes.

Finally, each node must also be endowed with a covariance matrix. For source nodes, such as $\vec{x}_1$ and $\vec{x}_2$, the associated covariance matrix reflects the prior knowledge on the variables in the node. For all other nodes, such as $\vec{y}_1$ and $\vec{y}_2$, the associated covariance matrix reflects the knowledge about values in the extra nodes $\vec{\varepsilon}_i$. In working with the *nucdataBaynet* package, these covariance matrix blocks are stored all together in a sparse matrix in compressed sparse row or column format, enabled by the *Matrix* package [12].

3 Establishing links

The second essential step is the definition of mappings between nodes. The definition of a mapping includes the specification of the source and the target variables, which is achieved by providing the associated indices given in the column *IDX* of the node data table.

Depending on the mapping type, additional information may be required. For instance, one type of mapping is linear interpolation. It can be used if the source variables are the cross sections on a computational mesh and the target variables measured cross sections associated with energies in-between the computational mesh points. Therefore, in order to establish a linear interpolation mapping, the energies associated with the cross sections of the computational mesh and those associated with the measured values are required.

In working with the *nucdataBaynet* package, the creation of a specific mapping is a two step procedure. First, a *named list* needs to be created. From the point of usage, they can be regarded as roughly equivalent to Python dictionaries, even though there are differences in terms of space and time complexity and field names in named lists are restricted to character strings. How this list creation can be done for a linear interpolation mapping is illustrated in Listing 1.

The *maptype* field indicates the type of mapping, here linear interpolation. The *mapname* field associates a name with this specific mapping, and can be helpful for debugging. The always required fields *src_idx* and *tar_idx* to denote the indices of the source and target variables, respectively, follow next. R data tables (as data frames in Python) provide powerful query functionality to select a

```
mesh_xs_to_exp <- list(
  maptype = "linearinterpol_map",
  mapname = "truexs_to_experiment",
  src_idx = dt[NODE=="truexs", IDX],
  tar_idx = dt[NODE=="exp", IDX],
  src_x = dt[NODE=="truexs", ENERGY],
  tar_x = dt[NODE=="exp", ENERGY],
)
```

Listing 1. Example R code for creating a list with the definition of a linear interpolation mapping

```
comp_map_def <- list(
  maptype = "compound_map",
  mapname = "my_cool_compound_map",
  maps = list(mesh_xs_to_exp, ...)
)
```

Listing 2. Example R code for creating a list with the definition of a compound mapping

subset of the table. Such queries are utilized in this example to retrieve the indices associated with the source node *truexs* and the target node *exp*, again analogous to how this could be achieved with *pandas* in Python. The last two fields, *src_x* and *tar_x*, are specific for a linear interpolation mapping and contain the x-values associated with the source and target variables. Query operations on the node data table are used again to extract the energies associated with the source and target variables. For other types of mappings, other fields may be present instead.

All required mappings for establishing the link structure of the Bayesian network can be constructed analogously to how it is sketched for a linear interpolation mapping above. In the example mapping definition above, the list was also assigned to the variable name *mesh_xs_to_exp* so that it can be used as an abbreviation in later definitions.

Once all required mappings are defined, they need to be bundled to a so-called compound mapping. An example of a compound mapping definition is given in Listing 2. This mapping only requires the specification of the *maptype* field with value *compound_map* and a field *maps*, which contains a list of all the mappings present in the Bayesian network.

Finally, the compound mapping definition is used to create a compound mapping object in the following way:

```
comp_map <- create_map(comp_map_def)
```

This mapping object provides the two functions *propagate* and *jacobian*. The former function allows to forward propagate values associated with nodes throughout the network. The latter function provides the Jacobian matrix, also known under the name sensitivity matrix.

These two functions can be used to perform sensitivity studies to assess the impact of changes in the Bayesian network. They are also leveraged for Baysian inference in the network as explained in the next section.

4 Bayesian inference

Bayesian networks rely on Bayesian inference. First a brief recapitulation of the Bayesian inference is presented and afterwards the connection to Bayesian networks established.

Bayesian inference is performed with the Bayesian update formula, e.g., [13],

$$\pi(\vec{z}\,|\,\vec{\sigma}) \propto \ell(\vec{\sigma}\,|\,\vec{z})\,\pi(\vec{z})\,, \tag{1}$$

which combines the knowledge expressed as prior distribution $\pi(\vec{z})$ for quantities of interest $\vec{z}$ with the likelihood $\pi(\vec{\sigma}\,|\,\vec{z})$ that connects probabilistically the quantities of interest with the observed values in $\vec{\sigma}$.

Bayesian modeling with the *nucdataBaynet* package makes the assumption that both the prior distribution and likelihood are multivariate normal. The likelihood is therefore given by $\vec{z} \sim \mathcal{N}(\mathcal{M}(\vec{z}), \boldsymbol{\Sigma}_{\mathrm{exp}})$, where $\mathcal{M}(\vec{z})$ is a (possibly non-linear) model to propagate values in $\vec{z}$ to values that are comparable with those in $\vec{\sigma}$ up to a difference explained by the uncertainties and correlations in the experimental covariance matrix $\boldsymbol{\Sigma}_{\mathrm{exp}}$. The prior distribution is given by $\pi(\vec{z}) \sim \mathcal{N}(\vec{z}_0, \boldsymbol{\Sigma}_0)$ with $\vec{z}_0$ being the prior best estimate and $\boldsymbol{\Sigma}_0$ the corresponding prior covariance matrix.

The GLS method addresses a special case where the model $\mathcal{M}(\vec{z})$ is linear. The Levenberg-Marquardt algorithm [14–16], an iterative algorithm for non-linear optimization, is employed to determine values $\vec{z} = \vec{z}_{\mathrm{MAP}}$ associated with the maximum of the posterior distribution $\pi(\vec{z}_{\mathrm{MAP}}\,|\,\vec{\sigma})$. This type of estimate is called a *Maximum A Posteriori* (MAP) estimate.

Following, we explain how the concept of nodes and links of a Bayesian network is connected to the Bayesian update formula in eq. (1). Let us consider again the Bayesian network illustrated in Figure 1. Because the links denote functional relationships between variables, we can express the structure of the network as a set of equations:

$$\vec{y}_1 = f_1(\vec{x}_1) + f_2(\vec{x}_2) + \vec{\varepsilon}_1 \tag{2}$$

$$\vec{y}_2 = f_2(\vec{y}_1) + \vec{\varepsilon}_2 \tag{3}$$

Each mapping in the compound mapping object corresponds to one function f_i. If the values of nodes without incoming links $\vec{x}_1, \vec{x}_2, \vec{\varepsilon}_1, \vec{\varepsilon}_2$ should be forward propagated through the network to obtain the values in $\vec{y}_1$ and $\vec{y}_2$, it is important to first apply the functions f_1 and f_2 and only afterwards f_3. If we are only interested in $\vec{y}_2$, e.g., because it corresponds to the observed quantity $\vec{\sigma}$, we may directly write

$$\vec{y}_2 = f_3(f_1(\vec{x}_1) + f_2(\vec{x}_2) + \vec{\varepsilon}_1) + \vec{\varepsilon}_2\,. \tag{4}$$

This equation clarifies that the structure of a Bayesian network defines a sequence of nested functions. If we are not interested in the nested structure of the functions, we can equally write

$$\vec{y}_2 = \mathcal{M}(\vec{x}_1, \vec{x}_2, \vec{\varepsilon}_1) + \vec{\varepsilon}_2 = \mathcal{M}(\vec{z}) + \vec{\varepsilon}_2\,, \tag{5}$$

where the mathematical model $\mathcal{M}$ summarizes the effect of the individual functions and the vector $\vec{z}$ is constructed by stacking together the vectors $\vec{x}_1$, $\vec{x}_2$ and $\vec{\varepsilon}_1$. The prior covariance matrix $\mathbf{\Sigma}_0$ in the Bayesian update equation is constructed by creating a block diagonal matrix from the covariance matrices associated with $\vec{x}_1$, $\vec{x}_2$ and $\vec{\varepsilon}_1$. The experimental covariance matrix $\mathbf{\Sigma}_{\text{exp}}$ is constructed by the extra nodes $\vec{\varepsilon}_i$ connected to the observed nodes, hence in the example by the covariance matrix associated with $\vec{\varepsilon}_2$.

The convenience from the user point of view lies in the fact that $\mathcal{M}(\vec{z})$ can be automatically constructed from the individual mappings. Similarly, the overall Jacobian matrix (sensitivity matrix) that encompasses all nodes at the same time can be obtained by applying the chain rule to combine the Jacobian matrices of the individual mappings. The logistics and bookkeeping involved are handled by the compound mapping object.

In the next section, an example is sketched that demonstrates the advantages of the Bayesian network framework in terms of computational and storage efficiency.

5 Benefits explained at an example

This section sketches a purely data-driven evaluation of the total cross section of ^{56}Fe between 1.0 and 1.1 MeV. A Gaussian process with a sparse structure will be assumed as mathematical fitting function. A good introduction to Gaussian processes is given in [17]. We consider one experimental data set of Cornelis et al. available in EXFOR [18, 19] under the accession number 22316001 and assume that the dataset is affected by statistical uncertainties and a normalization uncertainty. For the statistical uncertainties we adopt the values in the EXFOR entry and we assume a value of 0.05 barn for the normalization uncertainty. The code to reproduce the example analysis showcased in the following is also included in a tutorial, which is part of the *nucdataBaynet* package.

We may model this situation with the Bayesian network depicted in Figure 2. The node with the vector $\vec{\sigma}$ contains the cross section values on a computational mesh of incident energies. The mesh covers the range from 1.0 to 1.1 MeV with a spacing of 10 eV between mesh points. We can assign as prior best estimates a vector of zeros and a diagonal covariance matrix with very large diagonal elements (10^{30}) to implement a non-informative prior. If we are doing an evaluation in the fast energy region, we are expecting evaluated cross sections to have a certain degree of smoothness. Therefore, we introduce a new node

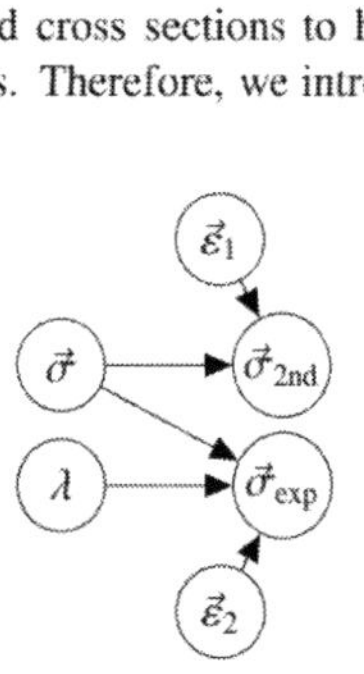

Figure 2. Bayesian network for the evaluation example. See the text for a description of the nodes and links.

$\vec{\sigma}_{\text{2nd}}$ containing the values of the second derivatives of the cross sections in $\vec{\sigma}$. The *nucdataBaynet* package already provides a mapping to go from a vector with values on an energy mesh to a vector with the second derivatives.

We assume that all the second derivates in $\vec{\sigma}_{\text{2nd}}$ were observed to be zero and that the associated covariance matrix (strictly speaking the prior covariance matrix associated with $\vec{\varepsilon}_2$) is a diagonal matrix with a single value α^2 repeated across the entire diagonal. We may call this observation a pseudo-observation because it does not correspond to any real measurement but is merely a mathematical trick to favor smooth solutions of the function encoded in a discretized way in the vector $\vec{\sigma}$. The value of α regulates the degree of smoothness. The construction of the prior knowledge for the cross sections in $\vec{\sigma}$ involving a pseudo-observation of the second derivatives in $\vec{\sigma}_{\text{2nd}}$ represents a Gaussian process that is in good approximation only opinionated on the degree of smoothness but not the range of the cross section values in $\vec{\sigma}$.

This construction is also attractive from a computational and storage efficiency point of view because all the covariance matrices associated with $\vec{\sigma}$ and $\vec{\sigma}_{\text{2nd}}$ are diagonal, which can be exploited in the Bayesian inference. For demonstration purposes, we will be doing the Bayesian inference two times, once with $\alpha = 10^4$ and another time with $\alpha = 10^7$ to show the impact on the evaluation result (the MAP estimate). As an aside, the *nucdataBaynet* package allows other Gaussian process constructions, e.g., based on pseudo-observations of first and third derivatives as well as based on user-provided covariance functions. At the time of writing, these options are however neither well documented in the package documentation nor demonstrated in the accompanying tutorials.

The measured values are summarized in the node $\vec{\sigma}_{\text{exp}}$. We assume that the prior best estimate of the values in $\vec{\varepsilon}_2$ is given by a zero vector and that the corresponding prior covariance matrix $\mathbf{D}$ is diagonal capturing the statistical uncertainties of the measured data points in $\vec{\sigma}_{\text{exp}}$. These statistical uncertainties are taken from the EXFOR entry and range from 0.04 to 0.06 barn. The normalization error λ is also associated with a node and linked to the measurement node $\vec{\sigma}_{\text{exp}}$ via a normalization mapping. The prior best estimate of λ is assumed to be zero and the corresponding prior variance η^2 (because a 1×1 covariance matrix is a single number—a variance) reflects the normalization uncertainty, which we assume to be 0.05 barn.

Noteworthy, the definition of the three nodes $\vec{\sigma}_{\text{exp}}$, λ and $\vec{\varepsilon}_2$ and the associated prior best estimates and covariance matrix blocks is fully equivalent to the specification that the experimental covariance matrix is given by $\mathbf{\Sigma}_{\text{exp}} = \mathbf{D} + \eta^2 \vec{1}\vec{1}^T$ where the diagonal matrix $\mathbf{D}$ reflects the statistical uncertainties and the second term the contribution of the normalization uncertainty. The notation $\vec{1}\vec{1}^T$ in the second term denotes a matrix of ones. This expression also shows that we are using an absolute normalization error—a shift—in this example rather than a relative one—a scaling factor—for the sake of simplicity. However, very often normalization errors are thought to be more adequately represented by scaling factors and this option is also supported by the *nucdataBaynet* package.

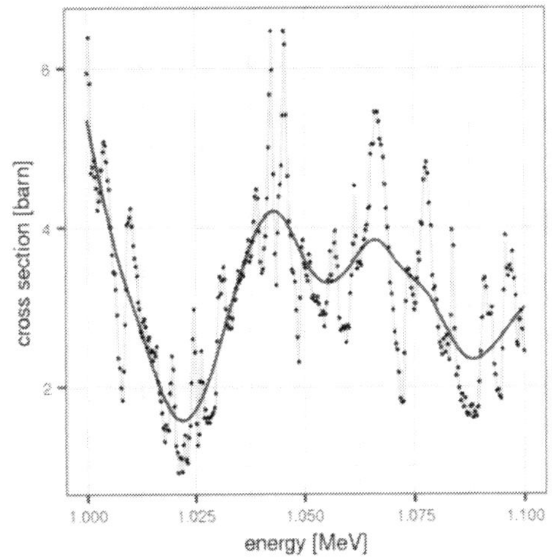

Figure 3. MAP estimate of the node $\vec{\sigma}$ resulting from two different prior uncertainty specifications of α. The blue line is the result for $\alpha = 10^4$ and the green line for $\alpha = 10^7$.

Regarding computational and storage efficiency, we can see here that the Bayesian network representation is very efficient. Instead of storing a dense matrix Σ_{exp} with all elements being non-zero, we can express identical uncertainty assumptions in the Bayesian network by relying on the nodes $\vec{\varepsilon}$ and λ with their diagonal prior covariance matrix blocks. This increased storage efficiency translates directly to an increased computational efficiency in the Bayesian inference because the inversion of a large and dense experimental covariance matrix can be avoided. The mathematical details regarding this statement can be found in [8].

Finally, we also need to connect the node $\vec{\sigma}$ to the measurement node $\vec{\sigma}_{\mathrm{exp}}$, which we will do here by an interpolation mapping as described in section 3. As an aside, the *nucdataBaynet* package also offers other mappings types that can take into account a finite experimental energy resolution or uncertainty in incident energy calibration.

The MAP estimates based on all the modeling assumption described in this section are shown in Figure 3, where it can also be clearly seen that the value of α acts as a regularization parameter to control the smoothness of the solution. As all relationships are linear in in this network, the MAP estimate for the variables in this network coincides with the solution of the GLS method. Noteworthy, even though the computational mesh contains 10000 mesh points, the calculation of the MAP estimate takes less than one tenth of a second on the notebook of the author thanks to the sparse structure of the Bayesian network.

6 Summary

This contribution elaborated on the construction of a Bayesian network with the *nucdataBaynet* package [9], which is a two step procedure. The first step consists in the definition of nodes and the second step in the creation of mappings between the nodes.

It was also explained how Bayesian inference as implemented by the *nucdataBaynet* package is linked to the Bayesian update formula. Finally, a Bayesian network was created for a schematic evaluation of the total cross section of ^{56}Fe between 1.0 and 1.1 MeV.

This contribution provided a glimpse on the capabilities of Bayesian networks as a framework to create Bayesian models. The author believes that the mental abstractions of nodes and links are very beneficial in the creation of Bayesian models that truly reflect the assumptions of the evaluator. For instance, some quantities are known to be positive but this assumption may not be always incorporated in the evaluation process due to limitations of some evaluation codes or the GLS method.

The *nucdataBaynet* R package is available on the GitHub account of the Nuclear Data Section at the IAEA. As a final remark, the GMA code [3] used since the early 1980s for the evaluation of the neutron data standards [20] has been translated to a Python package *gmapy* [21] with new evaluation features. It is planned that also the gmapy package will support the creation of Bayesian models for nuclear data evaluation using the Bayesian network paradigm.

References

[1] T. Kariya, H. Kurata, *Generalized Least Squares*, 1st edn. (Wiley, 2007)

[2] D.W. Muir, A. Trkov, I. Kodeli et al., *The Global Assessment of Nuclear Data, GANDR* (EDP Sciences, 2007)

[3] W.P. Poenitz, S.E. Aumeier, Tech. Rep. ANL/NDM-139, Argonne National Laboratory, Argonne, Illinois (1997)

[4] N.M. Larson, Tech. Rep. ORNL/TM-9179/R4, Oak Ridge National Laboratory, Oak Ridge (1998)

[5] D. Neudecker, R. Capote, H. Leeb, Nuclear Instruments and Methods in Physics Research Section A: Accelerators, Spectrometers, Detectors and Associated Equipment **723**, 163 (2013)

[6] P. Helgesson, H. Sjöstrand, Annals of Nuclear Energy **120**, 35 (2018)

[7] J. Pearl, *Probabilistic Reasoning in Intelligent Systems: Networks of Plausible Inference*, 1st edn. (Morgan Kaufmann, 2014)

[8] G. Schnabel, R. Capote, A. Koning et al., *Nuclear data evaluation with Bayesian networks* (2021), `arxiv:2110.10322`

[9] G. Schnabel, *IAEA-NDS/nucdataBaynet*, https://github.com/IAEA-NDS/nucdataBaynet (2021)

[10] M. Dowle, A. Srinivasan, *Data.table: Extension of 'data.frame'*, https://CRAN.R-project.org/package=data.table (2022)

[11] Wes McKinney, *Data Structures for Statistical Computing in Python*, in *Proceedings of the 9th Python in*

Science Conference, edited by S. van der Walt, Jarrod Millman (2010), pp. 56–61

[12] D. Bates, M. Maechler, *Matrix: Sparse and dense matrix classes and methods*, https://CRAN.R-project.org/package=Matrix (2021)

[13] D.S. Sivia, *Data Analysis: A Bayesian Tutorial* (Clarendon Press, 1996), ISBN 978-0-19-851889-1

[14] K. Levenberg, Quarterly of Applied Mathematics **2**, 164 (1944)

[15] D.W. Marquardt, Journal of the Society for Industrial and Applied Mathematics **11**, 431 (1963)

[16] P. Helgesson, H. Sjöstrand, Review of Scientific Instruments **88**, 115114 (2017)

[17] C.E. Rasmussen, C.K.I. Williams, *Gaussian Processes for Machine Learning* (MIT Press, Cambridge, Mass., 2006), ISBN 0-262-18253-X 978-0-262-18253-9

[18] V. Zerkin, B. Pritychenko, Nuclear Instruments and Methods in Physics Research Section A: Accelerators, Spectrometers, Detectors and Associated Equipment **888**, 31 (2018)

[19] N. Otuka, E. Dupont, V. Semkova et al., Nuclear Data Sheets **120**, 272 (2014)

[20] A. Carlson, V. Pronyaev, R. Capote et al., Nuclear Data Sheets **148**, 143 (2018)

[21] G. Schnabel, *GitHub - IAEA-NDS/gmapy: Gmapy: A Python package for nuclear data evaluation*, https://github.com/IAEA-NDS/gmapy

Covariances and parameter confidence intervals from light-element R-matrix evaluations

Mark Paris[1,*] and *Gerald* Hale[1,**]

[1]Theoretical Divsion, Los Alamos National Laboratory, MS B283, Los Alamos, New Mexico 87545

Abstract. R-matrix parameter covariances for light elements with $A \leq 16$ are calculated within the Wigner-Eisenbud multichannel unitary R-matrix theory. We review the theoretical foundation, numerical approach, and determination of the parameter covariances within this approach. We derive the relation between the parameter variance, as the diagonal elements of the covariance matrix, and the parameter confidence interval based upon the chi-squared distribution. Cross section covariances computed from the parameter confidence intervals are calculated for several compound systems and discussed.

1 Introduction

Nuclear reaction and scattering cross section information continues to be an indispensible component in design, modeling, simulation and refinement of complex systems in nuclear science, energy, and security. The ability to assess the validity of modeling and simulation capabilities, in particular, depends on the characterization and quantification of uncertainty associated with nuclear reaction and scattering observables. We should emphasize that we are not just referring to the energy-dependent, angle-integrated cross sections $\sigma(E)$, but the full, complete set of observables that comprise those associated with a particlar compound system ${}^A\mathscr{Z}$ that couples to all of the two- (and more) body partitions. For example, for the compund system of ^{5}Li, relevant two body partitions in order of increasing total partition mass, are

$$^5\mathrm{Li} \simeq p \otimes {}^4\mathrm{He} + d \otimes {}^3\mathrm{He} \cdots . \quad (1)$$

The observables for this system are the set of total, angle-integrated, and angular distributions of both unpolarized (spin-independent) and polarized or spin-dependent versions of these quantities. The advantage of being able to handle all of these types differential data has been discussed in Ref.[1]. These observables may be formed from the initial state spin density matrix elements $\rho^{(i)}$ and the transition T matrices for reactions and scattering (where the final partition composed of the same nuclides as the initial partition, but the masses may differ for inelastic scattering) as

$$\rho^{(f)} = T\rho^{(i)}T^{\dagger}. \quad (2)$$

This spin-density matrix approach, which follows closely along the lines of Ref.[2], constitutes a general methodology for computing model representations and confers a smooth (differentiable) representation of reaction and scattering data for general compound system.

Our focus in this present work is on light $A \leq 20$ systems; the R-matrix formalism is discussed in some detail in Ref.[1]. Observed data for all processes (total, elastic and reactions for both unpolarized and polarized data) is fit via least squares, the topic of the next section, Sec.2. The determination of the covariances from such a fit is the main focus of the present work. In section 3, we apply the method to the several compound systems. Conclusions are given in Sec.4.

2 Least squares fitting and covariance determination

Least squares fitting is accomplished via the R-matrix approach to reaction theory. The χ^2 function is formed as a sum over the included data for all experiments of processes that couple to a particular compound system. The general expression for the χ^2 function, which may be regarded as a log-likelihood function, is given as:

$$\chi^2_{EDA}(p, n_M) = \sum_{M,i_M}\left[\frac{n_M X_{i_M}(p) - R_{i_M}}{\Delta R_{i_M}}\right]^2 + \left[\frac{n_M S_M - 1}{\Delta S_M / S_M}\right]^2 . \quad (3)$$

The sum is over the measurement setups M or different experiments. Each experiment M is comprised of kinematical points and observable values denoted R_{i_M}, indexed by i_M. Predictions $X_{i_M}(p)$ are given generally as ratios of the elements of the spin-density matrix and their sums (such as the trace for the unpolarized angular distributions) and expressed as functions of the variational parameters p_α, which are N_p in number. Normalization factors n_M are varied for either relative or absolute measurements. In the case of relative measurements, the experimental normalization is chosen to eliminate the second term in the

*e-mail: mparis@lanl.gov
**e-mail: ghale@lanl.gov

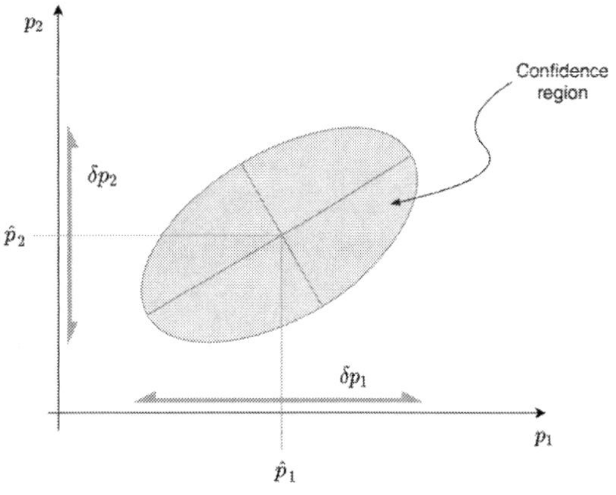

Figure 1. Confidence ellipse determined near the minimum of the function $\chi^2_{EDA}(\hat{p})$.

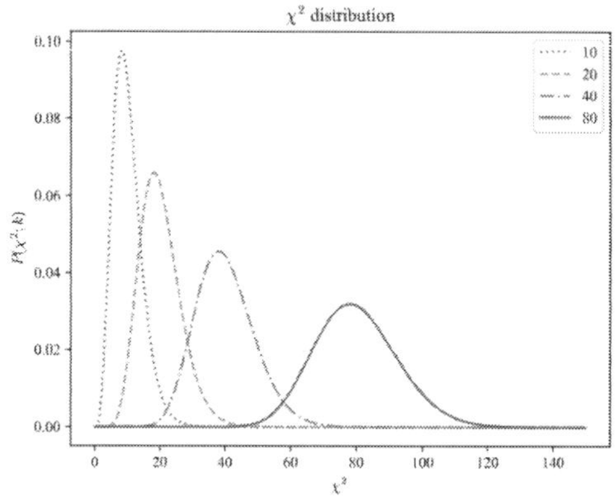

Figure 2. The χ^2 distribution for k degrees of freedom.

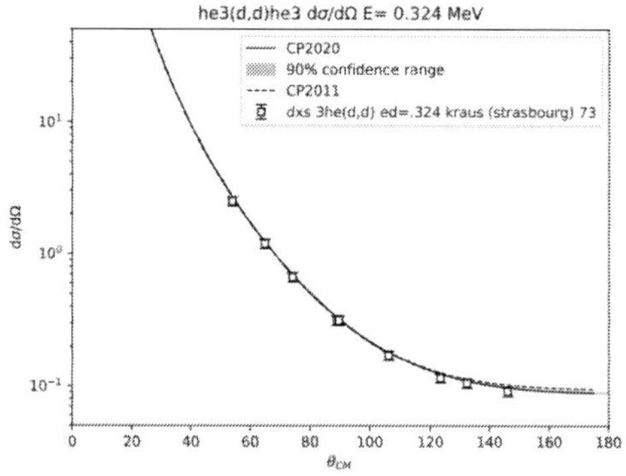

Figure 3. The angular distribution for ^{3}He$(d,d)^3$He for deuteron incident laboratory energy 324 keV of the unpolarized differential cross section in the center of mass as a function of the center of mass angle θ_{cm}. The evaluation is shown by the solid curve. The gray band is the 90% confidence interval determined as described in the text.

above equation [eq. (3)]. Absolute measurement normalizations are allowed to vary in accordance with either the quoted experimental normalization uncertainty ΔS_M or, if this uncertainty is underdetermined, by an assumed amount deemed to be reasonable, sometimes as large as a few 10's of percent but usually < 10%.

The above form for $\chi^2_{EDA}(p, n_M)$ should be compared to a *standard* form[3], which takes into account data covariance through the elements of a matrix V_{ij} as

$$\chi^2(p) = \sum_{ij} \left(X_i(p) - A_i\right) V^{-1}_{ij} \left(X_j(p) - A_j\right), \qquad (4)$$

in a simplified notation that keeps the experimental setup index M implied. (We also drop explicit reference to the normalizations, n_M, which may be considered to present.) Here absolute measurements are defined as $A_i = R_{i_M} S_M$, where the observable R_{i_M} and the experimental normalization S_M are defined above. The form $\chi^2_{EDA}(p)$ is recovered in the limit that the data covariance matrix has the form

$$\text{cov}(R_i, R_j) = \delta_{ij}(\delta R_i)^2, \qquad (5)$$
$$\text{cov}(S_i, S_j) = \delta_{ij}(\delta S_i)^2, \qquad (6)$$
$$\text{cov}(R_i, S_j) = 0. \qquad (7)$$

2.0.1 Parameter uncertainty scaling

The determination of uncertainties in the parameter values is given by the usual analysis[4] at a local (assumed to be global) minimum of the function $\chi^2_{EDA}(p)$ given by parameter values $p = \hat{p}$. About this minimum, the function $\chi^2_{EDA}(p)$ assumes the quadratic form

$$\delta\chi^2_{EDA}(p) \equiv \chi^2_{EDA}(p) - \chi^2_{EDA}(\hat{p}) = \delta p_\alpha C^{-1}_{\alpha\beta} \delta p_\beta, \qquad (8)$$

where $\delta p_\alpha \equiv p_\alpha - p_{0,\alpha}$ and C is the parameter covariance matrix (the summation convention on repeated indices is implied). The standard analysis, given in Arndt and MacGregor[4], at this local minimum, which considers the marginalization of a single parameter p_α, gives the uncertainty in terms of the change of the χ^2 function and the diagonal covariance element (the variance) $C_{\alpha\alpha}$ as

$$(\delta p_\alpha)^2 = \delta\chi^2 C_{\alpha\alpha}. \qquad (9)$$

Choosing $\delta\chi^2$ in the usual manner, $\delta\chi^2 \equiv 1$ however, leads to the following scaling behavior with N_p, the number of fit parameters:

$$\delta p_\alpha = (C_{\alpha\alpha})^{1/2} \sim \mathcal{O}(N_p^{-1/2}), \qquad (10)$$

as can readily be seen in fig. 1; fixing the χ^2 hypersurface as $\delta\chi^2 = 1$ results in the scaling relation above. This scaling behavior results in the uncomfortable situation that adding marginal or redundant parameters lowers the uncertainty of all the parameters.

An improvement in the scaling behavior in eq. (10) is obtained if we choose a measure of the χ^2 hypersur-

Channel	a_c (fm)	ℓ_{max}
$d+{}^3\text{He}(\frac{1}{2}^+)$	4.8	4
$p+{}^4\text{He}(0^+)$	2.9	4
$p+{}^4\text{He}^*(0^+; 20.2\text{ MeV})$	3.4	2
$d_0+{}^3\text{He}(\frac{1}{2}^+)$	5.1	0

Reaction	Energy Range (MeV)	# Data Points	Observables
${}^3\text{He}(d,d){}^3\text{He}$	$E_d = 0.32 - 10.0$	2,229	$\sigma(\theta), A_i, A_{ii}, C_{i,j}, C_{ij,k}, K_{i,j'k'}, K_{ij,k'l'}$
${}^3\text{He}(d,p){}^4\text{He}$	$E_d = 0.13 - 10.0$	3,839	$\sigma(E), \sigma(\theta), A_i, A_{ii}, C_{i,j}, K_{ij,k'}$
${}^3\text{He}(d,p){}^4\text{He}^*$	$E_d = 3.70 - 6.70$	28	$\sigma(\theta)$
${}^4\text{He}(p,p){}^4\text{He}$	$E_p = 0.92 - 34.3$	867	$\sigma(E), \sigma(\theta), A_y, P_y$
	Total:	6963	

Table 1. Channel configuration (top) and data summary (bottom) for the ^{5}Li system analysis. The column labeled "Observables" indicates the following data types: $\sigma(E)$, integrated cross section; $\sigma(\theta)$, unpolarized angular distributions (energy-dependence suppressed); A initial-state analyzing power; P final-state polarization; C spin correlation coefficients; K polarization transfer coefficients. (We have suppressed the indices $i, j, \ldots$ which take on values x, y, z for spins/polarization directions in configuration space.) All polarization and spin distributions are angular distributions, which depend on the angle of the outgoing particle. Chi-squared per degree of freedom for the analysis is $\chi^2/\text{dof} \simeq 2.7$ over 7,178 data points, 215 of which were discarded by eliminating individual data points which contribute to $\chi^2 > 40$.

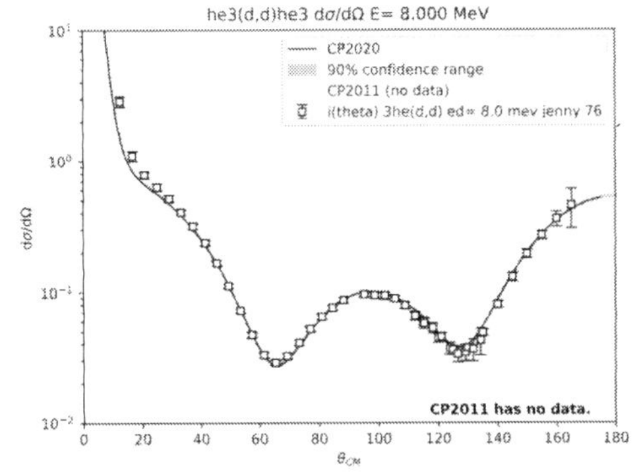

Figure 4. The angular distribution for ${}^3\text{He}(d,d){}^3\text{He}$ for deuteron incident laboratory energy 8.0 MeV of the unpolarized differential cross section in the center of mass as a function of the center of mass angle θ_{cm}. The evaluation is shown by the solid curve. The gray band is the 90% confidence interval determined as described in the text.

face that depends on the number of parameters N_p. Following Ref.[5], we recognize that, given the fact that the χ^2-function is a statistic distributed according to the χ^2-distribution. A detailed analysis[5] supports this observation and we arrive at a description of the parameter uncertainty in terms of the confidence interval, chosen as the upper limit of the cumulative χ^2 distribution function $P(\delta\chi^2; k)$ for k degrees of freedom as:

$$P(\delta\chi^2; k) = \left(2^{k/2}\Gamma(k/2)\right)^{-1} \int_0^{\delta\chi^2} dt\, t^{k/2-1} e^{-t/2}, \quad (11)$$

which is shown in fig. 2. Given that the mode and average of the χ^2 distribution are $k-2$ and k, respectively, we estimate using eq. (9), the scaling of the uncertainty of p_α to be

$$\delta p_\alpha \sim N_p^{1/2} C_{\alpha\alpha}^{1/2}. \quad (12)$$

The above analysis has been applied to the ^{5}Li compound system. The R-matrix configuration for this system is shown in the table.

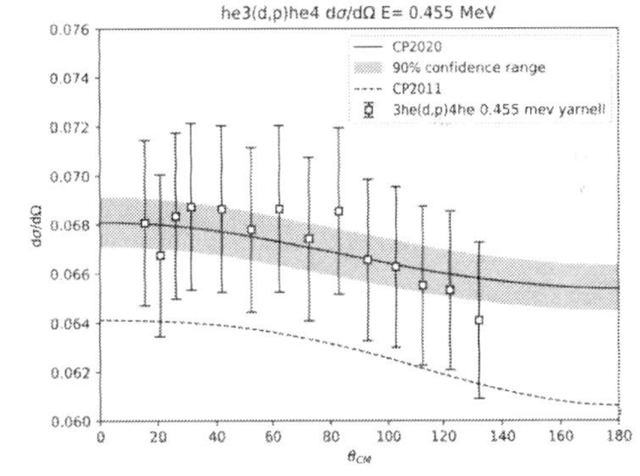

Figure 5. The angular distribution for ${}^3\text{He}(d,p){}^4\text{He}$ for deuteron incident laboratory energy 455 keV of the unpolarized differential cross section in the center of mass as a function of the center of mass angle θ_{cm}. The evaluation is shown by the solid curve. The gray band is the 90% confidence interval determined as described in the text.

3 Application to ^{5}Li compound system

We have applied the uncertainty quantification in terms of parameter confidence intervals described above to the ^{5}Li

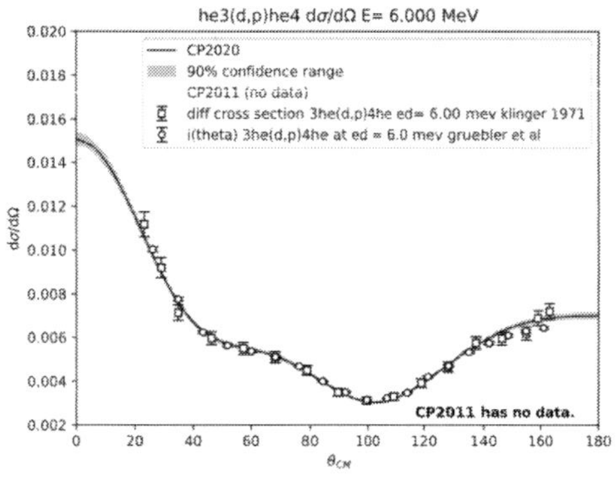

Figure 6. The angular distribution for ^{3}He$(d, p)^4$He for deuteron incident laboratory energy 6.0 MeV of the unpolarized differential cross section in the center of mass as a function of the center of mass angle θ_{cm}. The evaluation is shown by the solid curve. The gray band is the 90% confidence interval determined as described in the text.

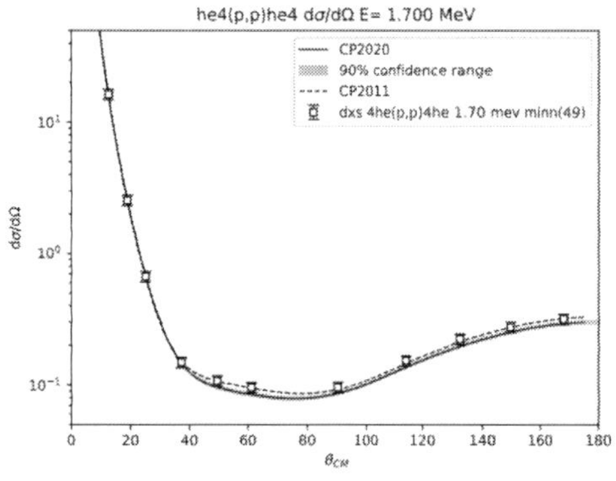

Figure 7. The angular distribution for ^{4}He$(p, p)^4$He for deuteron incident laboratory energy 1.7 MeV of the unpolarized differential cross section in the center of mass as a function of the center of mass angle θ_{cm}. The evaluation is shown by the solid curve. The gray band is the 90% confidence interval determined as described in the text.

compound system. The R-matrix configuration is shown in table 1. The two-particle partitions include the ground-state ^{3}He component, d +^{3}He$(\frac{1}{2}^+)$, p +^{4}He(0^+) – the proton ^{4}He ground-state component, its first excited state p +^{4}He$^*(0^+$; 20.2 MeV), and the scalar-isovector partner of the two-nucleon virtual bound state, d_0+^{3}He$(\frac{1}{2}^+)$. These are shown in the upper part of table 1, along with the channel radius a_c and the maximum value of the partition orbital angular momentum, ℓ_{max}. The fit data – processes (elastic, reaction, . . .), energy range, number and types of data – included in the ^{5}Li system evaluation is shown in the bottom table.

At the solution, set by an upper limit on the size of the gradient of 1 part in 10^6, we determine the parameter uncertainties as described in the previous section and propagate their error according to standard error propagation via the "sandwich" rule; see Ref.[1]. A few examples (out of hundreds generated by fitting the ~ 7k data points for the ^{5}Li system) are shown in figs. 3 to 8. Figures 3 and 4 show the elastic scattering of d from ^{3}He for a deuteron laboratory energy of 324 keV. Curves are shown in fig. 3 for a more recent evaluation (CP2020, which will be included in the ENDF/B-VIII.1 release in 2024) and the older evaluation (CP2011, which is equivalent to ENDF/B-VIII.0[6]) plotted against the data by Ref.[7]. Figure 4 shows data similar to the previous figure, plotted against the data from Ref.[8]. The (d, p) reaction data is shown in figs. 5 and 6 and finally the proton elastic data is displayed in figs. 7 and 8.

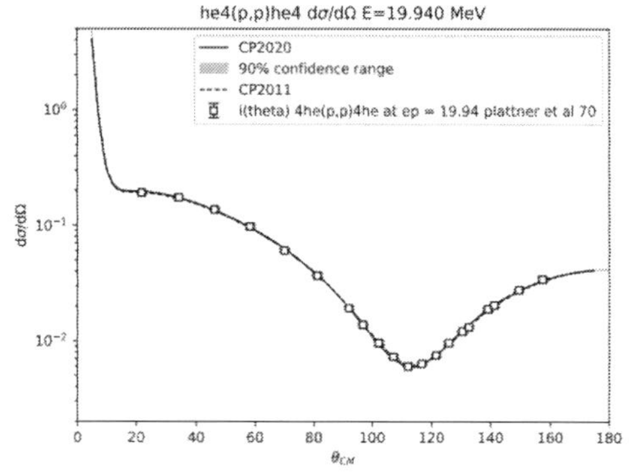

Figure 8. The angular distribution for ^{4}He$(p, p)^4$He for deuteron incident laboratory energy 19.94 MeV of the unpolarized differential cross section in the center of mass as a function of the center of mass angle θ_{cm}. The evaluation is shown by the solid curve. The gray band is the 90% confidence interval determined as described in the text.

4 Conclusions

Confidence intervals for R-matrix evaluations have been determined from the χ^2 distribution. We have motivated an improvement in the scaling of the parameter uncertainties with the number of parameters N_p in the fit. We applied this methodolgy to the ^{5}Li compound system in an evaluation of ~ 7k differential and integral data points. The improved scaling of the parameter uncertainties resolves some long standing issues associated with R-matrix evaluations yielding cross section uncertainties that are too small[9].

References

[1] G. Hale, Nucl. Data Sheets **109**, 2812 (2008)
[2] L. Wolfenstein, Annual Review of Nuclear Science **6**, 43 (1956)
[3] D. Smith, *Probability, Statistics, and Data Uncertainties in Nuclear Science and Technology*, Monograph Series (American Nuclear Society, 1991), ISBN 9780894480362

[4] R.A. Arndt, M.H. MacGregor, Methods in Computational Physics **6**, 253 (1966)
[5] Y. Avni, Astrophys. Journ. **210**, 642 (1976)
[6] D. Brown et al., Nucl. Data Sheets **148**, 1 (2018)
[7] L. Kraus (1971), PhD. Thesis
[8] B. Jenny, W. Gruebler, V. Konig, P.A. Schmelzbach, R. Risler, H.R. Burgi, D.O. Boerma, Nuclear Physics, Section A **324**, 99 (1979)
[9] A. Carlson et al., Nucl. Data Sheets **148**, 143 (2018)

Similarity evaluation between neutron multiplication factors and nuclide inventories during nuclear fuel burnup

Hiroki Harada[1,*] and *Go* Chiba[1,**]

[1]Hokkaido University, Faculty of Engineering, Kita 13 Nishi 8, Sapporo, Hokkaido 060-8628, Japan

Abstract. There are several different types of the integral data of nuclear fission systems, and the accuracy of the numerical predictions of them are related with each other via the nuclear data commonly used in the numerical simulations. Thus, it is sometimes possible that measurement data of one type of the integral data are used to improve the prediction accuracy of the different type of the integral data. To quantitatively evaluate such possibility, the similarity of the different types of the integral data is important. In this study, we quantitatively evaluate the similarity between neutron multiplication factors and nuclide inventories during nuclear fuel burnup from the viewpoint of the nuclear data uncertainties using the representativity factors. Using the Burner module of the CBZ reactor physics code system for fuel pin-cell burnup problems, we calculated the sensitivity of the neutron infinite multiplication factors and the inventories of the 17 actinoids at several fuel burnup points. The neutron multiplication factor during the fuel burnup was considered the target parameter for which the prediction accuracy is improved, and the degree of similarity of the nuclide inventory data during burnup to the target was quantitatively evaluated using the representative factor. Subsequently, multiple nuclide inventory data were combined using the concept of the extended bias factor method to create a fictitious parameter, and we investigated how much this fictitious parameter can increase the representativity factor for the target parameter. As a result, the representativity factor for the target parameters could be increased to more than 0.8 using some nuclide inventory data, and up to 0.92 depending on the burnup, even taking into consideration the measurement error.

1 Introduction

Measurement data on reactor core characteristics (integral data) are used for validation of evaluated nuclear data and for data assimilation to improve the accuracy of nuclear data. A typical example of integral data is the effective neutron multiplication factor (or criticality characteristics) of nuclear reactors. Measurement data on the effective neutron multiplication factor of various nuclear reactors is summarized in the ICSBEP Handbook, which is used worldwide. Data related to the nuclear fuel composition (or nuclide inventory) during/after the nuclear fuel burnup are likewise integral data and contribute to the validation and improvement of the accuracy of nuclear data. Recently, a project called EUCLID (Experiments Underpinned by Computational Learning for Improvements in nuclear Data) was started at Los Alamos National Laboratory[1, 2]. This project focuses on integral data other than effective neutron multiplication factors and actively uses them to verify and improve the accuracy of nuclear data. Nuclide inventory data are not explicitly referred in the paper of EUCLID, but these are also the integral data and are expected to have characteristics not found in other integral data. Therefore, it would be essential data for validating and improving the accuracy of nuclear data. However, there have been no quantitative examples demonstrating its features and uniqueness. For this reason, in this study, we examine the usefulness of nuclide inventory data.

The sensitivity coefficient of the integral data with respect to the nuclear data is useful, and sensitvities of parameters related to the fuel burnup, *burnup sensitivities*, can be calculated by the depletion perturbation theory[3, 4].The reactor physics code system CBZ, which is under development at Hokkaido University, would be the only code having the capability to calculate the burnup sensitivity of complex systems such as LWR (light water reactor) fuel assemblies[5]. By using CBZ, it is possible to quantitatively evaluate the characteristics of the sensitivity of the nuclide inventory data and the degree of its similarity to the neutron multiplication factor data.

2 Theory and procedure

In the present work, we evaluated the degree of similarity of the nuclide inventory data to the target parameter, the neutron infinite multiplication factor k_∞, of the fuel pin cell during burnup. More specifically, we calculated the sensitivity of k_∞ and the inventories of the 17 actinoids at concerning fuel burnup points using the Burner module of CBZ. Based on the calculated sensitivities, representativity factors (RFs) [6–9] between the target parameters and

*e-mail: hirokih0114bbpcf8@eis.hokudai.ac.jp
**e-mail: go_chiba@eng.hokudai.ac.jp

the nuclide inventory data were also calculated in order to evaluate the similarity. Then, we examined how much the RF could be increased by using multiple nuclide inventory data. The concept of the extended bias factor method[10] was used to combine multiple nuclide inventory data.

2.1 Representativity factor

Let $\mathbf{S}_i$ and $\mathbf{S}_j$ be the sensitivity vectors of the target parameter i and the mockup parameter j, and let $\mathbf{V}$ be the covariance matrix of the nuclear data. The RF r is defined as follows:

$$r = \frac{\mathbf{S}_i^T \mathbf{V} \mathbf{S}_j}{\sqrt{\mathbf{S}_i^T \mathbf{V} \mathbf{S}_i}\sqrt{\mathbf{S}_j^T \mathbf{V} \mathbf{S}_j}}. \tag{1}$$

The RF is the correlation between the two integral data including the nuclear data-induced uncertainties. The closer this absolute value is to unity, the more similar the data are in terms of nuclear data uncertainty. The degree of the similarity is important when the information on the mockup parameter is used to improve the prediction accuracy of the target parameter by the bias factor method[11]. Regarding the uncertainty (variance) of the prediction of the target parameters, let ΔR_0^2 be the uncertainty evaluated without the bias factor method and ΔR_1^2 with the bias factor method. There is the following relationship between them[6]:

$$\Delta R_1^2 = \Delta R_0^2 \left(1 - r^2\right). \tag{2}$$

In other words, the RF also represents how much the uncertainty of the target parameter is reduced by the bias factor method using information from the mockup parameter.

2.2 Extended bias factor method

While the classical bias factor method uses data from a single mockup parameter, the use of multiple parameters can further reduce uncertainty in target parameter predictions. One such method is the extended bias factor method[10]. In the bias correction, it is important to prepare mockup parameter whose sensitivity is similar to the sensitivity of the target parameter. The extended bias factor method attempts to reproduce the sensitivity of the target parameter by superimposing the sensitivities of multiple mockup parameters. The extended bias factor method deals with errors due to nuclear data, errors due to calculation methods, and measurement errors, but for the sake of simplicity, we focus only on errors due to nuclear data.

Let p_i denote multiple mockup parameters, and define the fictitious parameter p as

$$p = \prod_i p_i^{a_i}. \tag{3}$$

The partial differentiation of p with respect to p_i was performed as

$$\frac{\partial p}{\partial p_i} = a_i p_i^{a_i - 1} \prod_{j \neq i} p_j^{a_j} = \frac{a_i}{p_i} p_i^{a_i} \prod_{j \neq i} p_j^{a_j} = \frac{a_i}{p_i} p, \tag{4}$$

and the following equation is obtained:

$$\frac{\partial p}{\partial p_i} \cdot \frac{p_i}{p} = a_i. \tag{5}$$

Thus, the sensitivity of p to nuclear data σ, S_σ^p, can be represented as

$$S_\sigma^p = \frac{\partial p}{\partial \sigma} \cdot \frac{\sigma}{p} = \sum_i \frac{\partial p}{\partial p_i} \cdot \frac{p_i}{p} \cdot \frac{\partial p_i}{\partial \sigma} \cdot \frac{\sigma}{p_i} = \sum_i a_i S_\sigma^{p_i}. \tag{6}$$

This equation means that the sensitivity of the fictitious parameter p is expressed by the linear combination of the sensitivities of multiple mockup parameters. Next, let us evaluate the uncertainty of the predictions when using the bias factor for the fictitious parameter p. As mentioned above, we ignore uncertainties due to other than nuclear data for the sake of simplicity. The variance V_p for the calculated value of parameter p can be written as

$$\begin{aligned} V_p &= \sum_i \sum_j \left(\frac{\partial p}{\partial p_i} \cdot \frac{p_i}{p}\right)\left(\frac{\partial p}{\partial p_j} \cdot \frac{p_j}{p}\right) \mathrm{cov}\left(p_i, p_j\right) \\ &= \sum_i \sum_j a_i a_j \,\mathrm{cov}\left(p_i, p_j\right) = \sum_i \sum_j a_i a_j \mathbf{S}_i^T \mathbf{V} \mathbf{S}_j. \end{aligned} \tag{7}$$

Let p_R be the target parameter to be predicted. The variance for the calculated value of p_R can be expressed using the sensitivity vector of the target parameter, $\mathbf{S}_R$, as

$$V_{p_R} = \mathbf{S}_R^T \mathbf{V} \mathbf{S}_R. \tag{8}$$

Since the covariance of the calculated values of p and p_R is represented as

$$\mathrm{cov}\left(p, p_R\right) = \sum_i a_i \mathbf{S}_i^T \mathbf{V} \mathbf{S}_R, \tag{9}$$

the uncertainty of $\tilde{p}_R$ which is the predicted value of p_R using the bias factor method, $V_{\tilde{p}_R}$, is represented as

$$\begin{aligned} V_{\tilde{p}_R} &= V_p + V_{p_R} - 2\,\mathrm{cov}\left(p, p_R\right) \\ &= \sum_i \sum_j a_i a_j \mathbf{S}_i^T \mathbf{V} \mathbf{S}_j + \mathbf{S}_R^T \mathbf{V} \mathbf{S}_R - 2 \sum_i a_i \mathbf{S}_i^T \mathbf{V} \mathbf{S}_R \\ &= \left(\mathbf{S}_R - \sum_i a_i \mathbf{S}_i\right)^T \mathbf{V} \left(\mathbf{S}_R - \sum_i a_i \mathbf{S}_i\right). \end{aligned} \tag{10}$$

The exponents a_i in the fictitious parameter p are determined from the condition $\frac{\partial V_{\tilde{p}_R}}{\partial a_i} = 0$, which minimizes $V_{\tilde{p}_R}$:

$$\sum_j a_j \mathbf{S}_i^T \mathbf{V} \mathbf{S}_j - \mathbf{S}_i^T \mathbf{V} \mathbf{S}_R = 0, \quad (i = 1, \ldots, N), \tag{11}$$

where N is the number of mockup parameters. When we define $\mathbf{A} = (a_1, a_2, \ldots, a_N)^T$ and $\mathbf{S} = (\mathbf{S}_1 \quad \mathbf{S}_2 \ldots \mathbf{S}_N)$, then the above equation can be rewritten as follows:

$$\mathbf{S}^T \mathbf{V} \mathbf{S} \mathbf{A} = \mathbf{S}^T \mathbf{V} \mathbf{S}_R. \tag{12}$$

Thus, $\mathbf{A}$ is calculated as

$$\mathbf{A} = \left(\mathbf{S}^T \mathbf{V} \mathbf{S}\right)^{-1} \left(\mathbf{S}^T \mathbf{V} \mathbf{S}_R\right). \tag{13}$$

If the matrix $\mathbf{S}^T \mathbf{V} \mathbf{S}$ is singular, the pseudo-inverse matrix can be used instead of the normal inverse matrix.

3 Target system and burnup calculation conditions

The target system of the present study is a PWR-UO2 pin-cell with a U-235 enrichment of 4.1%. Line power density is set at 179 W/cm and fuel burnup up to 40 GWd/t is concerned. As nuclear data, JENDL-4.0 was used for neutron reaction cross sections, JENDL/FPD-2011 for decay data, and JENDL/FPY-2011 for fission yield data. For the burnup calculations, a simplified burnup (nuclide transmutation) chain consisting of 21 actinides and 138 fission products was used. Under the above burnup conditions, the sensitivity of k_∞ was calculated from 0 to 40 GWd/t in 5 GWd/t steps. For the same pin-cell, the sensitivity of the nuclide inventory was calculated from 5 to 40 GWd/t in 5 GWd/t steps. The mockup nuclide inventory data were for 17 nuclides excluding U-238 and the short-lived nuclides U-237, Np-239, and Am-242 from the heavy nuclides included in the chain used. Covariance data for nuclear data are required for the calculation of RFs. In this study, only covariance data for the reaction cross section of actinides were considered, and the JENDL-4.0 evaluation data were used.

4 Results

4.1 Maximum representativity factor calculation

The nuclear data-induced uncertainty of k_∞ of the target system is shown in **Figure 1**. The uncertainty at the beginning of the burnup is largely due to the capture cross section of U-238, and at the end it is largely due to the capture cross section of Pu-241. Additionally, the nuclear data-induced uncertainty of 17 nuclides inventory data for several burnup points is shown in **Figure 2**.

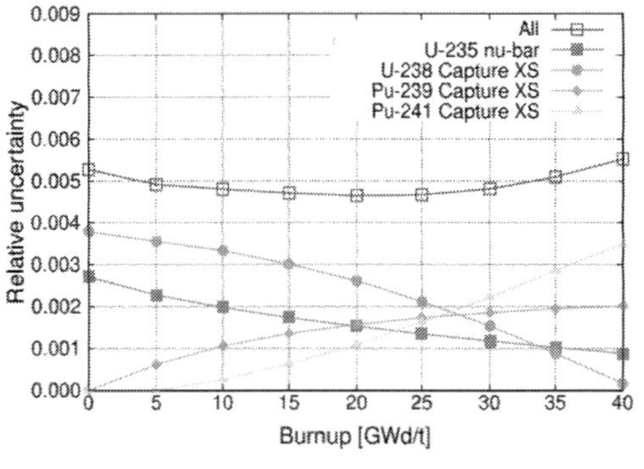

Figure 1. Total and component-wise nuclear data-induced uncertainty of k_∞

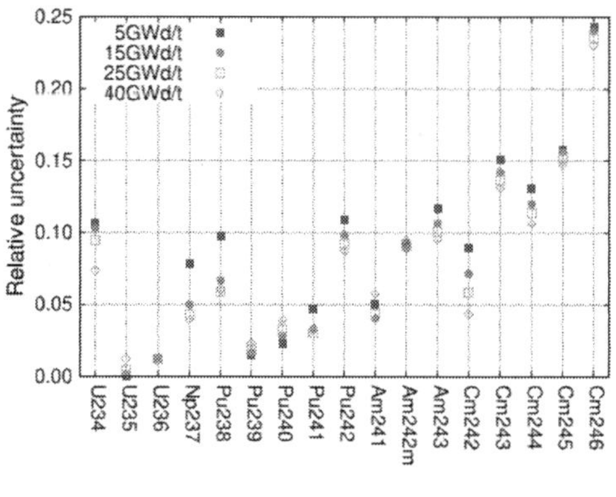

Figure 2. The Nuclear data-induced uncertainty of inventory data

The target parameters were set as k_∞ at 9 burnup points (0-40 GWd/t). We calculated RFs of all nuclide inventory data for all burnup points (5-40 GWd/t) for each target k_∞. The largest calculated RFs are shown in **Figure 3** and the corresponding mockup data are shown in **Table 1** as Case 1. The RF was about 0.75 at the beginning and end of the burnup, and about 0.55 at the middle. The mockup data yielding the largest RF was Pu-239 at the beginning and switched to Pu-242 as the burnup proceeds. Plutonium-239 and -242 are produced by the neutron capture reactions of U-238 and Pu-241, respectively, and these reaction cross sections are main contributors to the k_∞ uncertainties as shown in Figure 1.

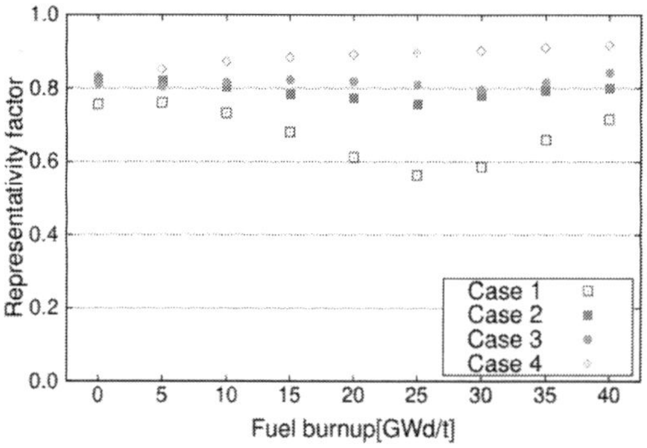

Figure 3. Maximum RFs of nuclide inventory data

Table 1. Mockup parameters and their combinations with which the maximum RFs are obtained

Target (GWd/t)	Case 1	Case 2		Case 3
0	Pu-239 5GWd/t	U-236 5GWd/t	Pu-239 5GWd/t	U-235
5	Pu-239 5GWd/t	U-235 5GWd/t	Pu-239 5GWd/t	U-235
10	Pu-239 5GWd/t	U-235 5GWd/t	Pu-239 40GWd/t	Pu-240
15	Pu-239 5GWd/t	Pu-240 15GWd/t	Pu-241 10GWd/t	Pu-240
20	Pu-239 40GWd/t	Pu-240 20GWd/t	Pu-241 5GWd/t	Pu-241
25	Pu-240 5GWd/t	Pu-240 5GWd/t	Pu-242 10GWd/t	Pu-241
30	Pu-242 10GWd/t	Pu-240 5GWd/t	Pu-242 15GWd/t	Pu-241
35	Pu-242 30GWd/t	Pu-240 5GWd/t	Pu-242 30GWd/t	Pu-242
40	Pu-242 40GWd/t	Pu-240 10GWd/t	Pu-242 35GWd/t	Pu-242

Next, in order to increase RF, a fictitious parameter was created using multiple nuclide inventory data by the extended bias factor method, and RFs for the targets were calculated. We considered three cases to combine multiple data: Cases 2, 3, and 4. In Case 2, we created fictitious parameters by combining two nuclide inventory data. In this case, the sensitivities of the same nuclide at different burnup points were not combined. In Case 3, we combined nuclide inventory data of unique nuclide at 8 burnup points from 5 to 40 GWd/t, and in Case 4 all the treated nuclide inventory data were combined.

The results are shown in Figure 3. For Cases 2 and 3, the maximum RFs among possible combinations are presented. The combinations yielding maximum RFs for Cases 2 and 3 are shown in Table 1. It is clear that the combination of multiple data can reduce the uncertainty of the target prediction.

In all of Cases 2, 3, and 4, the RF could be increased to 0.75 or higher by combining multiple data. In Case 2, the combination of data included Pu-239 at the beginning of the burnup and Pu-242 at the middle and end. In Case 3, it was found that, depending on the burnup of the target, RF could be increased to about 0.85 even with data for only one nuclide. Also here, it was found that the mockup inventory data yielding the maximum RF changed from U-235 to Pu-242 as the fuel burnup proceeds. In Case 4, since all data were utilized, the RF is theoretically increased to the maximum among all the cases. It can be seen that there is a limit to increasing the RF at the beginning of the burnup. This is due to the relatively large contribution of the number of neutrons produced per fission $\bar{\nu}$ of U-235. Since $\bar{\nu}$ has no significant effect on the nuclide inventory, it is difficult to reduce the uncertainty due to $\bar{\nu}$ only with the nuclide inventory data.

4.2 Identification of the important mockup parameters in the fictitious mockup parameters

Next, we attempt to identify the important mockup parameters in Case 4. First, we focus on the weights of the mockup parameters in the fictitious parameters determined by the extended bias factor method. The weights of each mockup parameter when targeting k_∞ at 40 GWd/t are shown in **Figure 4**. The range of the weights taken is quite wide and expands from -200 to 150. In this case, it is possible for nuclides having negligibly small sensitivities to have large weights relative to its importance. Therefore, by using weights alone, we cannot identify important mockup parameters. As another approach, weight of the mockup parameters was changed to zero individually and the amount of reduction in RF was quantified. Only the weights of the individual mockup data were reset to zero, and the other mockup data weights were not changed. It means that we did not perform the extended bias factor calculation where one mockup data was removed. The results are shown in **Figure 5**. There were some cases where the reduction in RF exceeded 0.9 (i.e., the RF became negative) when the weight of only one mockup parameter was set to zero.

We investigate this large reduction in RF. When k_∞ at 40 GWd/t is concerned, the main source of the uncertainty due to nuclear data is the capture cross section of Pu-241. The sensitivity of the target to the Pu-241 capture cross section and that reproduced using the extended bias factor method are shown in **Figure 6**. We can confirm that the sensitivity of the target is well reproduced. The fictitious parameter was created by linear combination of the sensitivities of the 17 nuclides × 8 burnup points as previously described. Among the mockup parameters, the sensitivities of the Cm-245 number density at 5 GWd/t and the Np-237 number density at 35 GWd/t to the Pu-241 capture cross section multiplied by their weights are shown in **Figure 7**. It can be found that the target sensitivity is reproduced by the cancellation of the large positive and negative sensitivities.

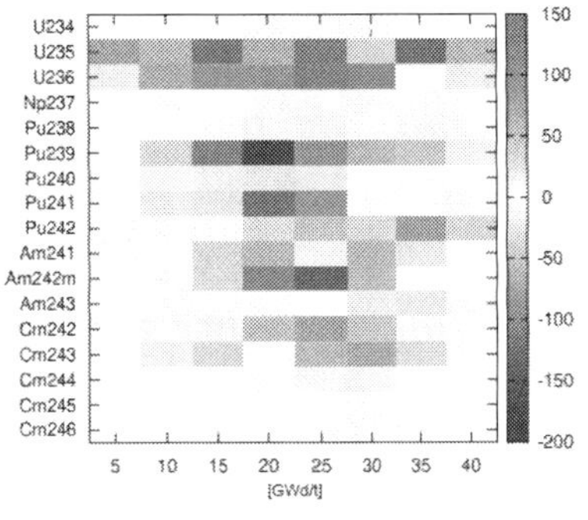

Figure 4. Weights of individual mockup parameters in the fictitious mockup parameters for k_∞ at 40 GWd/t (Case 4)

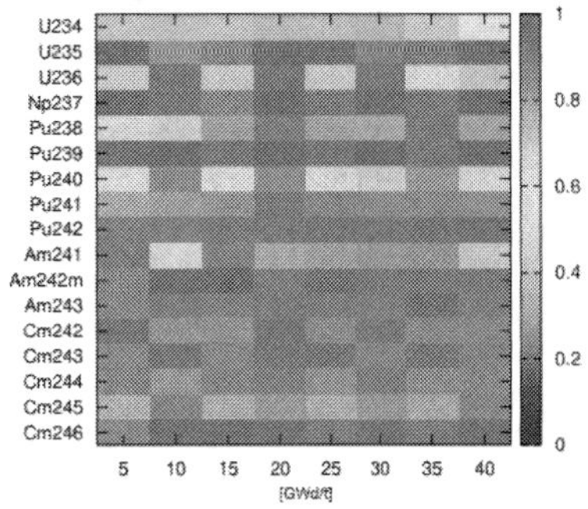

Figure 5. Reduction of RFs when setting individual weights zero (target: k_∞ at 40 GWd/t, Case 4)

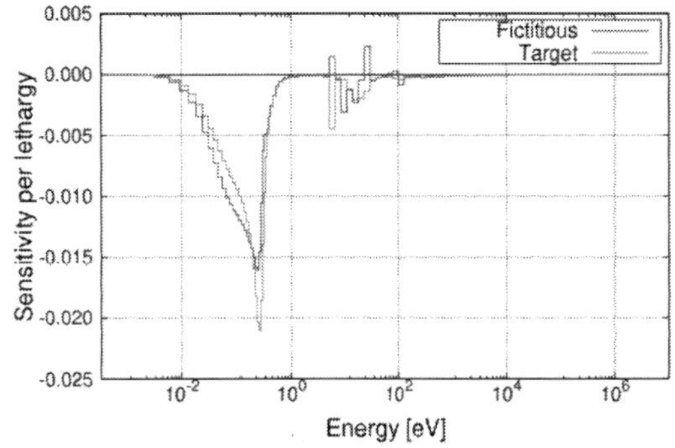

Figure 6. Sensitivities of target and fictitious parameters to Pu-241 capture cross section in Case 4

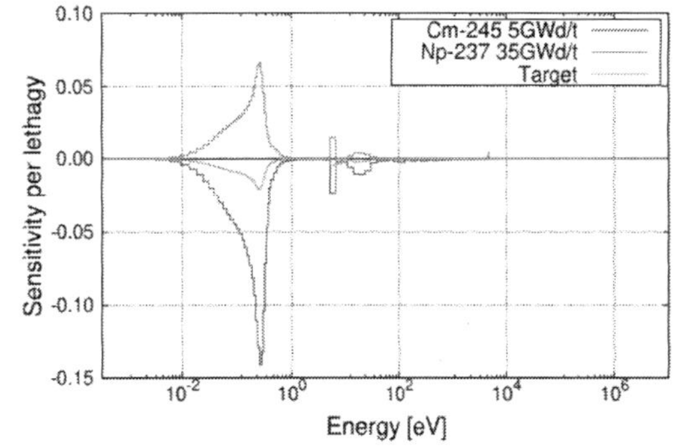

Figure 7. Decomposed sensitivities of fictitious parameters to Pu-241 capture cross section in Case 4

The above calculations use the extended bias factor method that deals only with errors due to nuclear data for simplicity. In this case, if measurement error is taken into consideration in some way, the uncertainty of the fictitious parameters becomes very large when large weights are given to individual mockup parameters. This makes the prediction error by the extended bias factor method worse significantly. Therefore, we reexamined the importance of mockup parameters by taking into consideration the measurement error of the mockup parameters. In the reference[11], RF that takes into consideration the measurement error of the mockup parameters is defined as follows:

$$\hat{r} = r \cdot \frac{\sqrt{V_c}}{\sqrt{V_c} + \sqrt{V_e}}, \tag{14}$$

where V_c represents the uncertainty (variance) due to nuclear data and V_e is the measurement error in the mockup parameters. Using this equation, the variance of the predictions by the bias factor method is modified as follows[11]:

$$\Delta R_1^2 = \Delta R_0^2 \left(1 - \hat{r}^2\right) = \Delta R_0^2 \left(1 - \frac{r^2}{1 + \frac{V_e}{V_c}}\right). \tag{15}$$

Thus, when the measurement error of the mockup parameters (V_e) is large, it is difficult to reduce the variance of the predictions even if r is close to unity. The weights of the extended bias factor method are determined by the following equation when measurement error is taken into consideration:

$$\mathbf{A} = \left(\mathbf{S}^T\mathbf{V}\mathbf{S} + \mathbf{V}_e\right)^{-1}\left(\mathbf{S}^T\mathbf{V}\mathbf{S}_R\right). \tag{16}$$

In the present work, the measurement error was assumed 0.5 % for all nuclide inventory data, and RF was recalculated in the same way as in Case 4. The results are shown in **Figure 8** as Case 4-2. The values of RF did not change significantly after the measurement error consideration. We attempted to identify the important mockup parameters from the same two perspectives as before. The results for k_∞ at 40 GWd/t are shown in **Figures 9** and **10**. Mockup parameters with extremely large weights do not appear when measurement errors are taken into consideration, and it is confirmed that many nuclides contribute universally to the reproduction of the target sensitivity. Compared to Case 4, the RF is no longer significantly reduced by setting only one weight to zero. The sensitivity of the target to the Pu-241 capture cross section and the sensitivity reproduced using the extended bias factor method to consider measurement error are shown in **Figure 11**. The sensitivity was found to be well reproduced in Case 4-2. The sensitivities of the Cm-244 number density at 10 and 40 GWd/t and the Am-243 number density at 5 GWd/t to the Pu-241 capture cross section multiplied by their weights are shown in **Figure 12**. Even in Case 4-2, cancellation of positive and negative sensitivities occurred. However, unlike Case 4, the norm of decomposed sensitivity vectors of the fictitious mockup parameter is no longer larger than the norm of sensitivity vector of the target.

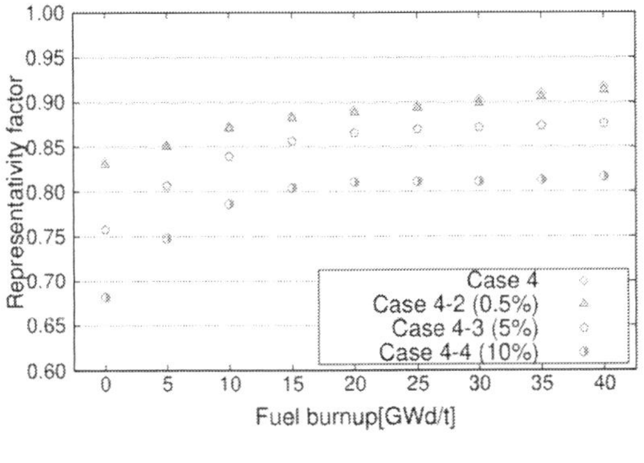

Figure 8. Maximum RFs of nuclide inventory data taking into consideration measurement error

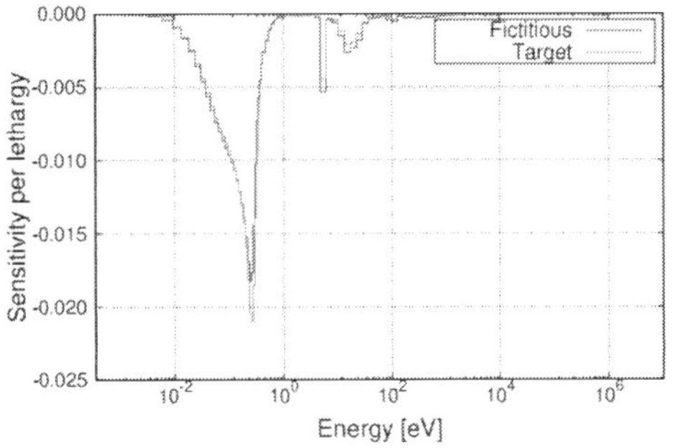

Figure 11. Sensitivity of target and fictitious parameter (Pu-241 capture,Case4-2)

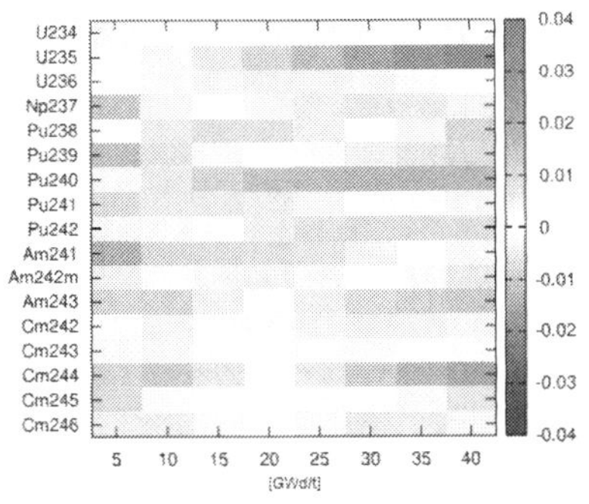

Figure 9. Weights of individual mockup parameters in the fictitious mockup parameters for k_∞ at 40 GWd/t (Case 4-2)

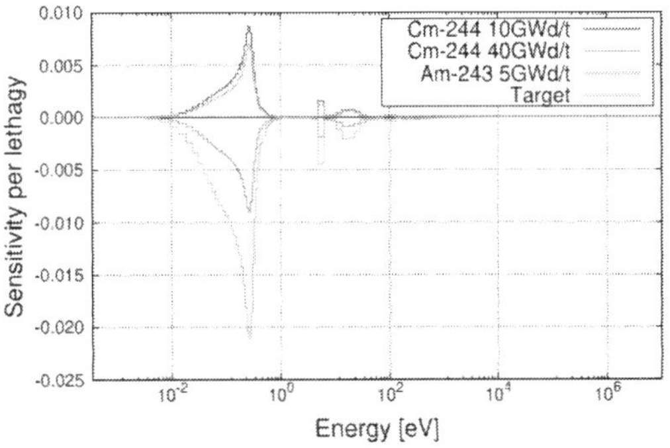

Figure 12. Decomposed sensitivities of fictitious parameters to Pu-241 capture cross section in Case4-2

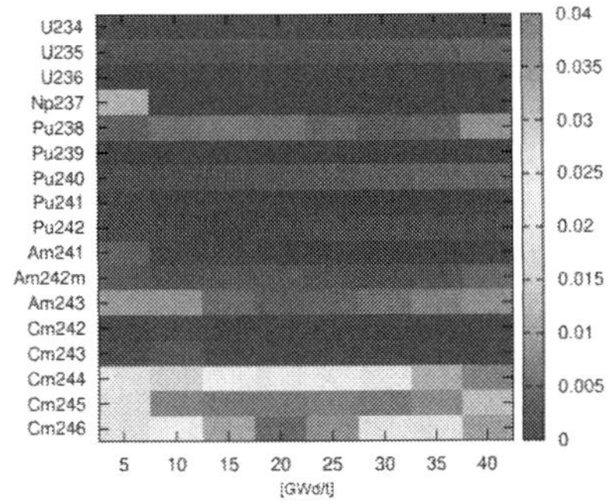

Figure 10. Reduction of the RFs when making individual weights zero (target: k_∞ at 40 GWd/t, Case 4-2)

In addition, we made the same calculations assuming a measurement error of 5 % or 10 % for all nuclide inventory data. The results are shown in **Figure 8** as Cases 4-3 and 4-4. In Case 4-3 the RF is about 0.75 at the beginning of the burnup and approached 0.88 as the burnup proceeds. In Case 4-4 the RF is about 0.68 at the beginning of the burnup and approached 0.82 as the burnup proceeds. We also attempted to identify important mockup parameters. The results for k_∞ at 40 GWd/t in Case 4-3 are shown in **Figures 13** and **14**. When the measurement error is set to 5%, the range of weights taken is smaller compared to 0.5%. The reduction of the RFs when making individual weights zero is also smaller. From the above, mockup parameters with extremely large weights do not appear when measurement errors are taken into consideration, and it is confirmed that many nuclides contribute universally to the reproduction of the target sensitivity. Compared to Case 4, the RF is no longer significantly reduced by setting only one weight to zero.

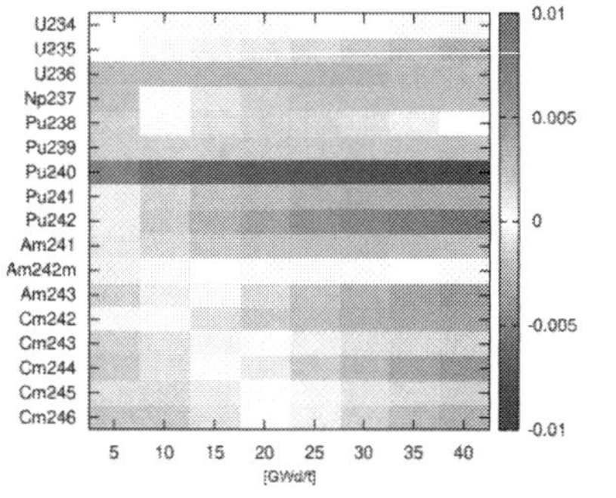

Figure 13. Weights of individual mockup parameters in the fictitious mockup parameters for k_∞ at 40 GWd/t (Case 4-3)

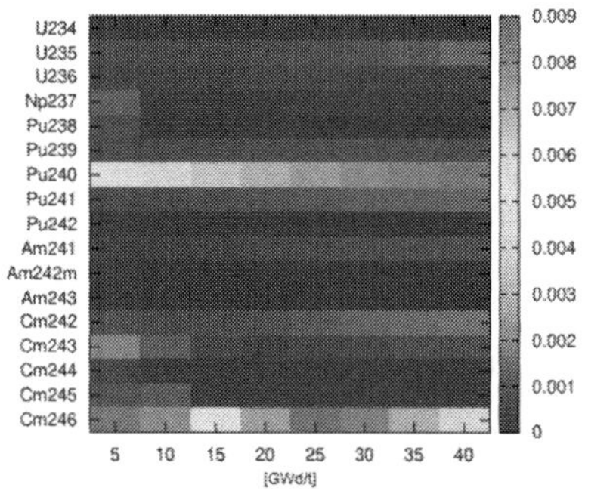

Figure 14. Reduction of the RFs when making individual weights zero (target: k_∞ at 40 GWd/t, Case 4-3)

5 Conclusion

The relationship between burnup quantity and criticality characteristics was quantified using RFs. The RFs for neutron infinite multiplication factor and nuclide inventory data were about 0.75 at the beginning and end of the burnup, and about 0.6 at the middle. We confirmed that the extended bias factor method can be used to create fictitious parameters which increase the RF with the target. The extended bias factor method here dealt only with uncertainties due to nuclear data. By combining multiple nuclide inventory data, we were able to increase RF to neutron infinite multiplication factor to about 0.92. By checking fictitious parameter combinations, it was found that Pu-239 and Pu-242 play important roles in reproducing the sensitivities of the target at the beginning and middle/end of the burnup.

When fictitious parameters were created using the extended bias factor method with a large number of mockup parameters, the weights for the mockup parameters became extremely large. We have only dealt with uncertainties due to nuclear data, but if the mockup parameters have measurement errors, the uncertainty of the fictitious parameter would become extremely large when large weights are given to individual mockup parameters. Therefore, the extended bias factor method was applied again assuming a measurement error of 0.5% or 5% for the mockup data. The weights of the individual mockup parameters were not extremely large, and we were able to create fictitious parameters in which many nuclides contributed universally. The RF with the target was about 0.8, even when considering a 10% measurement error in the inventory data. From the above, it can be said that burnup quantity data are useful for criticality characteristics data.

References

[1] M. Herman, et al., "Infrastructure for the new paradigm of nuclear reaction evaluation," Ann. Nucl. Energy, 163, 108494 (2021).

[2] J. Hutchinson, et al., "Investigation of delayed neutron sensitivities for several ICSBEP benchmarks using MCNP," Trans. Am. Nucl. Soc., 125, p.620 (2021).

[3] A. Gandini, "A method of correlation of burnup measurements for physics prediction of fast power-reactor life," Nucl. Sci. Eng., 38, p.1 (1969).

[4] M. L. Williams, "Development of depletion perturbation theory for coupled neutron/nuclide field," Nucl. Sci. Eng., 70, p.20 (1979).

[5] G.Chiba, H.Harada, "Validation of LWR fuel depletion calculation module of reactor physics code system CBZ," J. Nucl. Sci. Technol., (accepted).

[6] G. Palmiotti, M. Salvatores, "Use of integral experiments in the assessment of large liquid-metal fast breeder reactor basic design parameters," Nucl. Sci. Eng., 87, p.333 (1984).

[7] M. Plaschy, et al., "Importance of the MUSE experiments for emerging ADS concepts from the nuclear data viewpoint," Ann. Nucl. Energy, 32, p.843 (2005).

[8] P.T. Krishna Kumar, et al., "Maximization of representativity factors for experimental planning of cross-section measurements: an information theoretic approach," Ann. Nucl. Energy, 35, p.2243 (2008).

[9] J.F. Lebrat, et al., "The use of representativity theory in the depletion calculations of SFR blankets," Ann. Nucl. Energy, 101, p.429 (2017).

[10] T. Kugo, T. Mori, T. Takeda, "Theoretical study on new bias factor methods to effectively use critical experiments for improvement of prediction accuracy of neutronics characteristics", J. Nucl. Sci. Technol., 44, p.1509 (2007).

[11] A. Gandini, "Uncertainty analysis and experimental data transposition methods based on perturbation theory," in Y. Ronen (Ed.) "Uncertainty Analysis", CRC press, (1988)

Covariance analysis of ^{235}U(n$_{th}$,f) independent and cumulative fission yields : propositions for JEFF4

Grégoire Kessedjian[1*], *Sidi-Mohamed* Cheikh[1], *Oliver* Serot[1], *Abdelhazize* Chebboubi[1], *David* Bernard[1], *Vanessa* Vallet[1], *Robert* Mills[2] and *Luigi* Capponi[2]

[1]CEA, DES, IRESNE, DER, SPRC, LEPh, Cadarache center, F-13108 Saint Paul lez Durance, France
[2] National Nuclear Laboratory, Central Laboratory, Sellafield, Seascale CA20 1PG, England

Abstract. The study of fission yields has a major impact on the characterization and understanding of the fission process and its applications. Fission yield evaluations represent the synthesis of experimental and theoretical knowledge in order to perform the best estimation of independent fission yields. Today, the lack of correlations between the different fission observables induces several inconsistencies in the evaluations. Different works proposed to estimate the correlations of the independent fission yields satisfying the consistency to the chain yield evaluations. Nevertheless, none of them introduces a prior correlations of the independent and chain evaluations in the evaluation process. Covariance matrix of fission yields depends on the evaluation method used according to the kinds of existing measurements. The consistency is deeply entangled to the statistical agreement between each dataset considering the covariance of measurements. Moreover, covariance of model parameters does not represent the only contribution to the evaluation covariance matrix. Thus, a new evaluation process is crucial to provide a complete and coherent evaluation file. The LEPh Laboratory of CEA Cadarache is developing this program for the future version of the JEFF-library.

1 Status on correlation matrix of fission yields

1.1 Previous analysis

1.1.1 Context and notations

Fission yields correspond to several observables describing the particles produced during the fission process. Starting from the scission point up to the decay products, we can cite:

- Fission fragment mass yields (before neutron emission) $Y(A^*)$
- Fission fragment isobaric distributions $P(Z\,|\,A^*)$ (before neutron emission)
- Neutron multiplicity $\upsilon(A^*)$
- Fission product mass yields $Y(A)$ (after neutron emission)
- Fission product isobaric distribution $P(Z\,|\,A)$ (after neutron emission)
- Fission product independent yields $Y(A,Z,m)$
- Isomeric ratio $IR(A,Z,m)$
- Kinetic energy distributions $P(E_k\,|\,A,Z)$
- Fission product cumulated yields $C(A,Z,m)$
- Chain yields $C(A)$

Historically, JEFF evaluations are based on experimental data associated with phenomenological models (Brosa [1], Zp model [2], Madland-England [3]) to extend the range of fission products up to the isotopes of interest for the applications. Thus, we find the same classification in the sorting of the experimental data. Some data are also model-dependent as, for instance, several chain yields where Zp model correction are applied. Up to now, the evaluations of the independent yields (after neutron emission) and the cumulative yields respect the conservation laws. Nevertheless the uncertainty analysis of each observable is based on the experimental data where they exist and model dependent for the extrapolated yields.

1.1.2 Correlation matrix of mass yields:

During the last decades, to fill the lack of a covariance matrix respecting the difference between the large independent yield uncertainties and the precise chain yields, several methodologies were proposed.
C. Devillers [4] proposed in 1977 a sum method of covariance in order to respect the precise chain yield uncertainties. This method is applied by K. Tsubakihara et al. [5] for the JENDL library (**Fig. 1**), L. Fiorito et al. [6, 7] for the JEFF one and T. Kawano et al. [8] followed by M.T. Pigni et al. [9] for the ENDF/B-VII.1 library. This method consists to determine the covariance

* Corresponding author: gregoire.kessedjian@cea.fr

elements between Y(A,Z,m) to obtain the variance of C(A,Z,m). This method assumes a null correlation between cumulative yields. Then, this only allows the filling of the lacks between independent and cumulative yields in order to respect the compatibility of the uncertainties of the two evaluations. Nevertheless, the hierarchy between the observables and the propagation of conservation laws is not properly defined.

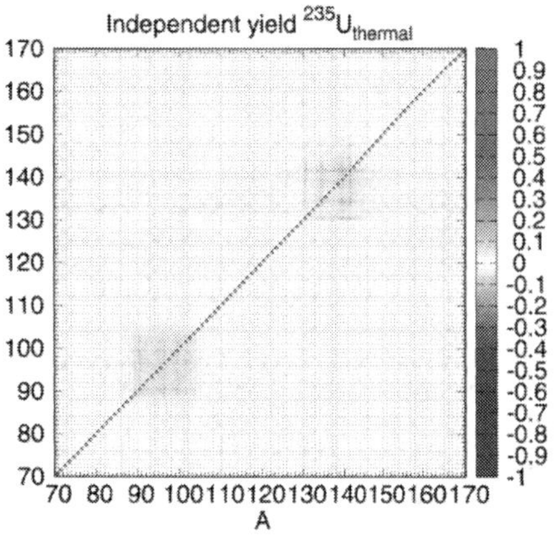

Fig. 1. Mass yield correlation matrix from K. Tsubakihara et al. (2021) [5].

In 2015, N. Terranova et al. [10,11] proposed a new approach considering the physical fission process to describe the covariance of the independent fission yields. Based on Brosa's model [1] and the Wahl parametrization of the Zp model [2] to describe the fission fragment distributions, the neutron emission yields $\upsilon(A^*)$ are adjusted in order to describe the JEFF-3.1.1 independent fission product yields [12]. The covariance matrix is obtained using the generalized perturbation theory. The cumulative yields are deduced from the previous calculations. Nevertheless, the calculated uncertainties of the chain yields are overestimated in comparison to those of JEFF-3.1.1. **Figs. 2 & 3** present the mass yield correlation matrix from the model parameter considering or not the prompt neutron yield uncertainties in order to observe the impacts of different model components.

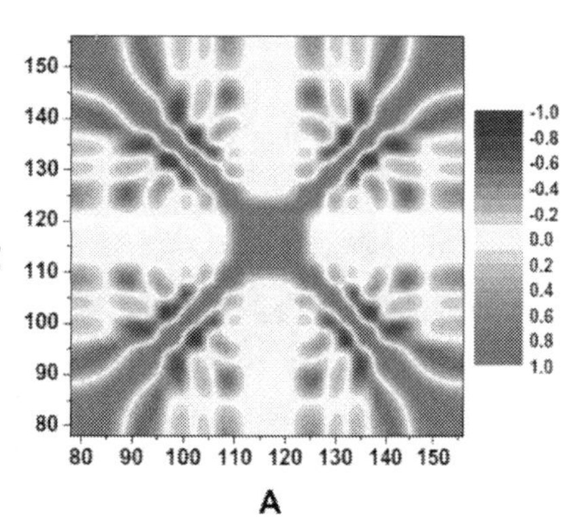

Fig. 2. Mass yield correlation matrix from CEA (2015) [10, 11] considering only the uncertainties of Brosa's fission mode parameters

The LOHENGRIN (ILL) collaboration on fission studies developed an experimental methodology to obtain absolute measurement of mass yields with associated experimental correlations (**Fig. 4**). There is not model input in this analysis and the correlation matrix come from the competition of the positive component of systematic uncertainties and the negative component due to the self-normalization [13].

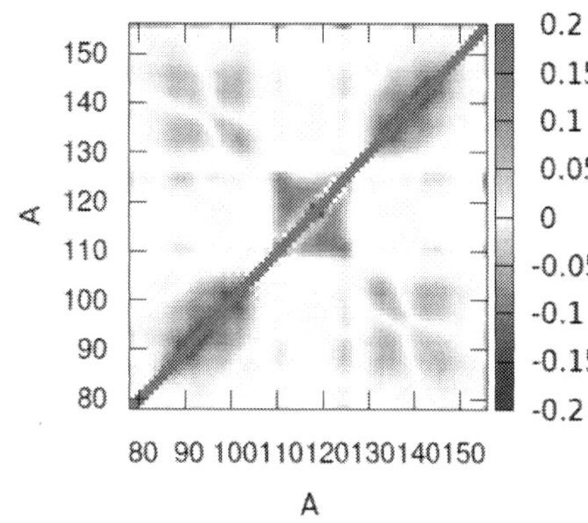

Fig. 3. Mass yield correlation matrix from CEA (2015) [10, 11] considering the uncertainties of Brosa's fission mode parameters and the prompt neutron yield uncertainties (saw tooth).

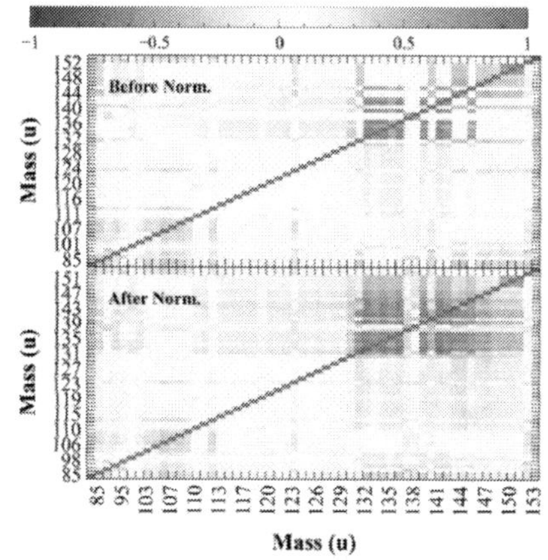

Fig. 4. (top) Experimental mass rate correlation of $^{233}U(n_{th},f)$ and (bottom) absolute yield correlation after self-normalization [13].

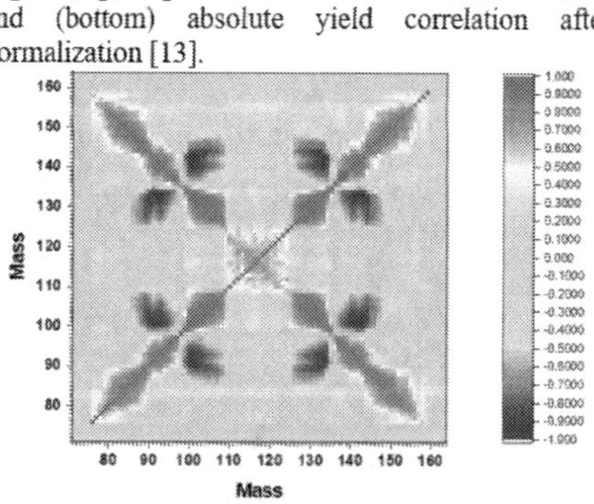

Fig. 5. K.H. Schmidt et al. GEF-2021/1.1 [14]

Another approach proposed by the NEA is the GEF modelling [14] based on a Brosa-like fission fragment description and many phenomenological parameters to describe a very large range of fissioning systems. A Monte Carlo code allows the generation of covariance of fission yields for one fissioning system (see **Fig. 5**) and between fissioning systems.

1.1.3 Correlation matrix of independent and cumulative yields:

A focus on the correlation matrices of independent fission yields according to the GEF model **(Fig.6)** and the Terranova's method **(Fig.7)** shows the same cross pattern but with important differences on the intensity of correlations. The analysis of L. Fiorito et al. [6, 7] is based on the sum rule to be consistent with the chain yield evaluation proposed by C. Devillers which structures the correlations matrix to go from about 5-10% independent yield uncertainty to 1-5% cumulative yield uncertainties (**Fig. 8 & 9**).

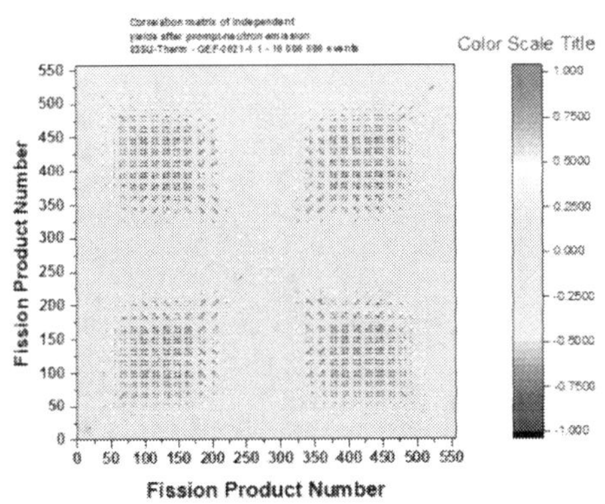

Fig. 6. Correlations of independent fission yields from the GEF-2021/1.1 code [14]

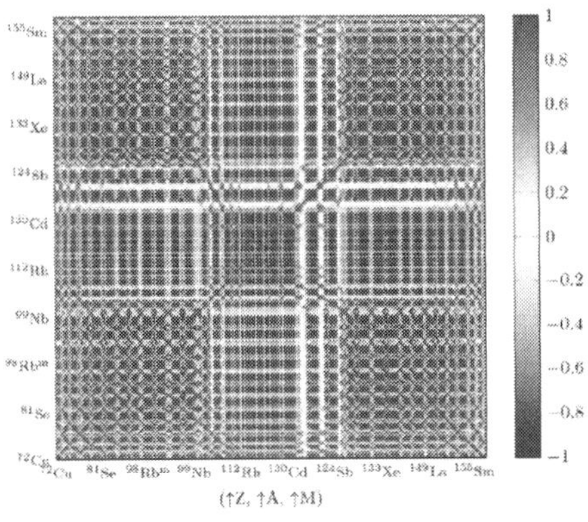

Fig. 7. Independent yield correlation matrix from CEA (2015) [10, 11]

1.1.4 discussion:

Thus, we observe that the structures are mainly due to the analysis methodologies. For the mass yields (post-neutron), a cross pattern comes from the mass modal model and the mass conservation law of the pre-neutron mass yields (see **Figs. 2 & 5**). When this stage of pre-neutron mass modeling is not introduce, the pattern of the correlation matrix presents only one diagonal as shown in **Figs. 1 & 4.**

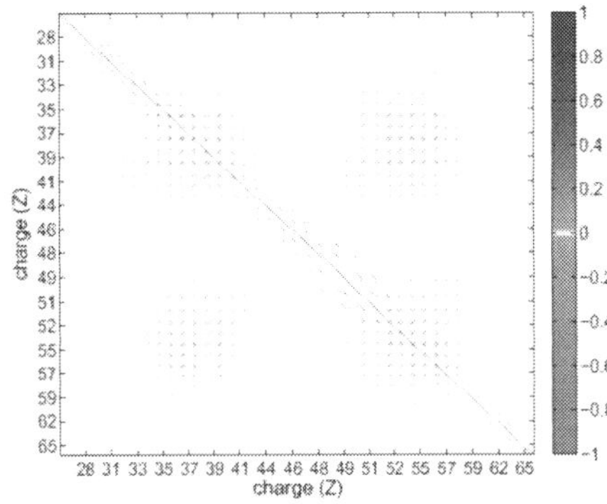

Fig. 8. Correlation matrix of independent yields proposed by L. Fiorito et al. [6, 7]

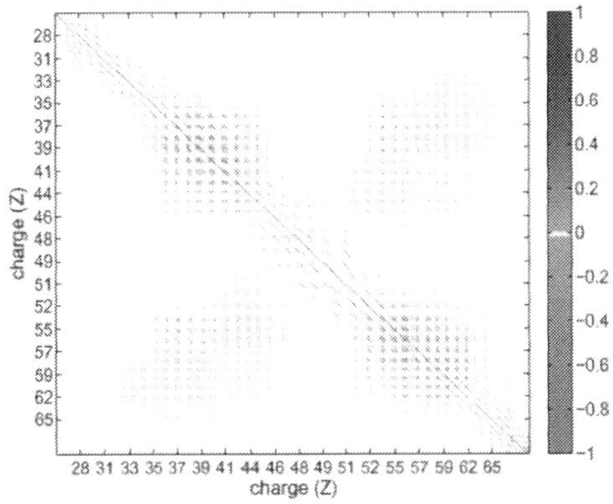

Fig. 9. Correlation matrix of cumulated yields proposed by L. Fiorito et al. [6, 7]

Adding the charge analysis, we observe a second cross pattern coming from the nuclear charge conservation law. This isotopic component of correlations can induce extreme correlations (see **Fig. 6 & 7**) when the component from Brosa's model (pre-neutron mass) and the components from conservation laws go in the same direction.

All of these works are based on partial evaluations of:

- mass, isobaric, isomeric (independent) yields,
- chain yields
- modeling of pre-neutron yields using Brosa's mode
- Wahl's systematics of charge polarization.

In all these studies, the weight of the model assumption is not necessarily clear.

1.2 New complete evaluation of ^{235}U(n$_{th}$,f) proposed to JEFF-4

In the recent work [15], we proposed a new consistent evaluation of independent and cumulative yields with minimal impact from models. This analysis is based on the following decomposition of independent fission yields:

$$Y(A,Z,m) = Y(A) \cdot P(Z|A) \cdot IR(A,Z,m) \quad (1)$$

Where $Y(A)$ is the mass yield (post-neutron), $P(Z|A)$ the charge distribution and $IR(A,Z,m)$ the probability populating the metastable of the $^{A}_{Z}X$ nucleus.

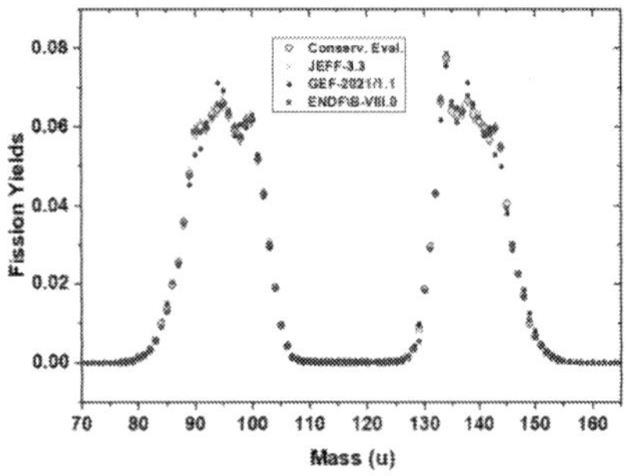

Fig. 10. Comparison of the conservative evaluation (see §2) of ^{235}U(n$_{th}$,f) [15] with the GEF model [14] and the JEFF-3.3 [12] and ENDF\B-VIII.0 evaluation [19].

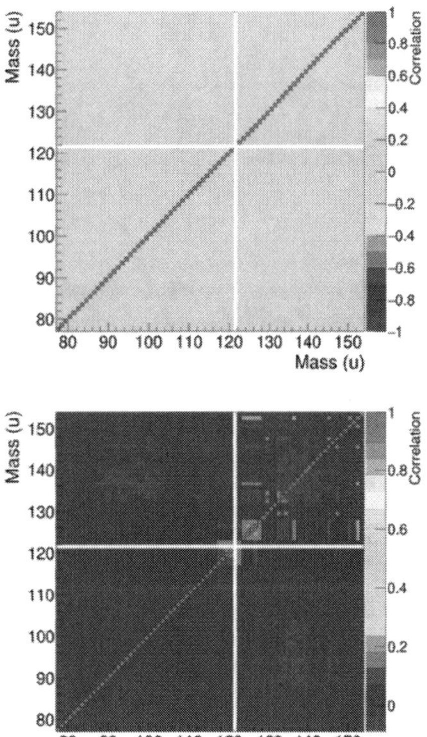

Fig. 11. (top) Mass yield correlations from of ^{235}U(n$_{th}$,f) [15] (bottom) zoom on the structures, principally on the heavy mass region.

For ^{235}U(n$_{th}$,f), the EXFOR database [16] is sufficiently complete to perform a mass yield evaluation without any model input. We obtained a new evaluation of mass yields which is in agreement to the JEFF-3.3 evaluation [12] (**Fig. 10**). The rigorous statistical analysis [15, 18] allows us to propose a synthesis of experimental data converging to quasi-null correlations of mass yields (see **Fig. 11 & §2**). Consistently to previous discussion (see §1.1.4), we observe a single diagonal pattern due to absence of mass number conservation law (at post-neutron stage) but a negative component due to the self-normalization step (**Fig. 11, bottom**).

For the isotopic yields, at this moment, we based our analysis on the JEFF-3.3 library considering this work as a synthesis of experimental isotopic rates completed by the GEF model in the tails of distributions. We apply the conservation laws and the self-normalization process in order to provide the correlation matrix of isotopic and isomeric distributions. As expected, the charge conservation law induces a cross pattern (no proton emission in thermal neutron induced fission). The self-normalization regulates the correlations in order to respect the sum rule:

$$Var[Y(A)] = \sum_{Z,m} Var\, Y(A,Z,m) + 2 \sum_{\substack{Z<Z' \\ m<m'}} Cov\big(Y(A,Z,m), Y(A,Z',m')\big) \quad (2)$$

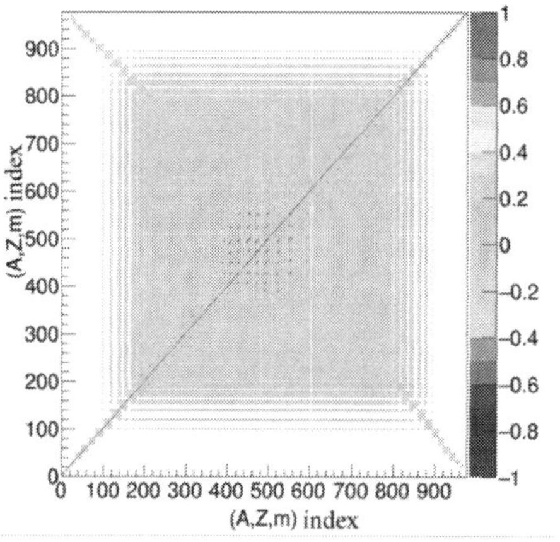

Fig. 12. Independent fission yield correlation matrix of ^{235}U(n$_{th}$,f) [15].

To take into account the chain yields in the mass yield analysis, we have to calculate the $C(A,Z,m)$ cumulative yields from the independent yields. These observables are calculated using the decay data defining the Q matrix:

$$C(A,Z,m) = \boldsymbol{Q}\, Y(A,Z,m) \quad (3)$$

With:

$$Q = (\boldsymbol{I} - \boldsymbol{B})^{-1} \quad (4)$$

Where $\boldsymbol{I}$ is the identity matrix and $\boldsymbol{B}$ is the matrix of decay corresponding to the list of (A,Z,m) isotopes. The $\boldsymbol{B}$ matrix is deduced from the JEFF-3.3 decay data file [12]. Thus, at the end of the process, we generate the evaluations of both independent and cumulative yields in a global process with complete and consistent correlation matrices (**Figs. 12 & 13**).

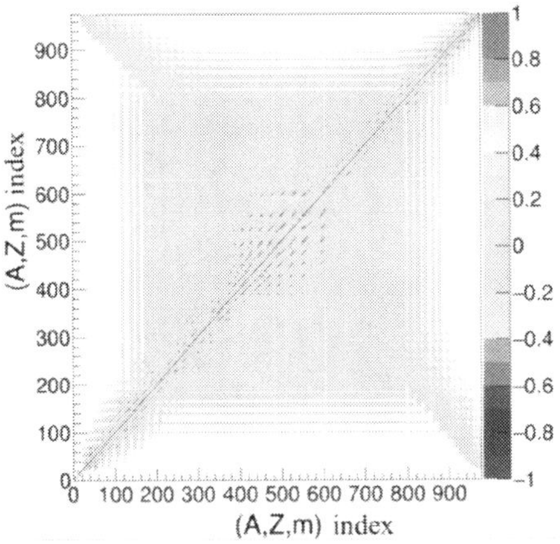

Fig. 13. Correlation matrix of cumulative yields of $^{235}U(n_{th},f)$ [15].

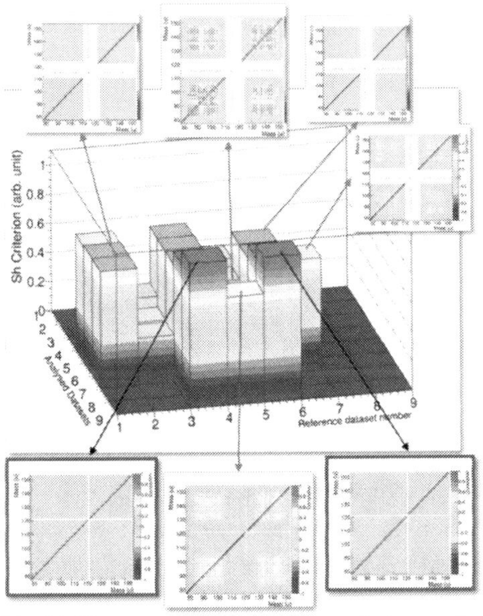

Fig. 14. Histogram of Shannon entropy criterion as a function of the reference dataset and the number of datasets used. The extrema are illustrated with different correlation matrices of evaluated mass yields (White bands correspond to the unevaluated mass yields).

2 Multiplicity of mass yield evaluations

The $^{235}U(n_{th},f)$ mass rate measurements are sufficiently complete to cover the fission product mass range (> 99.5%) to ensure the self-normalization of the fission yields. Nevertheless, the dataset provided by experimentalists are not in agreement and a cross-normalization of fission rates per mass is required [17]. We have shown that the use of one reference measurement to cross-normalized all the datasets (before absolute normalization) introduces different results according to the reference dataset. **Fig. 14** illustrates the dependence in the choice of reference dataset that maximizes or minimizes the information provided. The quantification of information is deduced from the Shannon entropy [18]

This result corresponds to non-ergodic analysis paths. The second limit of this process is that the tested mass rates are reduced to the overlapping mass rates (white cross). Thus, in order to solve this non-ergodic analysis, we have developed a new methodology.

2.1 Several analysis approaches

The test and the sorting of datasets are the key point to obtained mean values per mass representing the data selected. Considering that the experimental information is reduced to mean values and the standard deviations for the majority of datasets, we assume a Gaussian distribution for each mass yield (per experiment). Then, the maximum of likelihood method is equivalent to the minimum of the generalized least square χ_g^2.

The analysis is performed in two steps. First, we compute the mean value for each mass by using all the measurements. Then we test if the evaluation based on these mean values is compatible at 3σ with all the datasets used. If the test fails, a regularization is necessary. In this work, we have developed 4 different regularization. The two global ones are:

- Strict solutions: the sorting of mass measurements follows the main contributors χ_g^2 up to obtain an acceptable χ_g^2, i.e. with a $P - value > 1 - CL$ [20]. Different sorting of data are possible and we kept the two possible solutions, called "strict-1" and "strict-2";
- Conservative solutions: we apply the same statistical test after adding an uncertainty of 2.5% for all masses to obtain an acceptable $P - value$ (See [15]). It turns out that the measurements of the mass 153 have huge discrepancies. Therefore we decided to have two different evaluations with two values of the mass 153 with an additional uncertainty of 2.5%. The two solutions are called: "conservative 1" and "conservative 2";

The two local regularizations are:

- Conservative Student solution: for mass where measurements disagree, we consider additional uncertainty per mass based on Student law standard deviation.
  - Conservative Best Estimated (B.E.) solution: we add adjusted uncertainties per mass to obtain acceptable χ_g^2 test.

By definition, only the two last solutions preserve all the initial datasets. In the large majority of masses, the mean values are very insensitive to the regularization methods except for incoherent datasets (A= 153, see effect §3). Conversely, standard deviations of mean values are very sensitive to the different approaches. **Fig. 15** presents the relative uncertainties according to the regularization methods. In comparison to the results presented in [15], the two last regularizations are new additional calculations. We observe that the uncertainty of the Best Estimated (B.E.) method or the Student approach are very similar and included between the conservative and strict methods. These two local regularizations produce the most realistic solution considering all the data adjusted on the database.

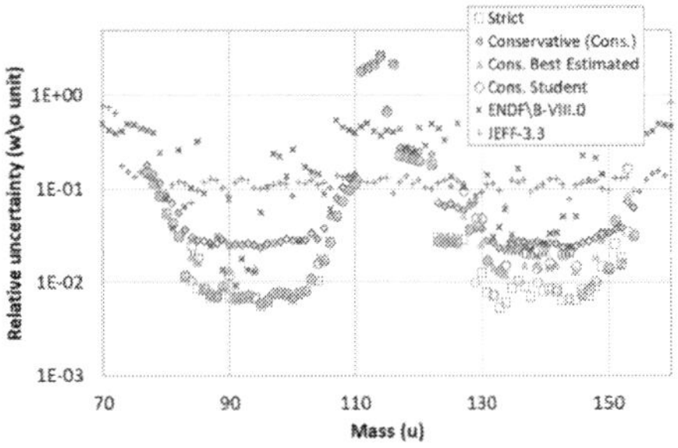

Fig. 15. Relative uncertainties of the different solutions as a function of mass in comparison to current evaluations.

2.2 Quantification of information provided by correlations

The output quantities of the evaluations are the mean values, the standard deviations and the correlations. We have seen that the mean values of different solutions are stable. Nevertheless, the standard deviations are dependent on the regularizations of datasets. In this part, we discuss the sensitivity of the correlations to the different evaluations.

Table 1. Shannon entropies associated to the different evaluations.

Regularizations	Shannon Entropy
Strict Eval.	6,222
Strict Eval. + Syst. 1%	6,228
Strict Eval. + Syst. 2%	6,233
Strict Eval. + Syst. 3%	6,236
Conservative Eval.	6,240
Cons. Eval. + Syst. 1%	6,238
Cons. Eval. + Syst. 2%	6,236
Cons. Eval. + Syst. 3%	6,233
Best estimated Eval.	6,222
Student Eval.	6,209

In our analysis tools, we are able to integrate the available experimental correlation matrices and to deduce the dependence of the evaluated correlation matrix to the lacks of experimental correlations, testing the effects of additional systematic uncertainties. We conclude that the evaluated uncertainties increase of 1% for 3% systematic uncertainty added. This effect is due to the cross-correlation between datasets in the renormalization process.

To compare the correlation matrices of all analyses, we have calculated the Shannon entropy associated to the different correlation matrices [18] by considering the previous kinds of regularizations. Table 1 presents the Shannon entropy calculations for the ten following cases. The information quantity distributed is very stable in contrary to the previous method using a reference dataset (**Table 1**).

2.3 Probability density functions (PDF) of fission rates

The evaluation of fission yields involves mixing of fission rates from different datasets assuming a Gaussian distribution for each measurement. The generalized least squares method (χ_g^2) is equivalent to the maximum likelihood method for Gaussian distributions associated to measurements. If there is no bias, this method leads to a minimal variance (Cramer Rao theorem) and the covariance matrix corresponds to the Fischer matrix associated to data [20].

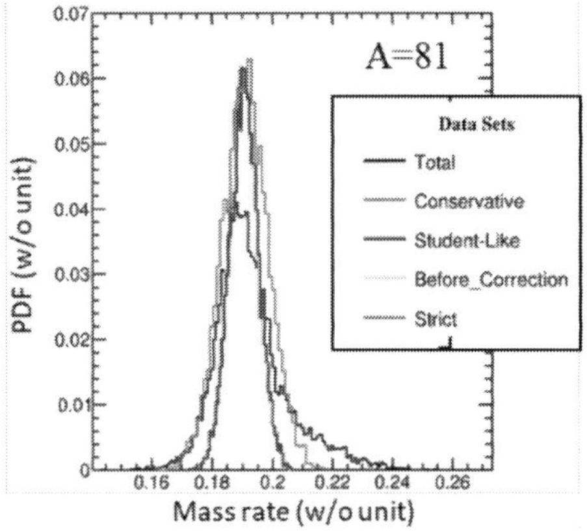

Fig. 16. PDFs of evaluated mass rates before absolute normalization for an acceptable $\chi_g^2(A)$ test.

Then, the mean value of data (maximum likelihood) is also distributed according to a Gaussian distribution. To test this property, we developed a Bayesian Monte Carlo (BMC) considering the different approaches and the empirical mean calculation to illustrate the PDFs of the evaluated values. **Fig. 16** presents the PDF for a regular dataset (A=81 u) and **Fig. 17** for an un-regular dataset (A=84 u). In the first case, the "Strict", "Student-like" and "without regularization" are similar. The empirical mean value is not the minimal variance estimator and the conservative assessment take into account a penalty to obtain an acceptable χ_g^2 test for all datasets. In the second case (**Fig. 17**), when the $\chi_g^2(A)$ test is not acceptable, we see clearly the bias in the mean values according to the regularization methods applied. In particular, we see that the mean value without any regularization underestimates the uncertainty for the dataset selected and presents a bias in comparison to the strict method. The Strict method applies a sorting on available data which generates two solutions for this mass A=84. The conservative method corresponds to a Gaussian reduction method for an incompatible initial dataset, hence the name used.

These BMC calculations illustrate that the most accurate assessment naturally corresponds to the selection of compatible data and leads to a Gaussian PDF associated to the mean rate.

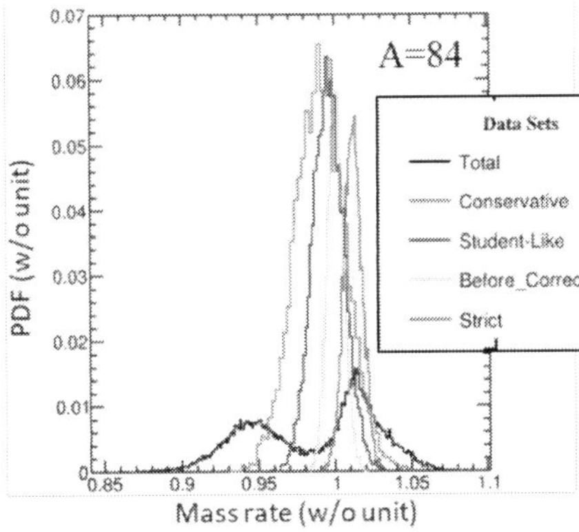

Fig. 17. PDF of evaluated mass rate before absolute normalization for a rejected $\chi_g^2(A)$ test.

2.4 Exclusion diagram

The comparison of the mean values of the different proposed evaluations is illustrated in **Figs. 18 & 19.** We see a general good agreement for the principal masses. Nevertheless, few mass yields of JEFF-3.3 and ENDF\B-VIII.0 evaluations are excluded for a confidence level of 99.7%. The limit of this plot is not to consider the correlation matrix. The P-values take into account the correlation matrices and have been presented in ref. [15].

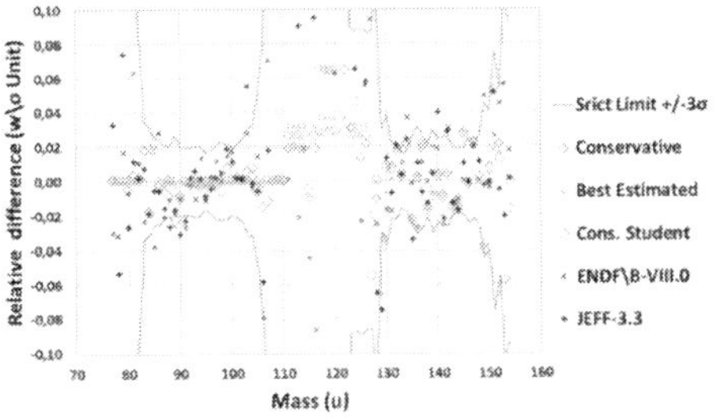

Fig. 18. Relative difference of available evaluations in comparison to the strict one. The green lines correspond to the limits for 99.7% confidence level.

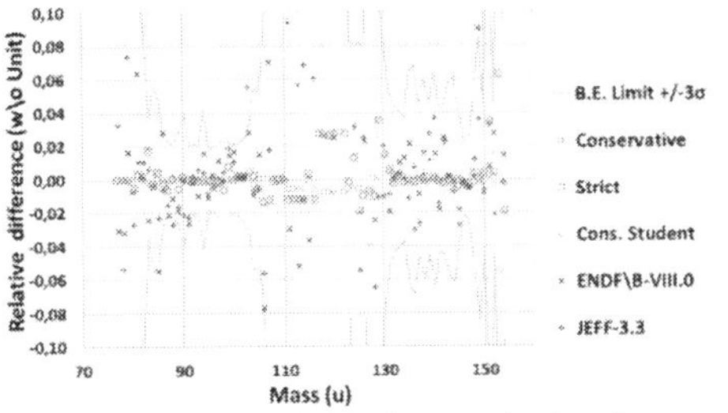

Fig. 19. Same plot as the previous one but in reference to the B.E. evaluation. The yellow lines correspond to the limits for 99.7% confidence level.

3 Cumulative yields

In this analysis, we determined the mass yields considering the experimental data on fission products and chain yields in the same analysis process (see § 1.2 and [15]). Thus, the correlation matrix of chain yields is coherent with those of independent yields.

For the conservative analysis, we obtain a limit of precision of 2.5 % in the best cases compared to about 1% accuracy in JEFF-3.3 (**Fig. 20**). Note that the independent yields and the chain yields of JEFF-3.3 are two different evaluations whereas in our analysis both observables are extracted from a single analysis. **Fig. 21** presents the relative uncertainties of the chain yields for the strict evaluation. We find the same precision level as JEFF-3.3. The process used in this work allows us to claim that the strict evaluation is the best estimation of the mass yields assuming no bias in the chain yields. **Figs. 22 & 23** present selected chain yields for current evaluations. The mass A=153 illustrates the impact of the sorting of data to obtain an acceptable statistical compatibility test of datasets. For this mass, two groups of two measurements exist. The recursive process of renormalization, test, sorting of data, renormalization, test (etc.) according to the major contributor of χ_g^2 generates several compatible data groups providing the different evaluations.

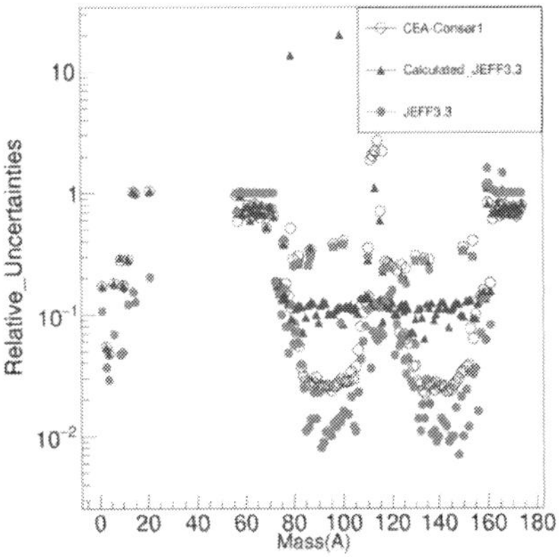

Fig. 20. Relative uncertainties of chain yields as a function of mass for the "conservative 1" evaluation.

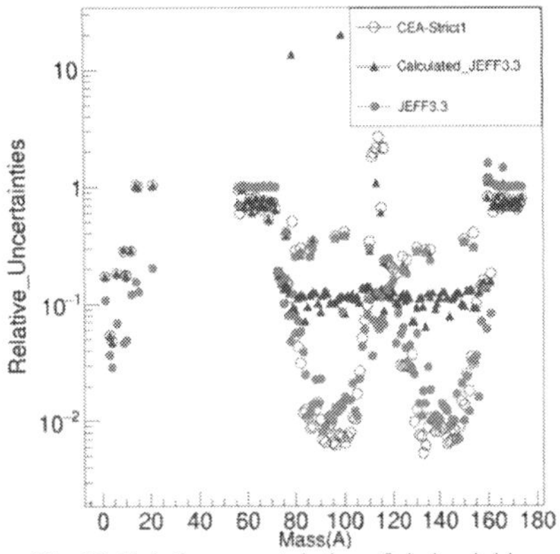

Fig. 21. Relative uncertainties of chain yields as a function of mass for the "Strict 1" evaluation.

4 Impact on reactivity calculations

A first calculation of the impact of the new fission yield evaluations concerns the reactivity effect on pin fuel calculations. Calculations are achieved with the Apollo2 [21, 22] code from CEA. Table 2 synthesises the reactivity calculations for two different burn up values. **Figs. 24 & 25** show the main contributors to the variation of reactivity in reference to the calculation with JEFF-3.3. These slight variations are due to the compatibility of our evaluations with the JEFF-3.3 one.

Table 2. Variation of reactivity in reference to JEFF-3.3 calculations.

	Δρ [pcm]	
Burn up (GWd/t)	1.5	50
CEA_Strict-1 /JEFF-3.3 -1	-58	24
CEA_Conservative-1 / JEFF-3.3 -1	-8	9

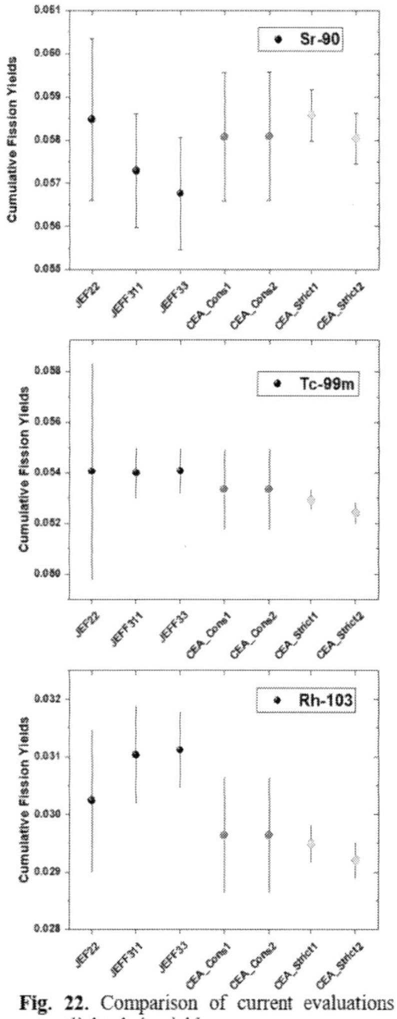

Fig. 22. Comparison of current evaluations for three important light chain yields.

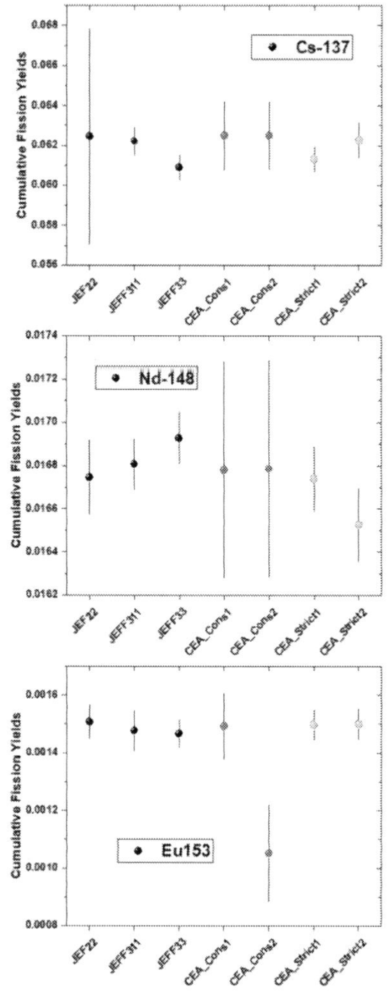

Fig. 23. Comparison of current evaluations for three important heavy chain yields.

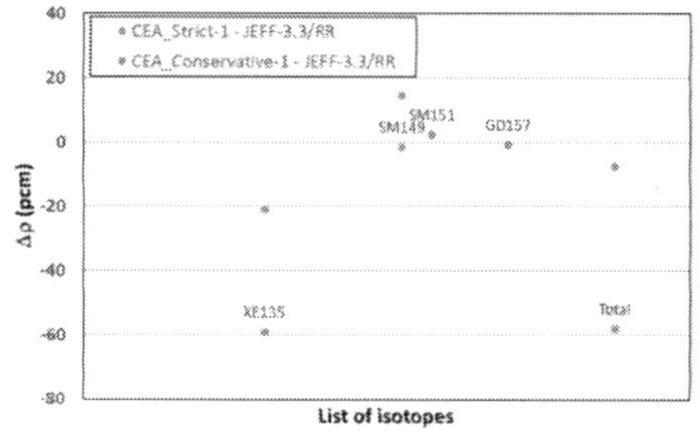

Fig. 24. (Preliminary results) Variation of reactivity for the "Strict 1" and "Conservative 1" FY evaluations in reference of the JEFF-3.3 for a burn up of 1.5 GWd/t.

https://doi.org/10.1051/epjconf/202328100022

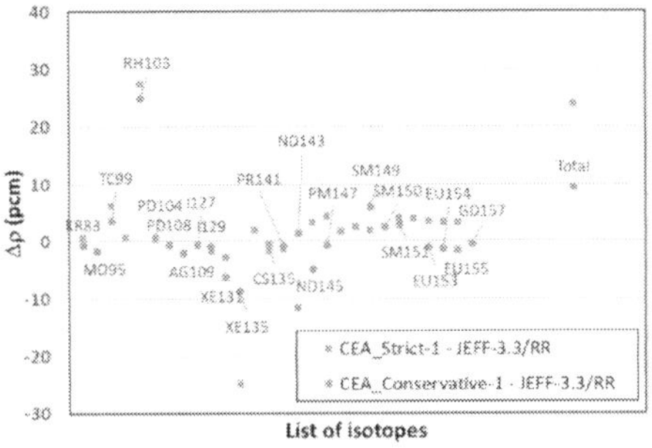

Fig. 25. (Preliminary results) Variation of reactivity for the "Strict 1" and "Conservative 1" FY evaluations in reference of the JEFF-3.3 for a burn up of 50 GWd/t.

5 Conclusion and Perspectives

The proposed analysis method of several fission observables allows us to determine a consistent evaluation of the independent and cumulative yields.
In this global approach, we conclude that the sorting of the data is an essential milestone to reach the best estimation of fission yields, limiting the bias and minimizing the uncertainties. If there is no bias in the very accurate measurements of chain yields, then the consistency of our work with the JEFF-3.3 library allows us to conclude that the strict evaluation corresponds to the best assessment of the $^{235}U(n_{th},f)$ mass yields. Nevertheless, this force us to exclude a part of the available measurements. On the other hand, to limit the potential bias of evaluation, the conservative method proposed a synthesis of the actual experimental knowledge taking into account the inconsistency of the datasets used. These two evaluations represent the upper and lower limits on the precision on these observables.

Acknowledgements
This work was supported by "le défi NEEDS" and the French I3P program.

References

1. U. Brosa, S. Grossmann and A. Müller, Nuclear Scission II Phys, vol. 197, pp. 167-262, 1990.
2. IAEA, «CRP 1991-1996» IAEA TECDOC-1168, 2000.
3. D. Madland and T. England, Nuclear Science Engineering, 64:859–865, 1977
4. C. Devillers, «The importance of fission product nuclear data in reactor design and operation,» IAEA Panel, Petten, 1977
5. K. Tsubakihara et al., J. Nucl. Sci and Technol., 2021, Vol. 58, No. 2, 151–165
6. L. Fiorito et al., Annals of Nuclear Energy, vol. 69, pp. 331-343, 2014.
7. L. Fiorito et al., Annals of Nuclear Energy, vol. 88, pp. 12-23, 2016.
8. T. Kawano and M. Chadwick, Journal of Nuclear Science and Technology, vol. 50, n° 110, pp. 1034-1042, 2013.
9. M. Pigni, M. Francis and I. Gauld, Nuclear Data Sheets, vol. 123, pp. 231-236, 2015.
10. N. Terranova, PhD thesis, Bologne University, Italia, 2016.
11. N. Terranova, O. Serot, P. Archier, C. De Saint Jean and M. Sumini, Annals of Nuclear Energy, vol. 109, pp. 469-489, 2017.
12. A. J. M. Plompen et al., Eur. Phys. J. A (2020) 56:181
13. A. Chebbouhi et al. Eur. Phys. J. A (2021) 57:335G.
14. K. H. Schmidt, B. Jurado, C. Amouroux and C. Schmitt, Nuclear Data Sheets, vol. 131, pp. 107-221, 2016.
15. S-M Cheikh, G. Kessedjian et al., ND2022, 15th International Conference on Nuclear Data for Science & Technology, Sacramento, California (2022), to be published.
16. EXFOR Database, https://www-nds.iaea.org/exfor/
17. Brieuc Voirin, Grégoire Kessedjian et al., EPJ Nuclear Sci. Technol. 4, 26 (2018)
18. G.Kessedjian, S-M Cheikh et al., EPJ Web of Conferences 242, 05001 (2020)
19. D.A.Brown et al., NDS, Vol. 148, February 2018, Pages 1-142
20. D.L. Smith, Probability, Statistics and Data Uncertainties in Nuclear Science and Technology, American, Nuclear Society, LaGrange Park, IL, USA (1991)
21. A. Santamarina et al., Advances in Nuclear Fuel Management IV (ANFM 2009), Hilton Head Island, South Carolina, USA, April 12-15, 2009,
22. D. Bernard et al., JEFDoc-2113, https://www.oecd-nea.org/dbdata/nds_jefdoc/jefdoc-2113.pdf

Test of models for $^{235}U(n_{th},f)$ charge distributions and their impacts on the covariance analysis

S.M. Cheikh[1], *G. Kessedjian*[1], *O. Serot*[1], *A. Chebboubi*[1] *and O. Litaize*[1],

[1]CEA, DES, IRESNE, DER, SPRC, LEPh, Cadarache center, F-13108 Saint Paul lez Durance, France

Abstract. Fission yields are of major interest for the nuclear industry and the study of the fission process. Fission yields are determined by evaluation process based on the analysis of experimental datasets and completed by phenomenological models. The aim is to provide the best estimation of independent yields of $^{235}U(n_{th},f)$ reaction. In this paper, we will focus on the methodology of assessment of the nuclear charge distribution per mass. In the JEFF-3.1.1, the Wahl's parametrization of the Z_p-model is used to complete the experimental knowledge. Nevertheless, the lack of this parametrization consists in the absence of uncertainty information. Consequently, it is impossible to generate the components of independent yield correlation matrix from the Z_p-model. In this context, we propose to use the Z_p model with a new parametrization to assess the charge distributions. Two different calculations have been developed: the first one is based on a Wahl-like linear approximation of the parameters as a function of mass. The second approach consists to apply a "direct-Z_p model" without linear assumptions. This analysis is achieved using the Markov-Chain Monte-Carlo (MCMC) method to adjust the model parameters and then to deduce the covariance of charge distributions. Moreover, with our development, we have directly and consistently described the pre-neutron and post-neutron charge distributions.

1 Introduction

The exploitation of nuclear reactors dedicated to the production of electricity by chain reactions requires the control of a large number of physical quantities, throughout the fuel cycle. This control is the basis of the operation and safety of the installation and is partly based on nuclear data. In this sense, several research projects in different fields are underway, in particular work aimed to improving the quality of nuclear data. Among these data, there are the fission yields, which represent the production rates of fission products. These data are a major interest for the nuclear industry and the study of the fission process. They are used in the calculation and quantification of several observables, in particular for the fuel inventory [1], the burn-up calculation [2] or still for the reactor decay heat after its shutdown.

Historically, very important efforts have been made to understand the fission process in a coherent theoretical framework [3]. Despite these efforts, the current understanding of fission process is patchy and incomplete, and then a consistent description of the fission observables (pre-neutron yields, post-neutron yields, cumulative yields, neutron emission) is not available. In particular, this consistency between pre and post neutron yields is the base of the description of the fission yields, which are very important for the applications and the test of the fission models. It is clear that the data resulting from those models are not sufficiently accurate to satisfy the requirements of the nuclear industry. However, it is undeniable that the best estimation of fission yields is provided by current evaluation process based on the analysis of experimental data completed by phenomenological models. Thus, the aim is to provide the best estimation of isobaric, isotopic and isomeric yields.

These phenomenological models such as Wahl's model [4] and Madland-England's one [5] have shown great success. They have been used for the development of several libraries of evaluations such as the JEFF-3.1.1 [6], ENDF/B-VIII.0 [7] and JENDL/FPY-2011 [8] libraries. In those libraries, the charge distribution of independent yields are completed using the empirical Z_p-model proposed by A. C. Wahl based on linear parameterization known as Wahl's systematics [4]. It is important to note that the Z_p-model is initially used to describe the fractional yields of pre-emission neutron and can be translated to the post-neutron using the prompt neutron multiplicity (so-called saw-tooth). But in libraries, it was used directly to calculate post-neutron charge distribution. This represents a limitation in those evaluated libraries. In addition, a limitation of Wahl's systematics parameterization is mainly related to the lack of information on the uncertainties and the correlations of the model parameters. Consequently, it is impossible to generate the component of correlation matrix of charge distribution from the Z_p model. In this context, we propose to use the Z_p model with new parameterizations to complete the charge distributions. The aim is to test and to improve the accuracy of the methodology in order to generate the covariance matrix of independent yields.

2 Description of Zp-model

The Z_p model consists in the modelling of the pre-neutron fractional yields with a Gaussian distribution of mean $Z_p(A^*)$ (representing the most probable charge) and the standard deviation $\sigma_Z(A^*)$. Two effects perturb the shape of the Gaussian distribution. The first is the charge polarization effect [4], which appears as a shift of $Z_p(A^*)$ from the charge Z_{UCD} based upon the assumption of the "unchanged charge density (UCD)" [4]. The UCD postulates that the charge-to-mass ratio of the fissioning nucleus, Z_F/A_F, is maintained for both fragments. The charge polarization effect is taken into account by adding a corrective term $\Delta Z(A^*) = Z_p(A^*) - Z_{UCD}$. The second is the pairing effect, which is related to the structure of the nucleus. To take into account this parity effect, A. C. Wahl propose two parameters, $F_Z(A^*)$ and $F_N(A^*)$ [4].
The pre-neutron emission charge distribution is therefore written:

$$P(Z|A^*) = \frac{F(A^*,Z)N(A^*)}{2}[erf(V) - erf(W)] \quad (1)$$

With:

$$V = \frac{Z - Z_p(A^*) + 0.5}{\sigma_Z(A^*)\sqrt{2}}$$

$$W = \frac{Z - Z_p(A^*) - 0.5}{\sigma_Z(A^*)\sqrt{2}}$$

Where A^* represents the pre-neutron mass of the fission fragment, $F(A^*, Z)$ is the term to include the pairing effect and $N(A^*)$ is the normalization coefficient. To complete the description of the Z_p-model, A. C. Wahl proposed a parameterization to determine $\Delta Z(A^*)$, $\sigma_Z(A^*)$, $F_Z(A^*)$ and $F_N(A^*)$ known as Wahl's systematics.

In the following, we will test the compatibility between the values from Wahl's systematics and the experimental data. We will also introduce two approaches to propose new parameterizations for the Z_p model. Then, a comparison of different types of calculations will be presented and the results on the pre-neutron charge distribution and post-neutron charge distribution will be discussed.

3 Evaluation method

3.1 Pre and Post-neutron transition

The Z_p model is developed to determine the charge distribution of the fission fragments before prompt neutron emission. To determine the parameters of this model, we need to adjust those parameters on the available experimental data. However, these data corresponds to fission products after prompt neutron emission. It is therefore necessary to formulate a connection between the pre-neutron and post-neutron data. This connection is established through neutron emission probabilities. Indeed, we know that a given nucleus of mass A and charge Z is formed either by cold fission (zero neutron emission) or by several emissions of neutrons as indicated in the **Fig. 1**. We can therefore calculate the post-neutron fission yields from the pre-neutron fission yields using the formula:

$$N(A,Z) = \sum_{\nu=0}^{\infty} Y(A^*)P(Z|A^*|\theta_{zp})P(\nu|A^*) \quad (2)$$

$$P(Z|A) = \frac{N(A,Z)}{\sum_Z N(A,Z)} \quad (3)$$

With:

$$A^* = A + \nu$$

Where A^* is the mass of fission fragment before neutron emission, A the mass of fission products, $P(\nu|A^*)$ represents the probability that a primary fission fragment of mass A^* emitting ν prompt neutrons, $Y(A^*)$ represents the mass yields, $P(Z|A^*|\theta_{zp})$ is the pre-neutron charge distribution, θ_{zp} is the parameters of Z_p model and P(Z|A) is the post-neutron charge distribution.

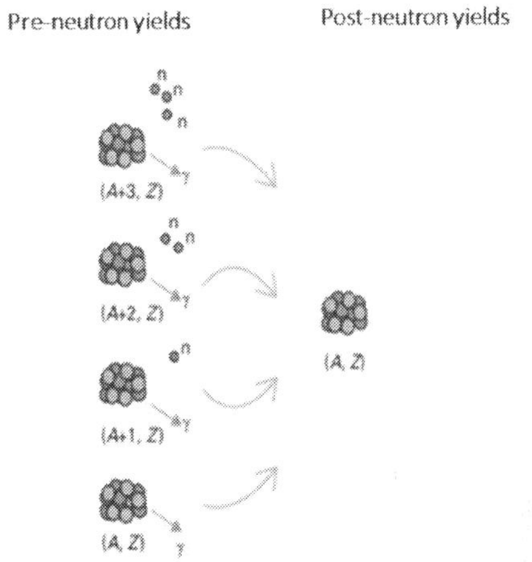

Fig. 1. Transition between pre-neutron data and post-neutron data [9].

For this work the mass yields $Y(A^*)$ are given by S. Zeynalov et al. [10], the $P(\nu|A^*)$ used is provided by the FIFRELIN code [11] and $P(Z|A^*)$ is determined by Z_p model. In the following, we will propose two approaches to provide the parameterization, but we limit our analysis to the calculation of $\Delta Z(A^*)$, $\sigma_Z(A^*)$. For $F_Z(A^*)$ and $F_N(A^*)$ we will use the Wahl's systematics for this first study. The objective of our analysis is therefore to adjust $\Delta Z(A^*)$, $\sigma_Z(A^*)$ to reproduce the post-neutron data (JEFF-3.1.1 or experimental data) as shown in **Fig. 2**.

FIFRELIN [11] is a Monte-Carlo code that simulates the de-excitation process of fission fragments and provides fission observables. In particular, we obtain the probability that a primary fission fragment of mass A^* emits ν prompt neutrons, prompt neutron spectrum, prompt gamma spectrum, neutron multiplicity etc. To determine these observables FIFRELIN needs a number of input data, such as the mass yields Y(A*), kinetic energy and spin distributions of the fission fragments. In the FIFRELIN calculations, we use the mass yields and kinetic energy distributions measured by S. Zeynalov et al. [10] to be consistent in our analysis.

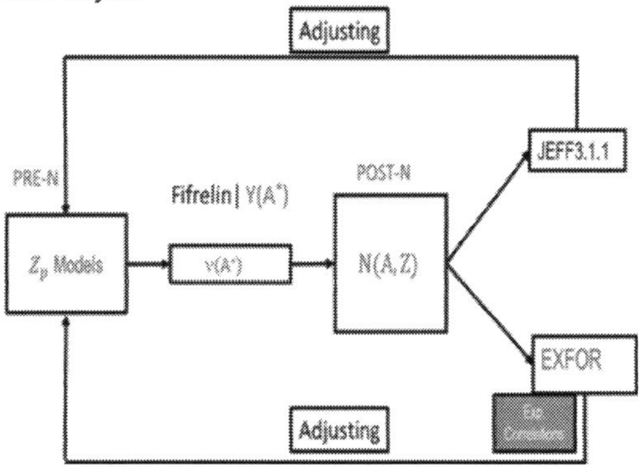

Fig. 2. The scheme for calculating and adjusting the Zp model parameters.

3.2 Adjustment of parameters

3.2.1 Linear parametrization (Wahl-like)

The principle of Wahl-like parameterization is to use the same assumptions as A.C. Wahl [4] i.e. that $\Delta Z(A^*)$ and $\sigma_Z(A^*)$ are linear in the pre-neutron mass region and then adjust the parameters of the lines to reproduce the post-neutron data. To determine $\Delta Z(A^*)$ for a mass region of [77-159] we need three parameters $\{\alpha_0, \alpha_1, \alpha_2\}$:

$$\begin{cases} \Delta Z(A^*) = \alpha_0 A^* + \alpha_1; \quad A^* \in [77-106] \\ \Delta Z(A^*) = \alpha_2(A^* - 106) + \Delta Z(106); \; [106-111] \end{cases} \tag{4}$$

By symmetry of the charge polarization, we determine the parameters in the complementary interval of [77-111] (i.e. [125-159]) and by continuity we determine the parameters in the interval of [111-125]. In a similar way we will use four parameters $\{\beta_0, \beta_1, \beta_2, \beta_3\}$ to determine $\sigma_Z(A^*)$:

$$\begin{cases} \sigma_Z(A^*) = \beta_0 A^* + \beta_1; \quad A^* \in [77,106] \\ \sigma_Z(A^*) = \beta_2; \; A^* \in [106,111] \\ \sigma_Z(A^*) = \beta_3; \; A^* \in [111,125] \end{cases} \tag{5}$$

By symmetry of $\sigma_Z(A^*)$, we calculate the parameters in the complementary interval of [77-111] (i.e. [125-159]).

To determine the best parameterization we will use an analysis based on a Classical Bayesian Monte-Carlo approach [12] :

$$p^{post}(\vec{\theta}|\vec{D}) = \frac{\mathcal{L}(\vec{D}|\vec{\theta})P(\vec{\theta})}{\int \mathcal{L}(\vec{D}|\vec{\theta})P(\vec{\theta})d\vec{\theta}} \tag{6}$$

$\vec{\theta} = \{\alpha_0, \alpha_1, \alpha_2, \beta_0, \beta_1, \beta_2, \beta_3\}$ is the parameter vector, $\vec{D}$ represents the data. The adjustment of the parameters is done with a classical Monte Carlo method by sampling the model parameters according to a prior distribution $P(\vec{\theta})$ (in our case a uniform distribution). The result is the posterior distribution of parameters given data used.

For the $\vec{D}$ experimental set associated with a given experimental covariance C, the likelihood function can be chosen as a multivariate Gaussian:

$$\mathcal{L}(\vec{D}|\vec{\theta}) = e^{\frac{-1}{2}(\vec{D}-\vec{t})^T C^{-1}(\vec{D}-\vec{t})}$$

Here, $\vec{t}$ represents the model predictions, which are functions of the model parameters $\vec{\theta}$.

Then we calculate the average values of $\vec{\theta}$ and the associate covariance matrix according to the following equations:

$$\langle\vec{\theta}\rangle = \sum_{i=1}^{n} \vec{\theta}_i * p^{post}(\vec{\theta}_i|\vec{D}) \tag{7}$$

$$C(\langle\vec{\theta}\rangle_k; \langle\vec{\theta}\rangle_l) = \sum_{i=1}^{n} [\vec{\theta}_i^k - \langle\vec{\theta}\rangle_k][\vec{\theta}_i^l - \langle\vec{\theta}\rangle_l] * p^{post}(\vec{\theta}_i|\vec{D}) \tag{8}$$

3.2.2 Direct-Z_p parametrization

The direct-Z_p approach consists in adjusting the parameters of the model without using any assumptions. This allows the suppression of the linear constraints on mass regions in the adjustment of free parameters. On the other hand, the complexity of implementing this approach is due to the very high number of parameters that must be adjusted. For example, in the mass region [77-159] the number of masses is 83. Each mass is represented by two parameters, which makes a total of (83*2=166) parameters. Thanks to the symmetry between the light and the heavy fragments, the mass interval is limited to [77-118] (42 masses) and the parameter number is reduced to 84. It is therefore difficult to use a classical Monte Carlo approach. To overcome this complexity, we used an adjustment based on Markov Chain Monte Carlo (MCMC) method. An important question about MCMC algorithms is how quickly they will converge to their stationary distribution (posterior distribution). In our case, using a Metropolis-algorithm [13], we were quickly confronted by the difficulty of convergence of the algorithm. The solution of this problem comes with the use of Weighted Likelihood Bootstrap [14]. The principle of this approach is relatively simple; it consists in sampling the experimental data. For each sample, we calculate with a Simulated Annealing algorithm [15] the vector of parameters guaranteeing a maximum likelihood between the sample and the data calculated from the model. This allows us to have the vector of parameters and the value of the corresponding likelihood function as shown in **Fig. 3**. At the end of this procedure, we calculate the posteriori distribution of parameters. This allows us to calculate the mean vector of the parameters and the associated covariance matrix based on equations (7) and (8).

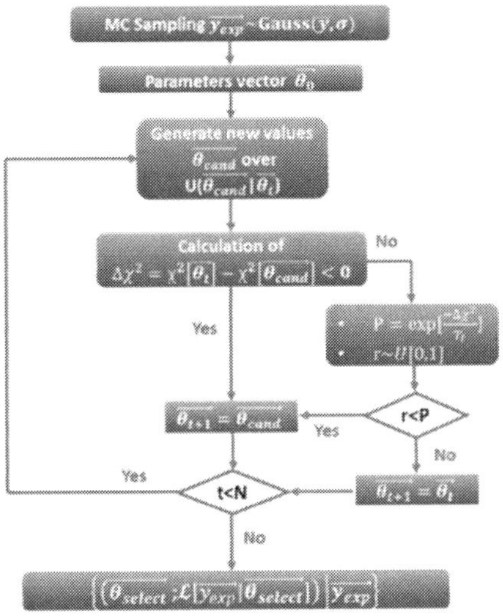

Fig. 3. Schematic diagram of the Bayesian Inference with the Bootstrap method and Simulated Annealing algorithm. The $\overrightarrow{y_{exp}}$ represents the sample of experimental data, $\overrightarrow{\theta_{cand}}$ represents the candidate parameter vector sampling according to a uniform distribution $U(\overrightarrow{\theta_{cand}}|\overrightarrow{\theta_t})$, $\overrightarrow{\theta_{select}}$ the selected parameter vector and N is the number of iterations.

4 Results

4.1 Adjustment based on JEFF-3.1.1

The adjustment of the parameters was carried out on the JEFF-3.1.1 library for ^{235}U(n$_{th}$, f). The JEFF-3.1.1 data were obtained from the evaluation process. They represent a synthesis of the experimental data completed by Wahl's model. However, we know that at the maximum only three or four charges Z per mass are experimentally determined. Then, we assume that the isotopes (A,Z) with probability P(Z|A) (defined in equation (3)) lower than an absolute value of 0.02 comes from the model. Therefore, we eliminate every isotope whose P(Z|A)<0.02. This allows us to avoid as much as possible using data from the Wahl's model in the JEFF-3.1.1 [6].

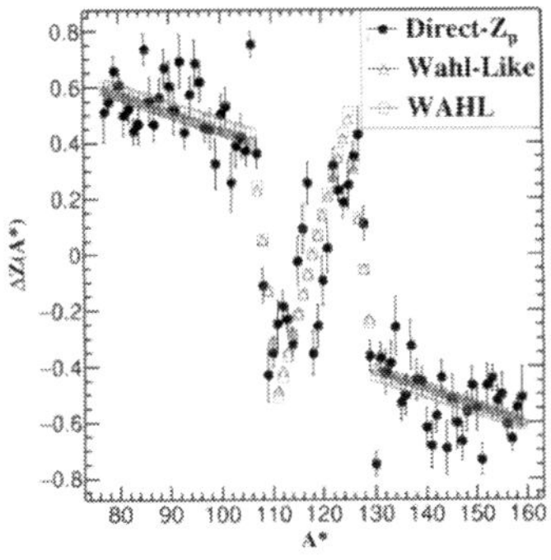

Fig. 4. Parameterization of the charge polarization based on Wahl's systematics (in red), Wahl-like approach with uncertainties (in blue) and direct-Z_p approach (in black).

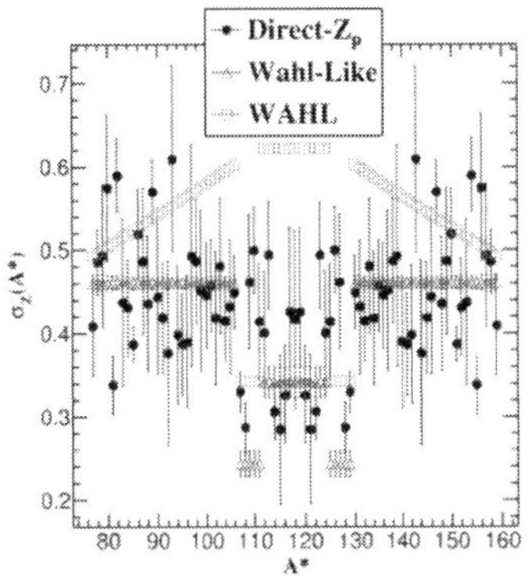

Fig. 5. Parameterization of $\sigma_Z(A^*)$ based on Wahl's systematics (in red), Wahl-like approach with uncertainties (in blue) and direct-Z_p approach (in black).

Fig. 4 and **Fig. 5** are comparing the Zp model parameters, $\Delta Z(A^*)$ and $\sigma_Z(A^*)$ as a function of fragment mass A^* obtained from the Wahl-like and Direct-Zp methods compared with the original Wahl's parameterization. We note that the charge polarization is very similar to those proposed by A. C. Wahl as indicated (**Fig. 4**). Nevertheless, the $\sigma_Z(A^*)$ trends are very different between Wahl-like and Wahl (**Fig. 5**) due to the fact that we determine directly the pre-neutron parameters in our calculation. Indeed, considering the neutron emission per mass (so-called saw-tooth) the charge distribution per (post-neutron emission) mass A is adjusted by several pre-neutron charge distributions at different pre-neutron masses ($A_0^* = A + 0$, $A_1^* = A + 1$, $A_2^* = A + 2$, $A_3^* = A + 3$...). The difference on the $\sigma_Z(A^*)$ trend for both calculations comes from the difference with de Wahl's methodology which determine the Zp parameters on the post-neutron charge distribution and apply a correction to determine the pre-neutrons polarization and its width as a function of pre-neutron mass (A^*).

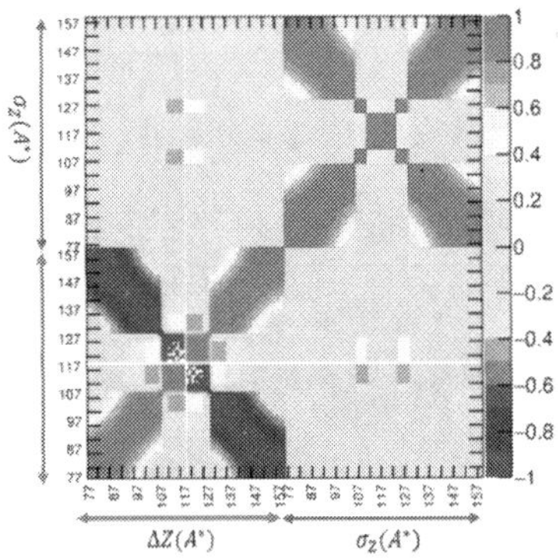

Fig. 6. The correlation matrix of the Wahl-like parameters.

Moreover, it is important to note that Wahl-like approach generates low uncertainties and extremely high correlations of the parameters. This is the result of the linearity assumption, which constrains the parameters and induces strong correlations (**Fig. 6**) and low uncertainties. The propagation of uncertainties and correlations of the parameters allows us to generate the correlation matrix of the pre-neutron charge distribution based on the equation (1) and illustrated in **Fig. 7** (left).

The calculation of the correlation matrix of the charge distribution allows us to generate the correlation matrix of the post-neutron charge distribution based on the equations (2-3) and presented **Fig. 8** (left). Note that the step of self-normalization and propagation of uncertainties of conservation laws to generate the absolute post-neutron charge distribution and the associated covariance matrix is not applied here. Then, **Fig. 8** (left) illustrates only the correlation component from Zp model used. For more details, see [16] and [17].

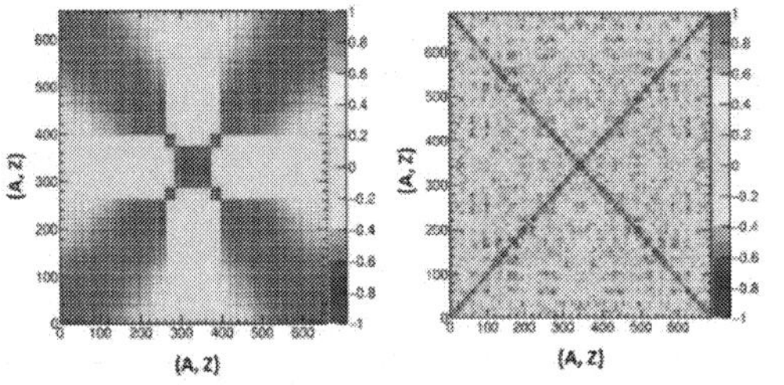

Fig. 7. The correlation matrices of the pre-neutron charge distribution based on Wahl-like approach (left) and direct-Z_p approach (right). The axes represent the isotope indexes.

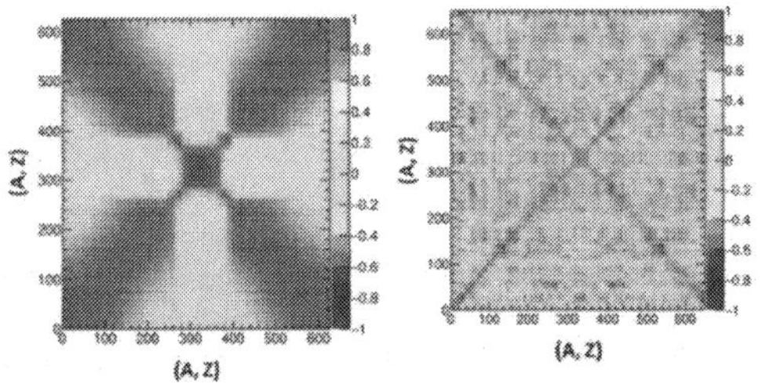

Fig. 8. The correlation matrices of post-neutron fission yields based on Wahl-like approach (left) and direct-Zp approach (right). The axes represent the isotope indexes.

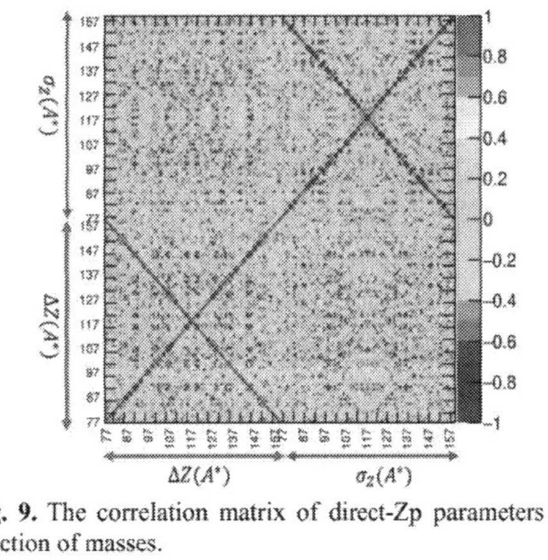

Fig. 9. The correlation matrix of direct-Zp parameters as a function of masses.

Similarly to the previous approach, we adjust the direct-Z_p parameterization to reproduce the major isotopes per mass of JEFF-3.1.1 data. The free adjustment reduces the correlation values between the parameters (**Fig. 9**) compared to Wahl-like parametrizations (**Fig. 6**), but it increases the uncertainties of these parameters as shown in **Fig. 4** and **Fig. 5**.

Based on the equation (1), we propagate the parameter uncertainties to generate the correlation of the pre-neutron charge distributions **Fig. 7** (right) and those of the post-neutron charge distribution **Fig. 8** (right) using the equation (2-3). The comparison of these distributions between calculations from Wahl's systematics and direct-Z_p approach shows that the Direct Z_p parameterization proposed better reproduces the data from JEFF-3.1.1 as illustrated in **Fig. 10**.

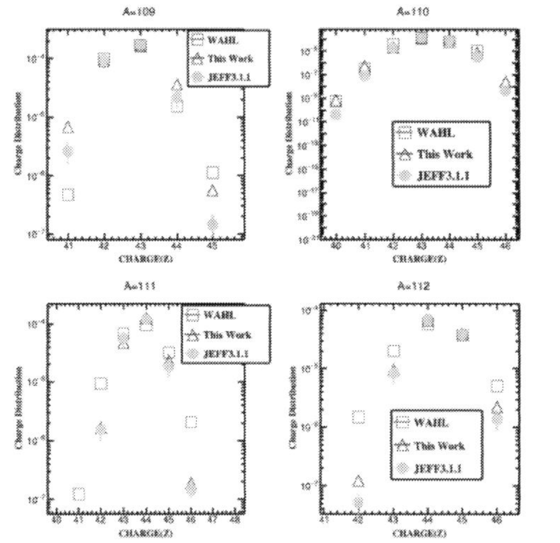

Fig. 10. The comparison of the charge distribution calculated using the Wahl's systematic (in red) and direct-Zp parameterization (in blue) with the JEFF-3.1.1 evaluation (in green) for the masses 109, 110, 111, 112.

A complementary comparison of the evaluations consists in the calculation of the empirical mean charges and the standard deviations of the post-neutron charge distributions according to the following equations:

$$\overline{Z(A)} = \sum_Z Z * \frac{N(A,Z)}{\sum_Z N(A,Z)}$$

$$\Delta Z(A) = \overline{Z(A)} - Z_{UCD}$$

$$\sigma_Z(A) = \sum_Z (Z - \overline{Z(A)})^2 * \frac{N(A,Z)}{\sum_Z N(A,Z)}$$

We compare the empirical mean values and standard deviations for the post-neutron charge distributions coming from the direct-Z_p approach. **Fig. 11** and **Fig. 12** show the evolution of the empirical mean values and the standard deviations as a function of the fission product mass obtained from the Direct-Z_p approach compared with JEFF-3.1.1. We observe a good agreement between both calculations.

Another check of consistency of the analysis is given by the generalized χ_g^2 statistical test. We obtain that the direct-Z_p approach generates a consistent evaluation (P-value > 0.003, for 99.7% confidence level) of data unlike the Wahl's systematic and Wahl-like data as indicated in **Table 1**.

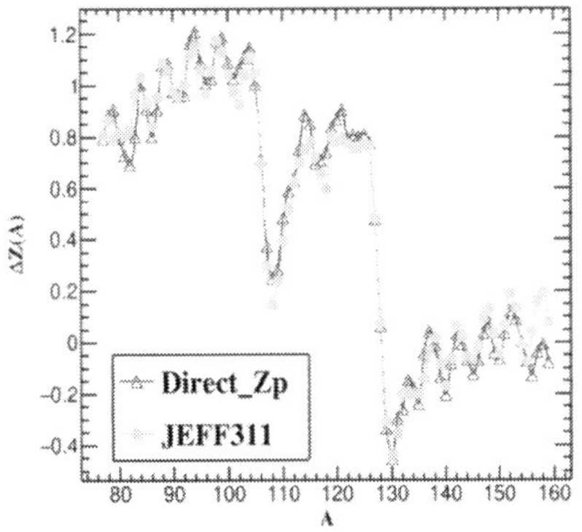

Fig. 11. Empirical calculation of post-neutron charge polarization from the JEFF-3.1.1 data (green) compared to the same calculation from charge distribution using the direct-Zp model (blue).

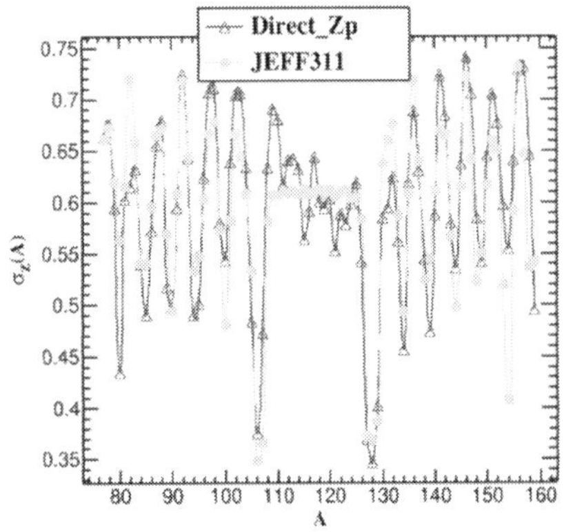

Fig. 12. Empirical standard deviation of post-neutron charge distributions calculated from the JEFF-3.1.1 library (green) compared to those from the direct-Zp approach (blue).

Table 1. Statistical test of charge distribution according to the different Zp parametrizations.

Approach	Wahl	Wahl-like	Direct Z_p
χ^2	1024	457.7	150.6
DOF	252	245	168
P-VALUE	10^{-93}	$9*10^{-15}$	0.83

4.2 Adjustment based on Lang's data

In this section, we will use the Direct-Z_p approach in order to compare the results of the adjustment of the parameters based on JEFF-3.1.1 [6] and on experimental data of W. Lang et al. [18]. Unlike the JEFF-3.1.1 data which are derived from the evaluation process. Lang's set are derived directly from the experimental process and have much smaller uncertainties, it covers a mass region between [80,110]. The **Fig. 13** and **Fig. 14** are comparing the results of adjusting the Z_p model parameters ($\Delta Z(A^*)$ and $\sigma_Z(A^*)$) using two different sources of data: JEFF-3.1.1 library and Lang's data. We observe many deviations, which show that a combination of all available experimental data is requested to assess the uncertainties on experimental knowledge. Moreover, the Lang's dataset remains limited and incomplete in the mass range of interest. In the near future, we plan to build a much larger experimental database containing an experiment data synthesis in order to use our approach on this database. This will allow us to improve the quality of the Z_p model parameters.

Nevertheless, we observe an important decrease of uncertainties of parameters with only one dataset. Thus, it will be interesting to introduce a similar analysis than mass yields (see [16] and [17]) of all available datasets to extract the experimental average dataset with realistic experimental standard deviations and correlation matrix.

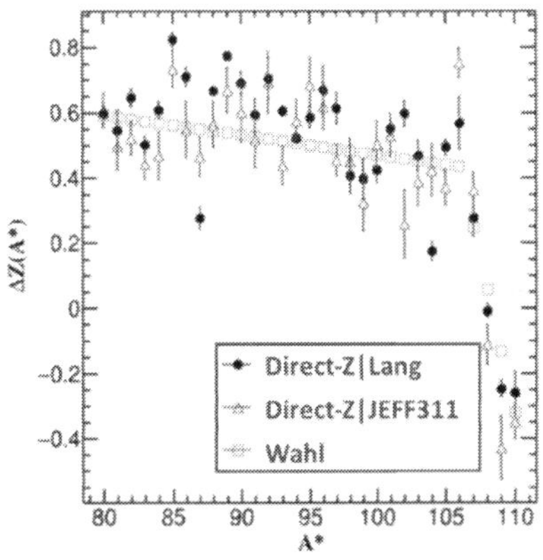

Fig. 13. The evolution of the pre-neutron emission charge polarization, in black the data determined via the adjustment based on Lang data, in blue the data determined via the adjustment based on JEFF-3.1.1 data and in red the data of Wahl's systematics parameterization.

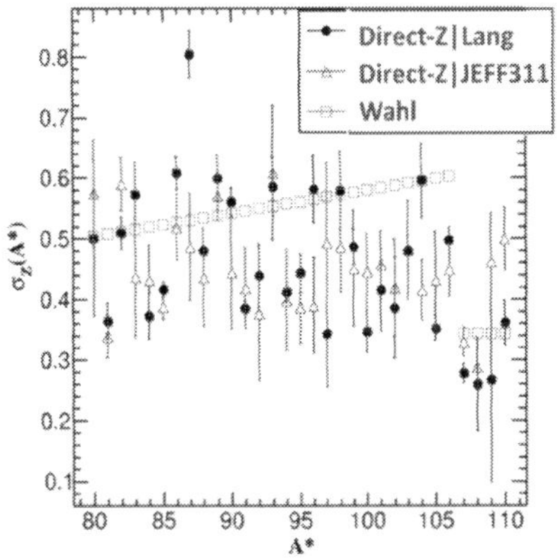

Fig. 14. The evolution of the pre-neutron emission $\sigma_Z(A^*)$, in black the data determined via the adjustment based on Lang results, in blue the data determined via the adjustment based on JEFF-3.1.1 library and in red the data of Wahl's systematics parameterization.

5 Conclusion

We have proposed two new approaches in order to determine Z_p model parameterization with their associated covariance matrix for $^{235}U(n_{th}, f)$ reaction. With these parameterizations, we propagated the parameter uncertainties to generate two covariance matrices of charge distribution associated to the JEFF library. These matrices will be used to fill the lack of charge distribution covariance information for the mixed analysis proposed for the fission yield JEFF-4 library [17]. For the direct-Z_p approach, an adjustment based on Lang's data was performed and the comparison with the adjustment based on JEFF-3.1.1 was made. This approach, starting from the pre-neutron distribution, allows the determination of Zp parameters the overall mass range using the charge conservation law. Thus, even if there is some lacks in the experimental data, the completeness of fission yield assessment is almost achievable for the $^{235}U(n_{th},f)$. With this method, we go back to the origin of the Z_p model but directly applied in the pre-neutron mass region. It will be interesting to apply this approach on the $^{252}Cf(sf)$ which is a standard reaction for fission yields.

References

[1] L. Fiorito, D. Piedra, O. Cabellos and C. Diez, "Inventory calculation and nuclear data uncertainty propagation on lightwater reactor fuel using ALEPH-2 and SCALE 6.2," *Annals of Nuclear Energy,* vol. 83, pp. 137-146, 2015.

[2] O. Cabellos, D. Piedra, C. Diez and L. Fiorito., "Impact of the fission yield nuclear data uncertainties in the pin-cell burn-up OECD/NEA UAM benchmark," in *presentation at the International Conference on the Role of Reactor Physics toward a Sustainable Future — PHYSOR*, Kyoto (Japan), 2014.

[3] N. Schunck and L. M. Robledo, "Microscopic theory of nuclear fission: a review," *Reports on Progress in Physics,* vol. 79, p. 116301, 2016.

[4] A. C. Wahl, "Nuclear-charge distribution and delayed-neutron yields for thermal-neutron-induced fission of 235U, 233U, and 239Pu and for spontaneous fission of 252Cf," *Atomic Data and Nuclear Data Tables,* vol. 39, pp. 1-156, 1988.

[5] D. Madland and T. England, "The Influence of Isomeric States on Independent Fission Product Yields," *Nuclear Science and Engineering,* vol. 64, p. 859–865, 1977.

[6] M. Kellet et al., "The JEFF-3.1/-3.1.1 radioactive decay data and fission yields sub-libraries," OECD NEA, JEFF Report 20 OECD NEA, 2009.

[7] D. A. Brown et al., "ENDF/B-VIII.0: The 8th Major Release of the Nuclear Reaction Data Librarywith CIELO-project Cross Sections, New Standards and Thermal Scattering Data," *Nuclear Data Sheets,* vol. 148, pp. 1-142, 2018.

[8] J. Katakura, "JENDL FP Decay data file 2011 and fission," in *JAEA-Data/Code 2011-025*, 2011.

[9] O. Serot, "Physique des réacteurs Modélisation et évaluation des observables de fission," *Techniques de l'ingénieur,* 2020.

[10] S. Zeynalov, W. Furman and F. Hambsch, "Investigation of Mass-TKE Distributions of Fission Fragments from the 235U(n,f)-Reaction in Resonances," in *Proceedings of the 13th International Seminar on Interaction of Neutrons with Nuclei (ISINN-13) - Neutron Spectroscopy, Nuclear Structure, Related Topics*, Dubna (Russia), 2006.

[11] O. Litaize, O. Serot and L. Berge, "Fission modelling with FIFRELIN," *The European Physical Journal A,* vol. 51, p. 177, 2015.

[12] A. J. Koning, "Bayesian Monte Carlo Method for Nuclear Data Evaluation," *Nuclear Data Scheets,* vol. 123, pp. 207-213, 2015.

[13] N. Metropolis, A. Rosenbluth, M. Rosenbluth, A. Teller and E. Teller, "Equations of state calculations by fast computing machines," *Journal Chemical Physics,* vol. 21, p. 1087–1092, 1953.

[14] A. Newton and E. Raftery, "Approximate Bayesian Inference with the Weighted Likelihood Bootstrap," *Journal of the Royal Statistical Society,* vol. 56, pp. 3-48, 1994.

[15] S. Kirkpatrick, C. D. Gelatt and M. P. Vecchi, "Optimization by Simulated Annealing," *American Association for the Advancement of Science,* vol. 220, no. 4598, pp. 671-680, 1983.

[16] G. Kessedjian et al., "to be published," in *CW 2022 conference*, 2022.

[17] S-M. Cheikh et al., "to be published," in *ND 2022 conference*, 2022.

[18] W. Lang, H.-G. Clerc and H. Wohlfarth, "Nuclear charge and mass yields for 235U(nth,f) as a function of the kinetic energy of the fission products," *Nuclear Physics,* vol. 345, pp. 34-71, 1980.

Marginalization methods for the production of conservative covariance on nuclear data

Pierre TAMAGNO[1,*]

[1]CEA, DES, IRESNE, DER, SPRC, LEPh,
Cadarache, F-13108 Saint Paul Les Durance, France

Abstract. The production of evaluated nuclear data consists not only in the determination of best estimate values for the quantities of interest but also on the estimation of the related uncertainties and correlations. When nuclear data are evaluated with underlying nuclear reaction models, model parameters are expected to synthesize all the information that is extracted from the experimental data they are adjusted on. When dealing with models with a small number of parameters compared to the number of experimental data points – *e.g.* in resonant cross section analysis – one sometimes faces excessively small evaluated uncertainty compared for instance with model/experimental data agreement. To solve this issue, an attempt was to propagate the uncertainty coming from experimental parameters involved in the data reduction process on the nuclear physics model parameters. It pushed experimentalists to separately supply random (statistical) and systematic uncertainties. It also pushed evaluators to include or mimic the data reduction process in the evaluation. In this way experimental parameters – also called nuisance parameters – could be used to increase evaluated parameter uncertainty through marginalization techniques. Two of these methods: Matrix and Bayesian marginalizations – respectively called sometimes Analytical and Monte-Carlo Marginalizations – that are currently used for evaluation will be discussed here and some limitations highlighted. A third alternative method, also based on a Bayesian approach but using the spectral decomposition of the correlation matrix, is also presented on a toy model, and on a a simple case of resonant cross section analysis.

1 Introduction

Since the discovery of the atomic nucleus, technological applications of nuclear energy have flourished. The design and modeling of such devices rely on technological properties (geometry, compositions, etc.) but also on fundamental constants describing the properties of spontaneous or induced nuclear reactions. These constants, called *nuclear data*, are however not known up to an arbitrarily high precision and special efforts must be brought to quantify related uncertainties, generally understood as *covariance data*. Nuclear data are obtained in a process called *evaluation* that combines experimental data and theoretical models. Once evaluated, these are usually further stored in files in ENDF format [1], these latter are then processed to create libraries specific to application codes, for instance to produce effective cross sections for medium at given temperature. The calculation scheme of the application codes can even internally imply some modifications of the nuclear data, for instance by the production of self-shielded energy-averaged cross sections. Each transformation of the nuclear data should imply a corresponding transformation of the related covariance data for a consistent propagation of nuclear data uncertainty to the quantities of interest for the applications. In practice, most of these transfers are handled in terms of best-estimate – or mean values – and *covariance matrices*, therefore focus is made here on these first two moments of the probability density function (PDF) describing nuclear data. The uncertainty propagation thus usually assumes mono-modale Gaussian distributions.

Quantifying uncertainties are a natural aspect of experimental science where for instance raw count rates follow Poisson distributions and detector response functions correlate eventual results. On the theoretical side, uncertainties can consists in "truncation errors" or in uncertain determination of the value of model parameters. The model parameter values are usually estimated from comparison with experimental data using least-square fit or within a more general Bayesian approach.

The production of "safe" covariance data has been a concern leading to several attempts to be conservative in the estimation of model parameters. A long-standing idea [2–6] has been to consider that the uncertainty on experimental setup characteristics (*i.e.* detector efficiency, target composition, etc.) – called *nuisance parameters* or *active variables*, written $\vec{\theta}$, involved in the setup response function – should be "transferred" to theoretical model parameters – called *physical parameters* or *passive variables*, written $\vec{x}$.

The present paper first reviews two techniques used to transfer uncertainty from $\vec{\theta}$ to $\vec{x}$. The first approach (hereafter called the matrix marginalization method) is shown

*e-mail: pierre.tamagno@cea.fr

to be ill-defined and does not guarantee to be conservative with respect to Bayesian posterior. The second approach (the Bayesian marginalization method) is shown to be possibly biased, namely when nuisance parameters are inaccurate. These methods have been tested on a previous study on a simple but realistic analysis of Prompt Neutron Fission Spectrum [7]. These approaches are tested here on a specifically designed toy model test-case, where they are shown to fail in producing consistent best-estimates and reliable propagated uncertainties. This test-case has inconsistent data, and is specifically design to exhibit the possible limits of the methods. The failure highlights that the two presented techniques – focusing on the transfer of nuisance parameter uncertainty – neglect the verification of agreement between posterior theory and experiment. An *ad hoc* procedure [7] is then presented in the paper, it aims at artificially increasing the estimated uncertainties but in a conservative manner with respect to the Bayesian approach and with consideration of agreement between posterior theory and experimental. This new technique is further exemplified on resonant cross section analysis.

2 Nuclear data evaluation with Bayesian inference

Evaluated nuclear data are often obtained from the synthesis of experimental results and theoretical models. The necessity of theoretical models may come from:

1. missing experimental data (comprehensiveness)
2. targeted uncertainties not met (precision)
3. different mixed-up phenomena blurred in recorded data (selectivity)

Theoretical models are not always fully predictive and sometimes require adjustment of model parameters. Note that the notion of "model parameters" can be understood in a larger sens. When one only aggregates several experimental values, the obtained mean value can also be seen as a model parameter; and the set of provided values supplied in an ENDF file can be seen as a piece-wise linear model reproducing the evaluated data.

In nuclear data evaluation, it is therefore general to consider parameter value estimation, and the Bayesian inference is an interesting mathematical framework for this task. The main idea is to treat parameter values $\vec{p}$ as random quantities described with related probability density functions (PDF). Starting with a possible initial *prior* PDF $P(\vec{p})$, information provided by $\vec{y}$ (a direct or indirect measurement of $\vec{p}$) yields an updated *posterior* PDF $P(\vec{p}|\vec{y})$. This relation is given by the Bayes' theorem [8]

$$P(\vec{p}|\vec{y}) = \frac{P(\vec{p})P(\vec{y}|\vec{p})}{\int d\vec{p}\,'P(\vec{p}\,')P(\vec{y}|\vec{p}\,')} \tag{1}$$

where $P(\vec{y}|\vec{p})$ is the likelihood describing the probability to observe $\vec{y}$ given a value of $\vec{p}$. The posterior distribution can further be used to obtain best-estimate values of a parameter p_i and related covariance matrix as the first and second moments of the posterior PDF:

$$\begin{aligned} \langle p_i \rangle &= \int d\vec{p} p_i P(\vec{p}|\vec{y}) \\ M_{ij} = \mathrm{Cov}(p_i, p_j) &= \int d\vec{p}(p_i - \langle p_i \rangle)(p_j - \langle p_j \rangle)P(\vec{p}|\vec{y}) \end{aligned} \tag{2}$$

Assuming no prior knowledge for $\vec{p}$ and if the likelihood is taken as

$$P(\vec{y}|\vec{p}) = \exp\left(-\frac{1}{2}(\vec{y} - \vec{t}(\vec{p}))^t M_y^{-1}(\vec{y} - \vec{t}(\vec{p}))\right) \tag{3}$$

the Bayesian approach reduces to a least-square fit but provides also a posterior covariance matrix. Here the likelihood is related to the usual χ^2, with M_y being the covariance matrix on the observation $\vec{y}$ and $\vec{t}(\vec{p})$ being the modeled values to be compared to $\vec{y}$.

To first order, meaning for small variations of $\vec{p}$ about $\langle \vec{p} \rangle$, the uncertainty on $\vec{p}$ can be propagated to modeled values $\vec{\sigma}(\vec{p})$ – which can be different from $\vec{t}(\vec{p})$ – using the sensitivity matrix G_{ij} defined as

$$G_{ij} = \frac{\partial \sigma_i}{\partial p_j} \tag{4}$$

The eventual covariance matrix on the $\vec{\sigma}$ quantities are obtained as

$$\begin{aligned} N &= GMG^t \\ \text{where} \quad N_{ij} &= \mathrm{Cov}(\sigma_i, \sigma_j) \end{aligned} \tag{5}$$

The model – providing theoretical values $\vec{t}_1$ to be compared with experimental values $\vec{y}$ – can be a composite model. It can depends on parameters $\vec{p}_1$ and on a sub-model $\vec{t}_2$ depending on its own parameters $\vec{p}_2$. This Russian dolls aspect can also apply for $\vec{t}_2$. The reason for this nested structure is that some experimental corrections can be handled during the Bayesian inference (described at the $\vec{t}$ level) by the evaluator. This can be for instance normalization or sample temperature for resonant cross section analysis. The parameters can also be frozen and fixed in the M_y matrix when treated by the experimentalist, for instance detector efficiency or dead time correction. However this latter approach may lead to so-called Peel's Pertinent Puzzles that are situation where the eventual χ^2 is minimized for values $\vec{t}$ that obviously mismatch $\vec{y}$. In general one can separate the "experimental" or "nuisance" parameters $\vec{\theta}$ to the theoretical parameters $\vec{x}$ that describe the fundamental nuclear data. The nested model will be written in short as $\vec{t}(\vec{\theta}, \vec{x})$ in the rest of the paper.

3 Uncertainty estimate – Traditional marginalization methods

The Bayesian inference described above produces best-estimates for $\vec{x}$ and $\vec{\theta}$ but also a covariance matrix between those parameters:

$$M_p = \begin{bmatrix} M_x & M_{x,\theta} \\ M_{\theta,x} & M_\theta \end{bmatrix} \tag{6}$$

where only the sub-matrix M_x will be used to propagate uncertainties on evaluated data $\sigma(\vec{x})$. A long-standing idea to produce conservative covariance data has been to modify M_x so that uncertainties propagated on $\vec{t}$ could be considered to originate from $\vec{x}$ alone, not from $\vec{x}$ *and* $\vec{\theta}$. Two methods will be reviewed in this section and exemplified on a toy model that will be presented first.

3.1 Toy model analysis

The toy model considered here consists in a linear model

$$\vec{t}(E, x, \theta) = \theta E + x \tag{7}$$

depending on a nuisance parameter θ (slope), a theoretical parameter x (value at origin) and is applied on an experimental condition parameter E. The parameters x and θ are inferred from "experimental" data points $y_i(E_i)$ generated as the following:

- E_i are sampled uniformly on $[0, 1]$
- $y_i = 0.55 \pm 0.1$ (with even probability of sign) are altered by a noise uniformly sampled in $[-0.01, 0.01]$
- A statistical uncertainty of 0.01 is associated to all y_i.

It is clear that expected final values should be

$$\theta_{\text{fin}} = 0 \quad \text{and} \quad x_{\text{fin}} = 0.55 \tag{8}$$

It is also clear that the model (7) is designed not to be able to reproduce the sampled data that follows a bi-modal distribution. This is done to emulate either model defects or experimental artifacts that could bias model parameter estimation. It is yet useful to have assimilation methods robust enough to yield realistic (mono-modal) mean values and standard deviations even in such cases. The sampled data are first analyzed with model (7) by Bayesian inference without consideration of prior knowledge for x and θ, which resumes to a least-square fit. The initial values used for the parameters are

$$\theta_{\text{ini}} = 0.055 \pm 0.055 \quad \text{and} \quad x_{\text{ini}} = 0.3 \tag{9}$$

The initial uncertainty on θ is provided as it will be used latter for marginalization techniques. One can either adjust θ along with x – these cases are called "informative" hereafter – or keeping θ constant, which is called here "non-informative". The data set is shown in Fig. 1 with only 2 sampled points analyzed. Of course with this small number of parameters, adjusting θ yields a perfect match of experimental data but terribly wrong values of x_{fin} and θ_{fin}. The simple model (7) allows to easily read the value of x : the model value for $E = 0$. The value of θ can also be estimated as the slope of the curves. The analysis for 20 sampled points is shown on Figure 2. It shows that increasing the amount of experimental information (increasing the number of data points) improves the estimation of final values in both cases (the final values are indicated in the figures). However either x or θ has a final uncertainty that is inconsistent with the expected values (8). It is striking however that the propagated uncertainty band (in green) does not cover the spread of the experimental data points. Even in the most favorable case (non-informative),

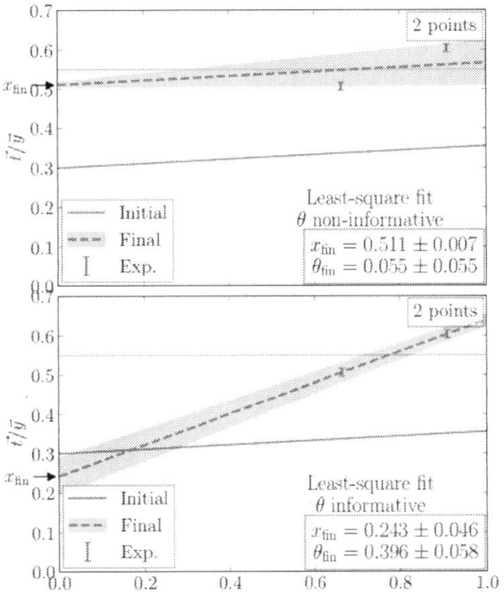

Figure 1. Analysis of the 2 points data set with the toy model data by Bayesian inference with and without fit of θ.

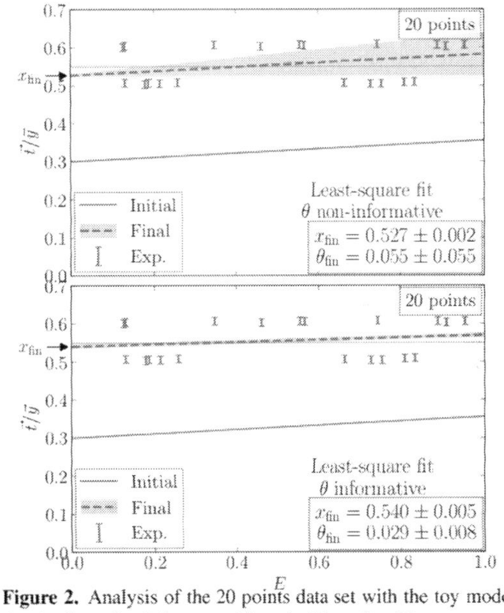

Figure 2. Analysis of the 20 points data set with the toy model data by Bayesian inference with and without fit of θ.

the large band is mostly due to the propagation of the initial uncertainty on θ. This can be seen as the uncertainty band becomes larger as E increases. Since only x is used for the applications $\sigma(\vec{x})$, the propagated uncertainty will be extremely small compared to the experimental knowledge of x. It is therefore misleading to propagate uncertainty on θ on such figures, for the rest of the paper only the uncertainty on x will be shown in the final uncertainty band. The initial relative uncertainty on θ (Eq. (9)) can be questioning since it stands for a calibration parameter that is in principle well known. It should be understood as θ is expected to be smaller than a given value – as for a background for instance – with an average propagated uncertainty on $\vec{t}$ of 5%. This can be high for a calibration parameter but for readability we restrict the toy model with a limited number of data points, which pushes for exaggerating the uncertainty on θ. This prior uncertainty of 0.055 for θ is rather consistent (for E between 0.5 and 1) with the bi-modal spread of experimental data points. However θ is not related to this bi-modal feature and thus cannot be expected to account for it. It will be shown latter that the uncertainty on θ is not necessarily the key for realistic uncertainty estimation.

3.2 "Matrix marginalization"

Several techniques have been designed to transfer uncertainty from θ on x. The first method presented here that is called matrix marginalization but is also called analytical marginalization [4, 6, 7] or zero-variance penalty marginalization [5] in other works. This method attempts to find a matrix M_x^{m} on x alone so that the propagated covariance matrix on $\vec{t}$ is the same than the one obtained with M_p of Eq. (6), namely

$$\underbrace{G_x M_x^{\mathrm{m}} G_x^{\mathrm{t}}}_{\text{expected matrix}} = \underbrace{G_p M_p G_p^{\mathrm{t}}}_{\text{(propagated) posterior matrix}} \tag{10}$$

where

$$[G_x]_{ij} = \frac{\partial t_i}{\partial x_j}, \qquad [G_\theta]_{ij} = \frac{\partial t_i}{\partial \theta_j}, \qquad G_p = [G_x\, G_\theta] \tag{11}$$

Multiplying Eq. (10) from the left by G_x^{t} and from the right by G_x one has

$$G_x^{\mathrm{t}} G_x M_x^{\mathrm{m}} G_x^{\mathrm{t}} G_x = G_x^{\mathrm{t}} G_p M_p G_p^{\mathrm{t}} G_x \tag{12}$$

After multiplication on both sides by $(G_x^{\mathrm{t}} G_x)^{-1}$ one obtains

$$M_x^{\mathrm{m}} = (G_x^{\mathrm{t}} G_x)^{-1} G_x^{\mathrm{t}} G_p M_p G_p^{\mathrm{t}} G_x (G_x^{\mathrm{t}} G_x)^{-1} \tag{13}$$

However inserting Eq. (13) in Eq. (10) one does no generally obtain the propagated posterior matrix $G_p M_p G_p^{\mathrm{t}}$, this could occur but in special cases where $G_x (G_x^{\mathrm{t}} G_x)^{-1} G_x^{\mathrm{t}} = \mathbb{1}$. This method does not yield the claimed properties, but it is still often used in practice.

Depending on how $\vec{\theta}$ are treated, three different cases can be considered with this method

1. *informative* case: mean value and standard deviation of $\vec{\theta}$ are updated by Bayesian inference
2. *non-informative* case: mean value and standard deviation of $\vec{\theta}$ are *not* updated by Bayesian inference
3. *conservative-informative* case: mean value of $\vec{\theta}$ is updated by Bayesian inference but its standard deviation is restored to the original value. One can replace M_θ in Eq. (6) by M_θ^{ini} and use the altered M_p' in Eq. (13):

$$M_p' = \begin{bmatrix} M_x & M_{x,\theta} \\ M_{\theta,x} & M_\theta^{\mathrm{ini}} \end{bmatrix} \tag{14}$$

where it is not clear if cross terms should also be restored to their original values. In the two latter cases a initial uncertainty on θ must be used. The results for the three cases are shown in Fig. 3.

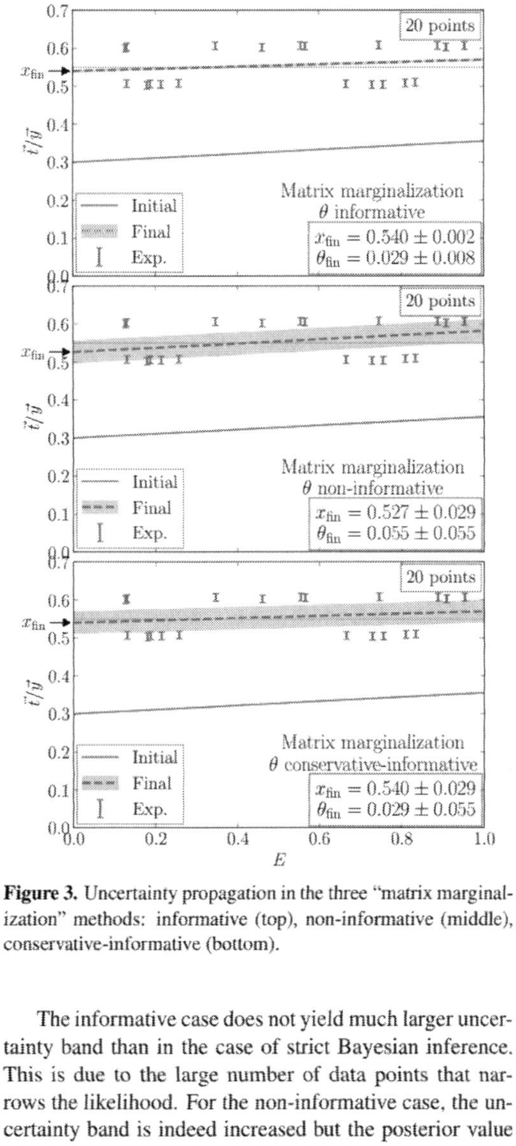

Figure 3. Uncertainty propagation in the three "matrix marginalization" methods: informative (top), non-informative (middle), conservative-informative (bottom).

The informative case does not yield much larger uncertainty band than in the case of strict Bayesian inference. This is due to the large number of data points that narrows the likelihood. For the non-informative case, the uncertainty band is indeed increased but the posterior value of x is less accurate than in the original Bayesian case (x should tend to 0.55). Finally the conservative-informative case exhibits the best behavior, x and θ tend both to good values and a larger uncertainty band is obtained. Two comments can yet be made:

1. M_x^{m} as given in Eq. (13) does not guarantee to obtained the desired propagated covariance matrix of Eq. (10).
2. The larger uncertainty band obtained here is only due to the prior θ uncertainty. In the sampled data points θ does not play any role and thus a much smaller initial uncertainty on θ would have been

plausible. As mentioned above, θ is not related to the bi-modal behavior of the sampled data. For instance in Fig. 4 a smaller initial value of θ is used (and a smaller prior uncertainty) on the same set. It yields this time a narrow uncertainty band that does not render the effect of the bi-modal distribution. This time the final parameter uncertainties do not overlap the true values.

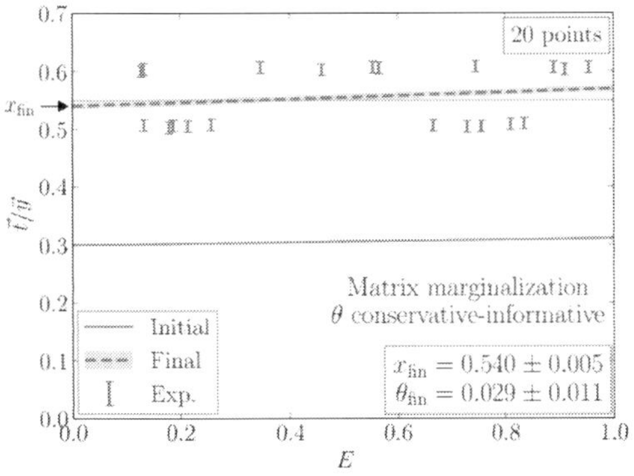

Figure 4. Uncertainty propagation with the conservative-informative "matrix marginalization" technique using a reduced initial uncertainty on θ: $\theta_{\text{ini}} = 0.011 \pm 0.011$.

3.3 "Bayesian marginalization"

Even if the matrix marginalization shows somehow reasonable behavior, in practice it was observed to be unstable in cases with larger number of parameters. This is possibly due to the improper connection from Eq. (13) to Eq. (10). Another traditional method has a stronger connection with the Bayesian framework [3]. This method is derived in few steps from the Bayesian posterior given in Eq. (1). First $\vec{p}$ is split into $\vec{x}$ and $\vec{\theta}$, then the conditional probability theorem is applied:

$$P(\vec{p}|\vec{y}) = P(\vec{x},\vec{\theta}|\vec{y}) = P(\vec{x}|\vec{\theta},\vec{y})P(\vec{\theta}|\vec{y}) \tag{15}$$

It is then assumed that $\vec{\theta}$ do not benefit from information from $\vec{y}$, so that

$$P(\vec{\theta}|\vec{y}) = P(\vec{\theta}) \tag{16}$$

The marginalized PDF of $\vec{x}$ is then defined as

$$P^{\text{m}}(\vec{x}) = \int \text{d}\vec{\theta} P(\vec{x},\vec{\theta}|\vec{y}) = \int \text{d}\vec{\theta} P(\vec{x}|\vec{\theta},\vec{y})P(\vec{\theta}) \tag{17}$$

The Bayesian inference is then applied on $P(\vec{x}|\vec{\theta},\vec{y})$ yielding

$$P(\vec{x}|\vec{\theta},\vec{y}) = \frac{P(\vec{x}|\vec{\theta})P(\vec{y}|\vec{\theta},\vec{x})}{\int \text{d}\vec{x}' P(\vec{x}'|\vec{\theta})P(\vec{y}|\vec{\theta},\vec{x}')} \tag{18}$$

Finally one assumes that there is no prior correlation between $\vec{x}$ and $\vec{\theta}$, namely

$$P(\vec{x}|\vec{\theta}) = P(\vec{x}) \tag{19}$$

so that the marginalized PDF of $\vec{x}$ is

$$P^{\text{m}}(\vec{x}) = \int \text{d}\vec{\theta} P(\vec{\theta}) \frac{P(\vec{x})P(\vec{y}|\vec{\theta},\vec{x})}{\int \text{d}\vec{x}' P(\vec{x}')P(\vec{y}|\vec{\theta},\vec{x}')} \tag{20}$$

From this distribution, one obtains the mean value and covariance matrix on $\vec{x}$. The corresponding results are shown in Fig. 5.

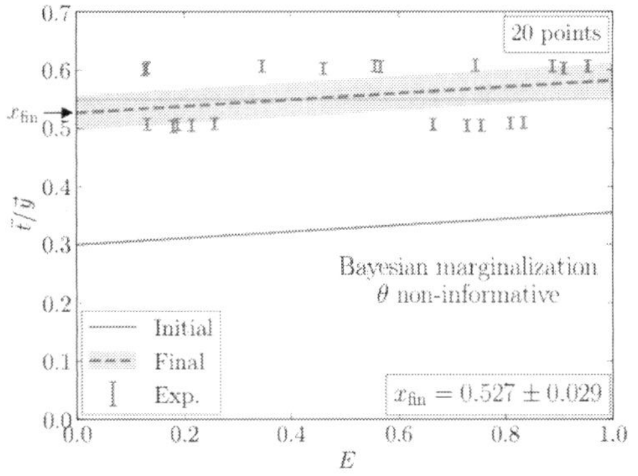

Figure 5. Uncertainty propagation in the Bayesian marginalization method.

One can see that the results of this method has the same behavior than the non-informative matrix marginalization (middle plot of Fig. 3). Namely since $\vec{\theta}$ is assumed not to benefit from information from $\vec{y}$ (Eq. (16)), the constant value of $\theta = \theta_{\text{ini}}$ bias the posterior estimate of x which is not as close to 0.55 as in the "informative" cases.

4 Spectral decomposition method

A third method [7] combines a good mathematical definition and the possibility to let the $\vec{\theta}$ parameters evolve if necessary. This method is based on the preservation of conservative uncertainty propagation for *any* quantity varying (linearly) with parameters $\vec{x}$. To demonstrate the method it is first recalled that covariance matrices must be positive (semi-)definite, namely all its eigenvalues must be positive or zero. This is true for the posterior covariance matrix on M_x, which can be decomposed as

$$M_x = Q\Lambda Q^{\text{t}} \tag{21}$$

where Q stores M_x eigenvectors and Λ is a diagonal matrix of its eigenvalues. The fact that $\forall i,\ \Lambda_i \geq 0$ implies that any quantity computed from $\vec{x}$ that varies linearly with $\vec{x}$ (possibly only locally), with sensitivity vector $\vec{S}$, has a positive propagated variance given by

$$\vec{S}^{\text{t}} M_x \vec{S} = (Q^{\text{t}}\vec{S})^{\text{t}} \Lambda Q^{\text{t}} \vec{S} = \sum_i \Lambda_i [Q^{\text{t}}\vec{S}]_i^2 \geq 0 \tag{22}$$

A conservative covariance matrix can thus be searched in the form $M_x^{\text{m}} = Q\Lambda^{\text{m}}Q^{\text{t}}$ where the Λ_i^{m} verify

$$\forall i,\ \Lambda_i^{\text{m}} \geq \Lambda_i \tag{23}$$

This parameterization, where the $\Lambda_i^{\rm m}$ parameters are adjusted to satisfy some criteria that will be defined below, ensures that the propagated variance will be larger than the posterior one. Note that in practice one can keep standard deviation outside the covariance matrix and work on the eigenvalue decomposition of the correlation matrix instead. This can improve numerical stability of the method.

It seems reasonable to search for a covariance matrix that would be conservative *just* as necessary. This requirement with condition Eq. (23) can be mathematically formulated as a minimization under constraints problem, for instance with the cost function

$$S = \frac{1}{2}\sum_i \left(\frac{\Lambda_i^m}{\Lambda_i} - 1\right)^2 \tag{24}$$

to be minimized and with constraints

$$\forall i,\; g_i = \frac{\Lambda_i^{\rm m}}{\Lambda_i} - 1 \geq 0 \tag{25}$$

Stated as such, the problem has a trivial solution: $\forall i,\; \Lambda_i^{\rm m} = \Lambda_i$. To be more conservative, additional constraints can be considered. In the same spirit as in the matrix marginalization, one can imposes the variance propagated from $M_x^{\rm m}$ on the t_n to be larger that the one obtained by propagation of M_p (or by M_p' given by Eq. (14)). This gives the constraints

$$\forall n,\; g_n = \frac{v_n^{\rm m}}{v_n} - 1 \geq 0, \quad \text{or} \quad g_n = \frac{v_n^{\rm m}}{v_n'} - 1 \geq 0 \tag{26}$$

with

$$\begin{aligned} v_n^{\rm m} &= [G_x M_x^{\rm m} G_x^{\rm t}]_{nn} \\ v_n &= [G_p M_p G_p^{\rm t}]_{nn} \\ v_n' &= [G_p M_p' G_p^{\rm t}]_{nn} \end{aligned} \tag{27}$$

Figure 6 shows the propagated results obtained with the informative approach (use of M_p and v_n) and in the conservative-informative approach (use of M_p' and v_n').

It can be seen that final expectation values of x and θ are as good as the best cases n the other methods because x and θ are adjusted during the Bayesian inference step and that the spectral method does not alter best-estimates. Even if a large uncertainty band is obtained, this method still suffers from the inconsistent definition of M_p' (Eq. (14)). Additionally if the initial uncertainty in θ does not account for the model defect, the uncertainty band will not be reliable (just like in Fig 4). To avoid these two latter points (to keep sound maths and robustness when initial uncertainties on $\vec{\theta}$ are small), additional constraints can be included.

The first additional set of constraints are called *local* constraints and enforce that the random variables $(t_n - y_n)/\sqrt{v_n^{\rm m} + [M_y]_{nn}}$ – considered as following a centered standard distribution – fall within an interval $[-\alpha, +\alpha]$. The value of α corresponding the a one–σ standard deviation is given by $\alpha = \sqrt{2}\,{\rm erf}^{-1}(68.27\%)$, where erf is the error function. With this choice $\alpha = 1$ but could be changed depending of the targeted confidence interval (here 68.27%). The corresponding constraints are thus

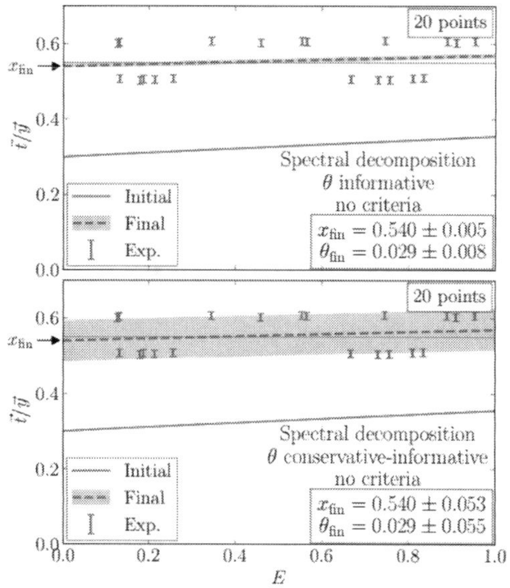

Figure 6. Uncertainty propagation in the informative (top) and conservative-informative (bottom) spectral methods.

given by

$$\forall n,\; g_n = \alpha^2 \frac{[M_y]_{nn} + v_n^{\rm m}}{(t_n - y_n)^2} \geq 0 \tag{28}$$

The results obtained with this approach is shown in Fig. 7.

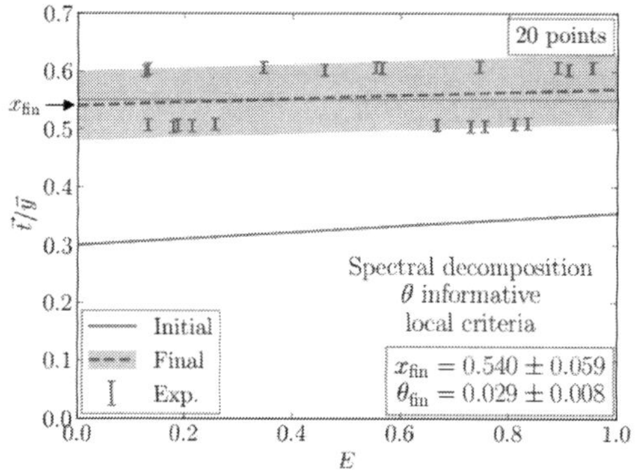

Figure 7. Uncertainty propagation in the spectral method with a local criteria constraints.

The local criteria constraint enforce that *all* points should fall within the experimental+propagated uncertainty band. Since the related constraints (Eq. (28)) are define for each point specifically, often a specific point will be responsible for the critical constraint ($g_n = 0$) limiting the minimization under constraint. This uncertainty propagation method can thus be used to identify possible outliers.

One may consider the local criteria as excessive. Indeed even with proper normally distributed random samples, it is not rare to have few points falling outside the one−σ interval. A global criteria [7] constraint can thus be given by

$$g = 1 - \frac{X^2}{\alpha^2} \geq 0 \tag{29}$$

with

$$X = \frac{1}{\sqrt{N_y}} \left(\sum_n \frac{(t_n - y_n)^2}{[M_y]_{nn} + v_n^{\mathrm{m}}} - N_y \right) \tag{30}$$

where N_y is the number of experimental data points. This latter criteria is valid if N_y is large because it assumes that an asymptotic behavior of a chi-square distribution with N_y degrees of freedom. The results obtained with this approach is shown in Fig. 8.

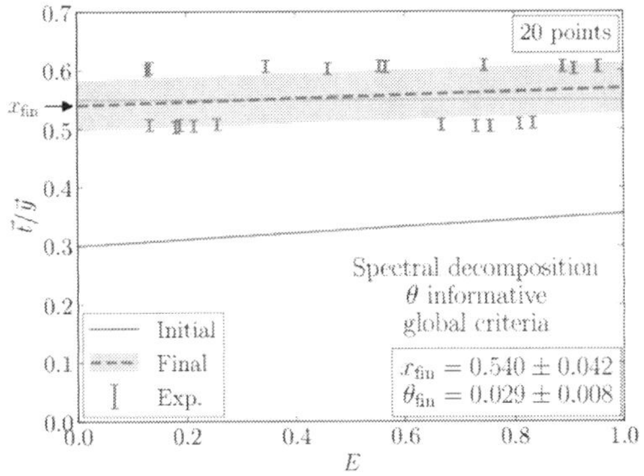

Figure 8. Uncertainty propagation in the spectral method with the global criteria constraint.

It is visible that the uncertainty band covers a large part of the data points but some points can lay outside since the global criteria is less strict than the local criteria. At first glance this latter method seems equivalent to the conservative-informative matrix marginalization results, here however:

1. the problem is mathematically well defined and is expected to scale better with increasing the number of parameters, namely because there is no consistency issues anymore between Eq. (13) and Eq. (10).
2. the large uncertainty band is longer driven by the prior uncertainty on θ but by the posterior agreement or disagreement between model values and experimental data.

5 Application to resonant cross section analysis and model defect

The spectral decomposition methods presented above can be used *even if there is no nuisance parameter considered in the analysis*. This application case of this marginalization technique is used to show its effects in case of realistic model defect. The spectral decomposition method is exemplified here in the case of an analysis of neutron total cross section measurement of sodium [9]. This measurement is reported to have a statistical uncertainty between 0.7% to 3.4% in the considered energy range, and 3% of systematic uncertainty, usually assumed to correspond to a normalization factor – not considered here. In Fig. 9 the experimental values are shown with the evaluated data provided in the JEFF-3.2 international evaluated nuclear data library [10] that is used here as a prior.

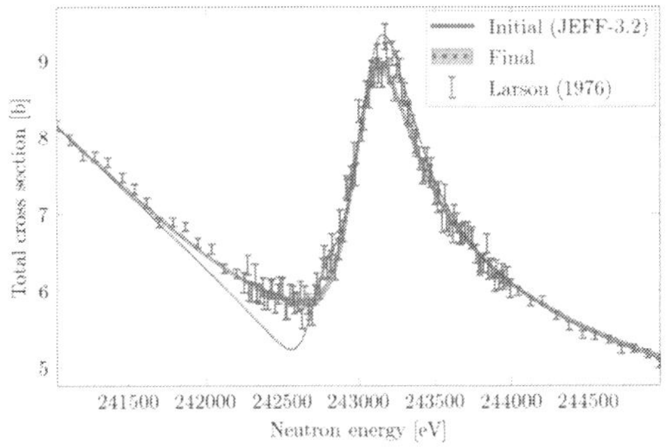

Figure 9. Resonance analysis using the JEFF-3.2 parameters.

The experimental data in the range [241 keV, 245 keV] of incident neutron kinetic energy have been used to adjust parameters provided in JEFF-3.2. These parameters are provided to an R-Matrix model [11] (with Reich-Moore approximation) used to reproduce cross sections in a large energy range. Here the setup response function is not considered, no resolution function and no Doppler broadening were included in the analysis, it was however tested that including them does not modify the presented conclusions. In the prior value of Fig. 9, an experimented eye can see that the structure is composed of two resonances, corresponding to two energies given in JEFF-3.2: E_λ = 242.7913 keV and E_λ = 243.0485 keV. For each of these two resonances, two other parameters (resonance partial widths) are given for the radiative capture and the elastic scattering channels. To perform consistent comparisons, these six parameters were adjusted on the experimental data set and yielded the green line in Fig. 9 where it is no longer possible to distinguish the contribution of the two resonances. To make sure of the necessity of having two resonances in this energy range, the three parameters related to each resonance (energy, gamma and neutron widths), have been, in turn, removed while those for the remaining resonance have been re-adjusted. The resulting χ^2 are reported in Tab. 1 along the χ^2 obtained while keeping both resonances.

Table 1. Summary of adjustments on ^{23}Na(n, tot) measurement

nb. of parameters	χ^2_{fin}	nb. iterations
6 (two res.)	1.63	15
3 (first res. kept)	2.96	7
3 (second res. kept)	2.82	3
4	1.06	4

The significant reduction of χ^2 while keeping two sets of resonance parameters can be seen as a justification of the inclusion of a second resonance. However it was tested to keep only the second resonance but using a generalized version of the R-matrix model where several neutron widths can be used. Two partial widths were thus used corresponding to orbital angular momentum $\ell = 0$ and $\ell = 2$ that are coupled to the $J^\pi = 0^+$ (second) resonance. In this case only the second resonance is kept but an extra parameter is included. The corresponding χ^2 is also reported in the last line of Tab. 1. The large reduction of χ^2 with a reduced number of adjusted parameters with respect to JEFF-3.2 allows to conclude that *there is a model defect* in the precise energy range in JEFF-3.2. Note that the number of iterations required to converge as reported in Tab. 1 also argues in the same direction.

We now consider only the two models

- the 6-parameters model, *i.e.* JEFF-3.2 refitted – the model with a model defect
- the 4-parameters model, *i.e.* the extend reference R-matrix model

And we will see how they behave with the spectral marginalization methods. The results obtained with the local criteria are shown in Fig. 10.

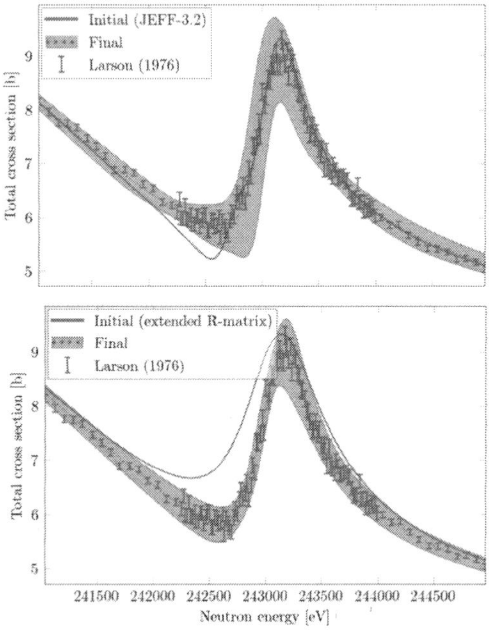

Figure 10. Uncertainty band obtained with the local criteria with the JEFF-3.2 prior (top) and with the extended model (bottom).

While for the extended R-matrix case the local criteria yields a smooth uncertainty band following nicely experimental "error bars", results for the uncertainty band obtained from the JEFF-3.2 model extends much farther, namely at the top and bottom of the resonance, highlighting the model defect penalty. This could indicate that even if a reasonable χ^2 is obtained with the addition of a second resonance (χ^2 divided by about a factor of two, *cf.* Tab. 1), the model has trouble overlapping the spread of experimental points within their reported uncertainties. The uncertainty band thus gives an idea of the model flexibility allowing to reproduce or not the observed shape. This also indicates that a second measurement with consistent – but different – experimental data points could be difficult to assimilated in the Generalized Least-Square approach. Indeed in the 6-parameters models, parameters must vary a lot to accommodate experimental data spread and uncertainties, this would correspond to a strong penalty is successive data assimilation.

If the *global* criteria is used instead of the *local* ones, the width of the uncertainty band width reduces but remains larger than the band obtained with the reference model. This latter case is not shown here, instead focus is made on the 6-parameters model. Indeed in most cases we are not aware if there is a model defect or not. The tested situation presented here consists in comparing the uncertainty band obtained with the Bayesian inference only and the one obtained after applying the global criteria. The result is shown in Fig. 11, where the uncertainty band obtained with the global criteria is 20% to 50% larger than one given by the Bayesian posterior. Both methods share the same expectation values since the spectral decomposition method only deals with second order moments.

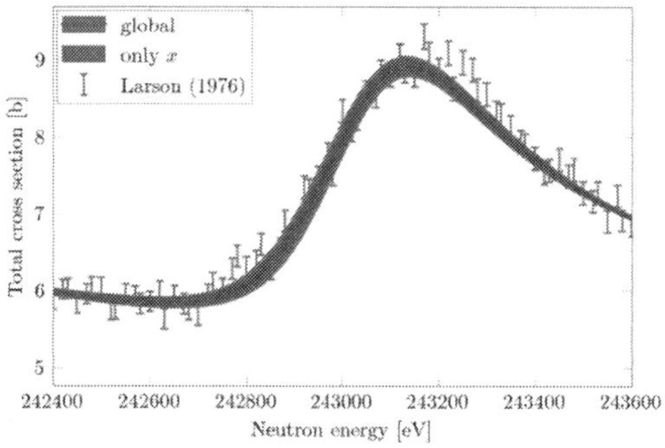

Figure 11. Uncertainty band obtained with the global criteria compared to the one obtained after the Bayesian analysis.

6 Conclusion

The necessity to estimate realistic uncertainties led to the development of marginalization techniques used to transfer uncertainties from nuisance (or experimental) parameters to evaluated parameters that are used to produce nuclear data. Two main methods have been reviewed here but shown to suffer from mathematical weakness or from lack of flexibility while keeping nuisance parameter values fixed. An alternative method based on the spectral decomposition of the covariance (or correlation) matrix has been presented. Here not only nuisances parameter values are free to be updated by data assimilation, but the produced uncertainties are guaranteed to be larger than the one propagated after Bayesian assimilation. This method allows to transfer uncertainty from nuisance parameters to model parameters of interest but can also be applied without nuisance parameters. In this latter cases, it is the compatibility of post-fit theoretical values and provided experimental data that is used to condition estimated uncertainties. This method has been applied here on a toy model experiment analysis but also on a real-case resonant cross section analysis. It has been shown that the spectral method (using the local criteria) could be used to indicate the presence of a not-obvious model defect. The same method (with global criteria) is shown to increase the propagated uncertainty by 50% with respect to original Bayesian assimilation. In all cases, the spectral method is applied *after* the Bayesian (or least-square) fit, it is therefore very fast as no extra model computations or derivatives are required. It is also rather easy to implement in existing codes.

References

[1] A. Trkov, D.A. Brown, Tech. Rep. BNL-203218-2018-INRE, Brookhaven National Lab.(BNL), Upton, NY (United States) (2018), `https://www.osti.gov/biblio/1425114`

[2] F. Fröhner, Nuclear Science and Engineering **126**, 1 (1997)

[3] C. De Saint Jean, G. Noguère, B. Habert, B. Iooss, Nuclear Science and Engineering **161**, 363 (2009)

[4] B. Habert, C. De Saint Jean, G. Noguère, L.C. Leal, Y. Rugama, Nuclear Science and Engineering **166**, 276 (2010)

[5] G. Noguère, P. Archier, C. De Saint Jean, B. Habert, Nuclear Science and Engineering **172**, 164 (2012)

[6] C. De Saint Jean, P. Archier, E. Privas, G. Noguère, B. Habert, P. Tamagno, Nuclear Data Sheets **148**, 383 (2018), special Issue on Nuclear Reaction Data

[7] P. Tamagno, European Physical Journal A **57**, 61 (2021)

[8] T. Bayes, Philosophical Transactions **53**, 370 (1763)

[9] D.C. Larson, J.A. Harvey, N.W. Hill, Tech. Rep. ORNL/TM-5614, Oak Ridge National Lab., Tenn. (USA) (1976), `https://www.osti.gov/scitech/servlets/purl/7342187`

[10] A.J.M. Plompen, O. Cabellos, C. De Saint Jean, M. Fleming, A. Algora, M. Angelone, P. Archier, E. Bauge, O. Bersillon, A. Blokhin et al., The European Physical Journal A **56** (2020)

[11] A.M. Lane, R.G. Thomas, Rev. Mod. Phys. **30**, 257 (1958)

Uncertainty estimation in neutron TOF measurements with ANNRI

Atsushi Kimura[1]*, *Shoji* Nakamura[1], and *Shunsuke* Endo[1]

[1]Japan Atomic Energy Agency, 2-4 Shirakata, Tokai-mura, Ibaraki 319-1195, Japan

Abstract. To improve data accuracy of neutron-capture and total cross sections of minor actinides and long-lived fission products, the time-of-flight experimental instrument named "Accurate Neutron-Nucleus Reaction measurement Instrument" was constructed, and neutron time-of-flight experiments have been performed with an intense pulsed-neutron beam at MLF in J-PARC. Detailed analysis methods and deduced uncertainties for capture and total cross-section measurements are presented using an example of the experiments of ^{243}Am. The deduced uncertainties are categorized into five types.

1 Introduction

For detailed engineering designs and safety evaluations of nuclear reactor systems, accurate cross-section data are crucial. Especially, neutron-capture cross sections of minor actinides (MAs) and long-lived fission products (LLFPs) are significant for evaluating and transmuting radioactive wastes and designing various innovative reactor systems[1–3]. However, accurate measurements of these cross sections are particularly challenging because of their high specific radioactivity. As a result, highly accurate data are not available for all materials.

To address this issue, Hokkaido University, Tokyo Institute of Technology, and JAEA collaborated to build the Accurate Neutron-Nucleus Reaction measurement Instrument (ANNRI). The ANNRI is installed at Beam Line No. 04 of the Materials and Life science experimental Facility (MLF) of the Japan Proton Accelerator Research Complex (J-PARC), which provides the world's most intense pulsed neutron source. A high flux neutrons supplied by the 1-MW pulsed spallation neutron source in the J-PARC allows better conditions for radioactive sample background reduction. Since 2008, high-intensity pulsed neutron measurements of the neutron capture and total cross sections of MAs, LLFPs, and some stable isotopes have been conducted [4].

This paper presents detailed information about the analysis methods and estimated uncertainties in TOF experiments in the ANNRI by using capture, and total cross-section measurements of ^{243}Am as examples [5].

2 Facilities

2.1 ANNRI Beamline

Pulsed neutrons are produced by the spallation reaction in the mercury target in MLF, by injecting 3-GeV pulsed protons from the rapid cycling synchrotron in the J-PARC [6]. Produced neutrons are slowed down in three types of supercritical hydrogen moderators and transported to beamlines in MLF. A Coupled Moderator, which produces the most intense neutron beam, supplies neutrons to the ANNRI.

Figure 1 shows the vertical cross-sectional view of ANNRI. To provide a high-quality neutron beam to detector systems, neutron collimators, neutron resonance filters, and disk choppers have been installed.

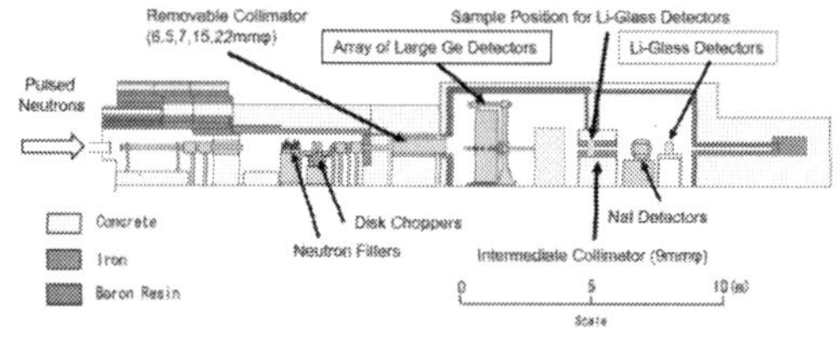

Figure 1. Vertical cross-sectional view of ANNRI.

2.2 Detector systems in ANNRI

The ANNRI has three detector systems. An array of large Ge detectors and a NaI(Tl) spectrometer were installed at flight lengths of 21.5 and 28-m, respectively. Measurements of capture cross-section are performed using these two gamma-ray spectrometers. Li-glass detectors are installed at a flight length of 28.5 m and are used for total cross-section measurements.

The array of Ge detectors consists of two cluster-type Ge detectors, eight coaxial-type Ge detectors, and anti-coincidence shields around each Ge detector [7]. There are seven Ge crystals in each cluster-type Ge detector. The array of Ge detectors consequently consists of 22 Ge crystals. The photo-peak efficiency of the array of Ge detectors is 2.28 ± 0.11 % for 1.33 MeV rays. Signals from the Ge

*e-mail: kimura.atsushi04@jaea.go.jp

detectors are fed into CAEN V1724 (14 bit, 100 MHz) ADC boards [8]. The TOF and pulse height(PH) are recorded in an event-by-event mode.

The NaI(Tl) spectrometer comprises two NaI(Tl) detectors with neutron and γ-ray shields [4]. The cylindrical NaI(Tl) detectors, 330 mm in diameter and 203 mm in length and 203 mm in diameter and 203 mm in length, are located at 90 degrees and 125 degrees to the neutron-beam line, respectively. If necessary, the detector at 125 degree is used to reduce the effects of angular distributions of capture γ rays from resonances with $l > 0$. Signals from NaI(Tl) detectors are analyzed in the DAQ with a CAEN 1720 (12bit 250MHz), and TOF and PH are recorded in the same manner as the array of Ge detectors.

Two types of Li-glass detectors are installed in ANNRI [9]. A ^{6}Li-glass detector is used to obtain the neutron TOF transmission spectra. The scintillator of the ^{6}Li-glass detector is GS-20 produced by Saint-Gobain with a dimension of 50 mm × 50 mm and a thickness of 1 mm using enriched ^{6}Li (> 95%). Neutrons are detected via the ^{6}Li(n,α)^{3}H reaction. A ^{7}Li-glass detector is used to determine the background shape due to γ rays. The scintillator of the ^{7}Li-glass detector is GS-30 (^{7}Li enrichment higher than 99.99%) produced by Saint-Gobain with the same size and chemical component as the GS-20 scintillator. Thus, the neutron sensitivity of the ^{7}Li-glass detector is significantly smaller than that of the ^{6}Li-glass detector due to the small neutron absorption cross section. However, both detectors have almost the same sensitivities to gamma rays. The ^{7}Li-glass detector is located on the upstream side of the ^{6}Li-glass detector. Signals from the detectors are digitized in a CAEN 1720, and TOF and PH are recorded.

2.3 Neutron beam at ANNRI

The neutron intensity at the 21.5 m sample position of the ANNRI under proton beam power of 17.5 kW is compared to those of DANCE at LANSCE and n TOF at CERN in Figiure 2 [10]. The current proton beam power is 800 kW in July 2022. The deduced present neutron intensity and the expected neutron intensity under the future 1-MW operation are also presented in Figiure 2. One finds that the present and future neutron intensities of ANNRI are much higher than those of the other facilities.

3 Types of uncertainties in TOF experiments

3.1 Categolization

In an analysis of TOF experiments, uncertainties are deduced step by step. Here, the deduced uncertainties are categorized into the following five types:

- Statistical uncertainty
- Type-1: proportional to the deduced cross section (normalization uncertainty),
- Type-2: proportional to other cross sections (impurities),

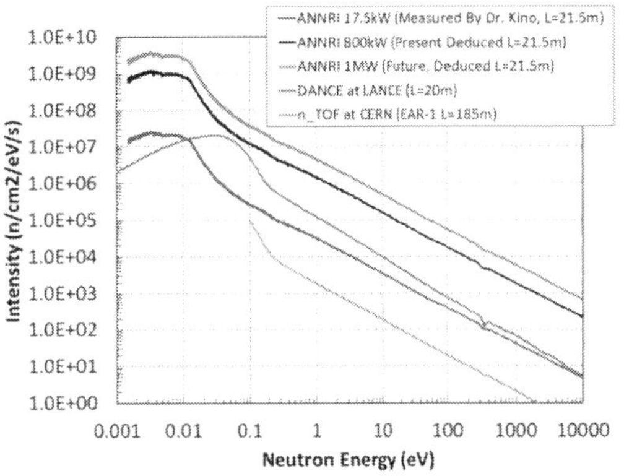

Figure 2. Neutron intensities per second at the 21.5-m sample position of ANNRI under 17.5-kW operation compared to that of DANCE at LANSCE and n TOF at CERN [10]. The deduced present neutron intensity under 800-kW operation and the expected neutron intensity under the future 1-MW operation are also shown.

- Type-3: constant on the deduced cross section,
- Type-4: the combination of some uncertainties.

One is statistical uncertainty. Uncertainties due to dead time correction and background subtraction include statistical uncertainty. The others are systematic uncertainties categorized into four types. Type-1 is the uncertainty proportional to the deduced cross section; for example, uncertainties due to normalization and pulse height weighting technique are categorized in Type-1 uncertainties. Type-2 is the uncertainty proportional to other cross sections; uncertainty due to impurity subtraction (sample composition) is categorized in this uncertainty. Type-3 is the uncertainty proportional to the transmission ratio and makes constant uncertainty in the deduced cross section; only neutron intensity normalization at the total cross-section measurement belongs to this type. Type-4 is the uncertainty due to a combination of some uncertainties; for example, the uncertainty due to self-shielding and multiple scattering belongs to this type. This uncertainty is deduced from sample shape uncertainty with Monte Carlo simulations.

3.2 Uncertainties in Resonance analysis

In the resonance region, resonance analysis is typically applied with resonance analysis codes, e.g., REFIT and SAMMY codes. Statistical uncertainty and Type-2 (impurities) are considered in the code. For Type-1 (normalization) and Type-3 uncertainties, some methods were proposed by P. Schillebeeckx et al.[11], and a new approach was presented by S. Endo et al. in this workshop. However, only a few studies have been reported. Unfortunately, for Type-4 uncertainty, there is still no fixed way to evaluate it. Further research is required.

4 Analysis methods and uncertainties in TOF experiments in ANNRI

This section presents the analysis and uncertainty estimation methods in TOF experiments in ANNRI using the ^{243}Am capture and total cross-section measurements as examples. Detailed information about the ^{243}Am experiments is described in Ref. [5].

4.1 Setup of ^{243}Am experiments

An ^{243}Am sample with a nominal activity of 240 MBq was prepared for the experiments. Americium dioxide (AmO_2) powder was mixed with Y_2O_3 powder and pressed into a pellet with a diameter of 10.0 ± 0.1 mm and a thickness of 0.5 ± 0.1 mm. The pellet was encapsulated in an Al container with a diameter of 22 mm and a thickness of 0.1 mm. Similarly, the dummy sample was prepared, which contained only a Y_2O_3 pellet. The calorimetric method was applied to determine the activity of the ^{243}Am sample. The determined activity was 281.8 ± 0.3 MBq. The isotopic composition of the Am sample was analyzed by thermal ionization mass spectrometry and alpha-ray spectroscopy. The isotopic purity of ^{243}Am was determined to be 97.67 ± 0.02 %.

The array of Ge detectors was used for the neutron capture cross-section measurement. The ^{243}Am sample was put in a bag of fluorinated ethylene propylene (FEP) films and attached to a sample holder. The measurement time was about 13 hours. For the background estimation, measurements were carried out with the dummy sample, the FEP bag without any sample (Blank), and an enriched ^{208}Pb sample. Measurements with an enriched ^{10}B sample with a diameter of 5.0 ± 0.1 mm and a weight of 170.2 ± 0.1 mg were also made to determine the energy dependence of the incident neutron flux.

The ^{6}Li-glass and ^{7}Li-glass detectors were utilized for transmission measurements. The ^{7}Li-glass detector was installed on the upstream side of the ^{6}Li-glass detector. A lead filter with a thickness of 87.5 mm was installed in the neutron beamline to reduce the background gamma ray on the beamline and to control the event rate of the Li-glass detectors. The background was estimated with the black resonance technique using neutron notch filters of ^{nat}Mn, ^{nat}Co, ^{nat}In and ^{nat}Ag. The measuring times of the ^{243}Am and the dummy samples were 16 h and 8 h, respectively.

4.2 Capture cross-section measurement

Figure 3 shows the data analysis procedure for the capture cross-section measurements in ANNRI. In the analysis of the ^{243}Am experiment, uncertainties were deduced step by step.

The Pulse Height Weighting Technique (PHWT) was usually applied to capture measurements [12]. The response functions of the array of Ge detectors were calculated by Monte Carlo simulation. The weighting functions for the sample were calculated from the response functions. Figure 4 shows weighted TOF spectra of the ^{243}Am and dummy samples. In PHWT, the most significant uncertainty originated from correcting undetected events below the threshold level. The correction was done by extrapolating the PH spectrum to under the threshold. In the analysis, an exponential function fitted the spectrum and the extrapolation of the spectrum below the threshold level. The extrapolation's uncertainty was deduced from the difference between the flat and exponential extrapolation. Consequently, the deduced uncertainty was about 3%. The deduced uncertainties are proportional to the final cross section. In short, this uncertainty belongs to Type-1.

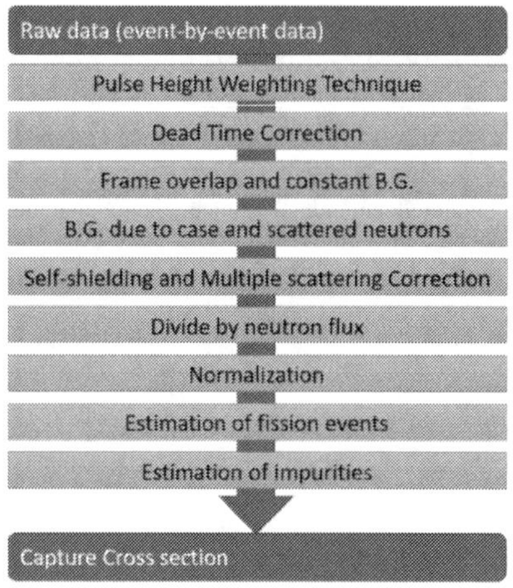

Figure 3. Flow chart of the data analysis procedure for the capture cross-section measurement.

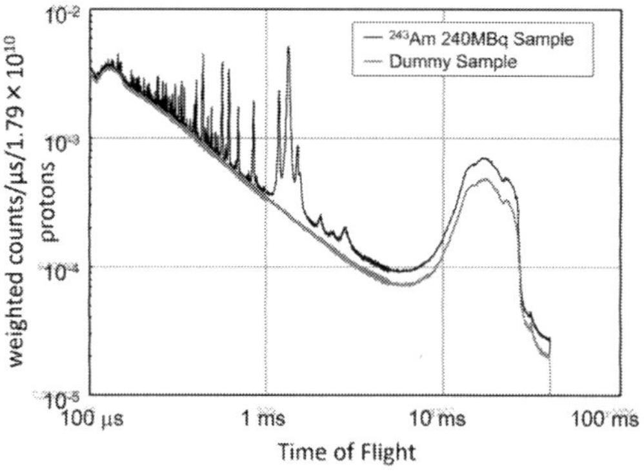

Figure 4. Weighted TOF spectra of the ^{243}Am and dummy samples.

The second step is the dead-time correction. In the analysis, the dead-time correction was applied with the extended dead-time model using a dead time of 5.67 ± 0.02 μs per event. This dead time uncertainty was small enough to be negligible. The remaining uncertainty was statistical uncertainty. Figure 5 shows the weighted

TOF spectrum of ^{243}Am sample with the dead time correction and the deduced uncertainty.

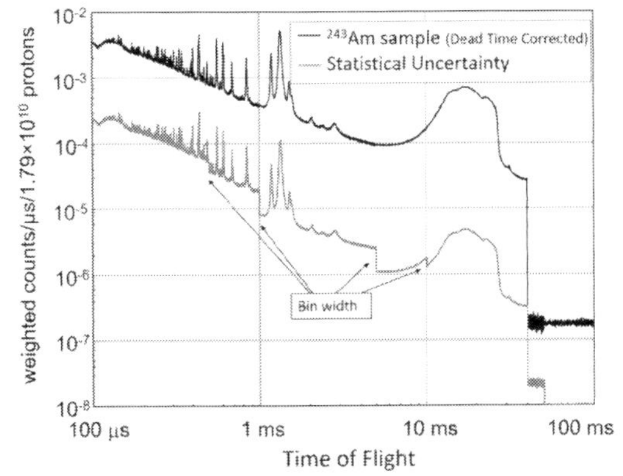

Figure 5. Weighted TOF spectrum of ^{243}Am sample with the dead time correction and the deduced uncertainty.

The third step is overlap and constant B.G. estimation and subtraction. In the experiment, the first 58 of every 62 shots were provided to MLF, and the last four were delivered to the other facilities. Thus, every 2.48 s, a broad TOF spectrum to 200 ms was obtained. The frame overlap background and constant background in the normal TOF range earlier than 25 ms were deduced by fitting an exponential decay plus constant function to this broad range TOF spectrum in the TOF region from 40 to 65 ms[13]. In this step, uncertainties were deduced from the uncertainties of the fitted parameters. The deduced uncertainties are energy-dependent, and the value of each TOF bin is correlated. Therefore, the uncertainty belongs to Type-4.

The background due to the aluminum container and Y_2O_3 included in the sample was derived from the spectrum of the dummy sample. The background due to scattered neutrons by ^{243}Am was estimated by the spectrum of the ^{208}Pb sample by considering the differences in the elastic-scattering cross-section of ^{243}Am and ^{208}Pb. The elastic-scattering cross-section was taken from the JENDL-4.0 [14]. Figure 6 shows the TOF spectra of ^{243}Am sample before and after the background subtraction. In this step, uncertainties were deduced from the uncertainties of the evaluated elastic cross-sections in JENDL-4.0 and the statistical uncertainties of the blank, dummy, and lead sample measurements. Therefore, the uncertainty is a sum of statistical uncertainty and Type-4 uncertainty.

Self-shielding and multiple scattering correction factors were calculated with the Monte Carlo simulation code. The uncertainties of the areal densities mainly come from the uncertainty of the pellet diameter. Therefore, uncertainties of the correction factors were derived from the uncertainty of the pellet diameter. To derive the uncertainties, the ideal sample with area densities larger by the 1-σ uncertainty than those in the actual sample was considered. The difference in the correction factors between the

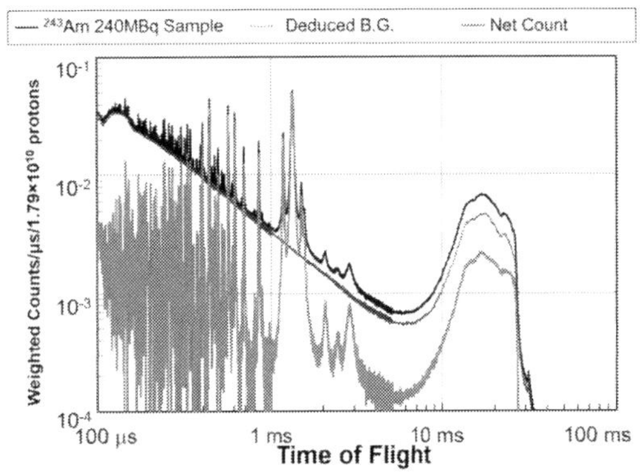

Figure 6. Deduced background and TOF spectrum of ^{243}Am sample before and after the background subtraction.

actual and ideal samples was used as the uncertainty. The uncertainty depends on the uncertainty of the areal density of ^{243}Am and the evaluated cross section. The uncertainty is categorized into Type-4.

The relative incident neutron energy spectrum was determined by measuring 478-keV gamma-rays from the ^{10}B(n, $\alpha\gamma$)^{7}Li reaction. The ^{10}B(n, $\alpha\gamma$) cross section was taken from JENDL-4.0. The dead-time, neutron self-shielding, and multiple scattering corrections were applied. The uncertainty of the dead-time correction was negligible. The uncertainty of the neutron self-shielding and multiple scattering correction was deduced from the uncertainties of the ^{10}B sample mass and diameter. The obtained relative neutron flux and uncertainties are shown in Figure 7. The uncertainty is a sum of statistical uncertainty and Type-4 uncertainty. The relative capture cross section of the ^{243}Am sample was calculated from the deduced net spectrum of the ^{243}Am sample and the relative neutron flux.

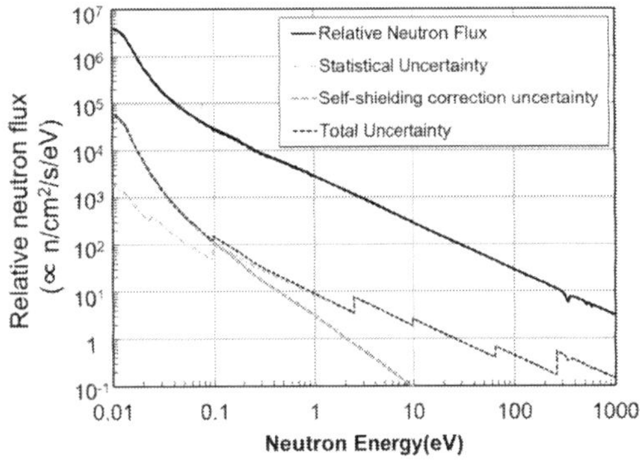

Figure 7. Relative incident neutron energy spectrum determined by measuring the 478 keV gamma rays from the ^{10}B(n, $\alpha\gamma$)^{7}Li reaction.

In the analysis, the absolute capture cross section at the 1.356 eV resonance was derived with another analysis method [5]. The relative capture cross section of the ^{243}Am sample was normalized to the derived absolute cross section. The normalization uncertainty was derived from the uncertainty of the absolute cross section. The deduced uncertainty was proportional to the capture cross section of the ^{243}Am. The uncertainty is categorized into Type-1.

In the last step, contributions from the influences of the impurities and fission events were subtracted. The contributions were deduced using the resolution function of ANNRI, the capture and fission cross sections in JENDL-4.0, and the sensitivity of fission events to the capture events (2.3 ± 1.0). The capture cross sections of ^{243}Am were derived by subtracting the contributions. The uncertainties were derived from the uncertainties of the abundances of the impurities, JENDL-4.0, and the ratio of the sensitivity. The deduced uncertainties were proportional to the cross-sections of other reactions. The uncertainty is categorized into Type-2.

Figure 8 shows a part of the deduced capture cross section of ^{243}Am and the uncertainties in the neutron energy range from 1 to 10 eV.

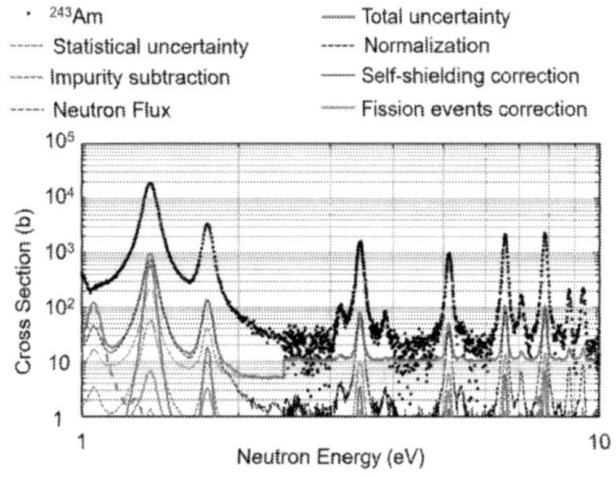

Figure 8. Deduced capture cross section of ^{243}Am and with a comparison to the deduced uncertainties in the neutron energy range from 1 to 10 eV.

4.3 Total cross-section measurement

The analysis procedure for the total cross-section measurements in ANNRI is shown in Figure 9.

Figure 10 shows TOF spectra of the ^{243}Am and dummy samples measured with the ^{6}Li-glass and ^{7}Li-glass detectors. The gamma-ray backgrounds deduced with the ^{7}Li-glass detector were much smaller than the neutron TOF spectra with the ^{6}Li-glass detector.

The dead-time correction was applied to the TOF spectra using a fixed dead time of 1264 ± 4 ns (316 ± 1 ch.). In this experiment, the event rate is less than 1.6 events per 100 μs and the dead time ratio was less than 2 %. The uncertainty of the dead time correction depends not on the uncertainty of the dead time (1264 ± 4ns) but on the statistical uncertainty.

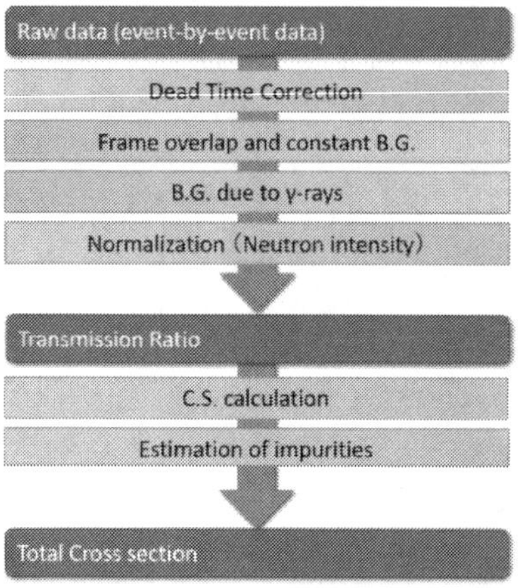

Figure 9. Flow chart of the data analysis procedure for total cross-section measurements in ANNRI.

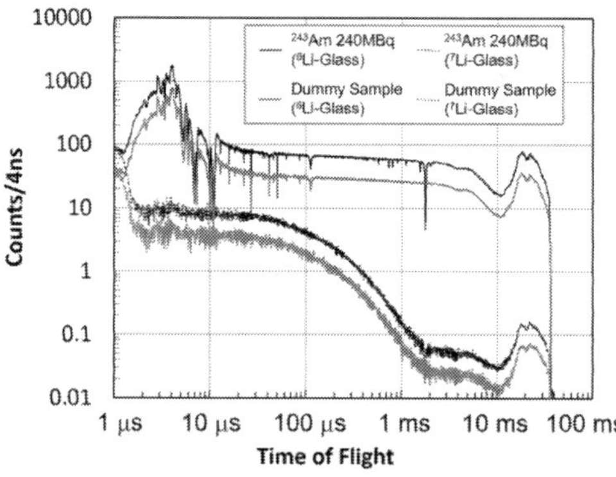

Figure 10. TOF spectra of the ^{243}Am and dummy samples measured with the ^{6}Li-glass and ^{7}Li-glass detectors.

Frame overlap and constant B.G. estimation are the same as those in the capture measurements. The deduced background is shown in Figure 11 in comparison with the broad and normal range TOF spectra. In this step, uncertainties were deduced from the uncertainties of the fitted parameters. Therefore, the uncertainty belongs to Type-4.

The TOF spectra obtained using the ^{7}Li-glass detector were used to determine the observed background events due to gamma rays. Gamma-ray efficiencies were slightly different between the ^{7}Li-glass and ^{6}Li-glass detectors. To correct the difference, the black resonance technique was applied. The TOF spectrum measured with the ^{7}Li-glass detector was normalized to that of the ^{6}Li-glass detector at the four dips by ^{55}Mn, ^{59}Co, ^{115}In and ^{109}Ag. The deduced normalization factor was 1.07 ± 0.03. Uncertainties due to this background subtraction were deduced from the

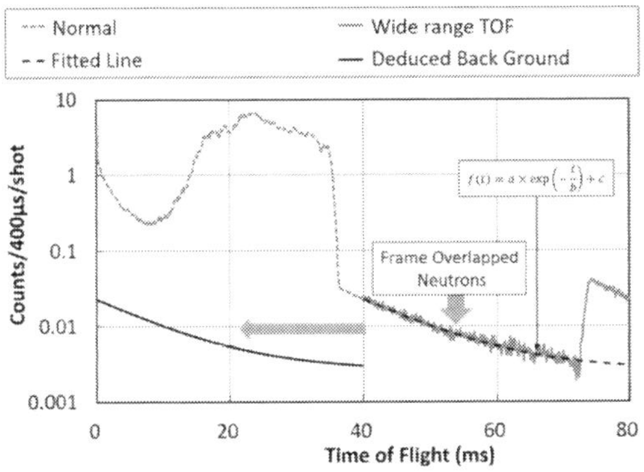

Figure 11. Broad range TOF spectrum after 40 ms and normal TOF spectrum in comparison with the fitted result and the deduced frame overlap and constant backgrounds.

uncertainties of the normalization factor (Type-4) and the statistical uncertainty.

In the experiment, an Al foil was set at the sample position of the array of large Ge detectors. Prompt gamma-rays from the ^{27}Al(n,γ) reaction were measured simultaneously with the Ge detector array to monitor the relative neutron intensities. The relative neutron intensity per proton beam shot for each sample was derived by dividing the number of gamma rays by the number of proton beam shots in the measurement. The factor was 1.0068 ± 0.0004 for the Am sample. Uncertainties were calculated from the uncertainty of this factor. The deduced uncertainties were proportional to the transmission ratio, i.e., constant uncertainty in the cross section (Type-3). The transmission ratio was obtained by dividing the TOF spectrum of the sample by that of the dummy. The deduced transmission ratio is shown in Figure 12.

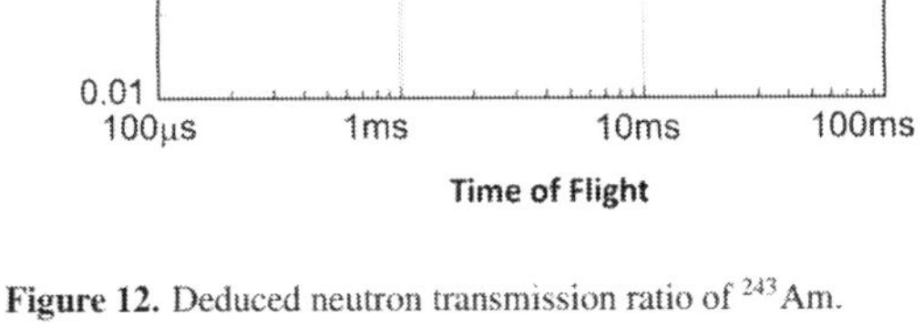

Figure 12. Deduced neutron transmission ratio of ^{243}Am.

The total cross section of the ^{243}Am sample was derived using the transmission ratios and the areal density of ^{243}Am. Uncertainty was calculated from the uncertainty of the areal density. The deduced uncertainty was proportional to the cross section, and this uncertainty belongs to Type-1.

In the last step, contributions from the influences of the impurities were subtracted. This procedure is the same as that in the capture measurements. The uncertainties were derived from the uncertainties of the abundances, JENDL-4.0, and the ratio of the sensitivity. The deduced uncertainties were proportional to the cross-sections of other reactions. The uncertainties are categorized into Type-2.

Deduced total cross section of ^{243}Am in the neutron energy range from 0.01 to 10 eV is shown in Figure 13 with a comparison to the deduced uncertainties.

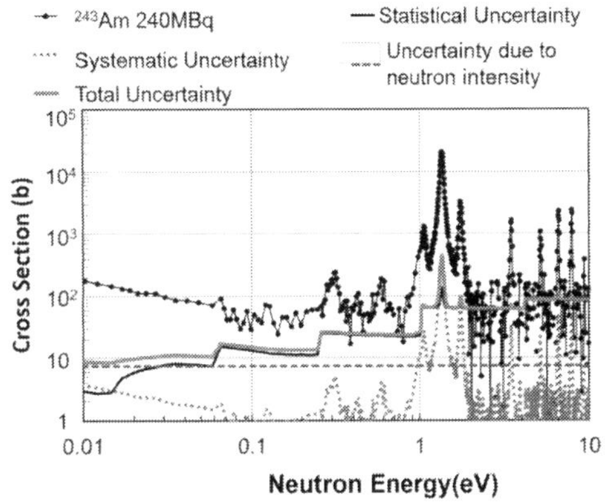

Figure 13. Deduced total cross section of ^{243}Am and with a comparison to the deduced uncertainties in the neutron energy range from 0.01 to 10 eV.

5 Conclusion

In the analysis of TOF experiments in ANNRI, uncertainties are deduced step by step. In experiments in ANNRI, each deduced uncertainty is reported separately. The deduced uncertainties are categorized into the five types.

Experimentalists have been trying to estimate all of the uncertainties. However, there is a possibility of missing some other uncertainties. The missing always results in an underestimation.

References

[1] M. Salvatores, A Report by the Working Party on International Evaluation Co-operation of the Nuclear Science Committee (2008)

[2] T. Sugawara, K. Nishihara, K. Tsujimoto, T. Sasa, H. Oigawa, Journal of Nuclear Science and Technology **47**, 521 (2010)

[3] H. Iwamoto, K. Nishihara, T. Sugawara, K. Tsujimoto, Journal of Nuclear Science and Technology **50**, 856 (2013)

[4] M. Igashira, Y. Kiyanagi, M. Oshima, Nuclear Instruments and Methods in Physics Research Section

A: Accelerators, Spectrometers, Detectors and Associated Equipment **600**, 332 (2009)

[5] A. Kimura, S. Nakamura, K. Terada, T. Nakao, K. Mizuyama, N. Iwamoto, O. Iwamoto, H. Harada, T. Katabuchi, M. Igashira et al., Journal of Nuclear Science and Technology **56**, 479 (2019)

[6] F. Maekawa, M. Harada, K. Oikawa, M. Teshigawara, T. Kai, S. ichiro Meigo, M. Ooi, S. Sakamoto, H. Takada, M. Futakawa et al., Nuclear Instruments and Methods in Physics Research Section A: Accelerators, Spectrometers, Detectors and Associated Equipment **620**, 159 (2010)

[7] T. Kin, K. Furutaka, S. Goko, H. Harada, A. Kimura, F. Kitatani, S. Nakamura, M. Ohta, M. Oshima, Y. Toh et al., Journal of the Korean Physical Society **59**, 1769 (2011)

[8] T. Nakao, K. Terada, A. Kimura, S. Nakamura, O. Iwamoto, H. Harada, T. Katabuchi, M. Igashira, J. Hori, *Developments of a new data acquisition system at ANNRI*, in *EPJ Web of Conferences* (EDP Sciences, 2017), Vol. 146, p. 03021, `https://doi.org/10.1051/epjconf/201714603021`

[9] K. Terada, A. Kimura, T. Nakao, S. Nakamura, K. Mizuyama, N. Iwamoto, O. Iwamoto, H. Harada, T. Katabuchi, M. Igashira et al., Journal of Nuclear Science and Technology **55**, 1198 (2018)

[10] K. Kino, M. Furusaka, F. Hiraga, T. Kamiyama, Y. Kiyanagi, K. Furutaka, S. Goko, H. Harada, M. Harada, T. Kai et al., Nuclear Instruments and Methods in Physics Research Section A: Accelerators, Spectrometers, Detectors and Associated Equipment **626-627**, 58 (2011)

[11] P. Schillebeeckx, B. Becker, Y. Danon, K. Guber, H. Harada, J. Heyse, A. Junghans, S. Kopecky, C. Massimi, M. Moxon et al., Nuclear Data Sheets **113**, 3054 (2012), special Issue on Nuclear Reaction Data

[12] R.L. Macklin, J.H. Gibbons, Phys. Rev. **159**, 1007 (1967)

[13] A. Kimura, T. Fujii, S. Fukutani, K. Furutaka, S. Goko, K.Y. Hara, H. Harada, K. Hirose, J.i. Hori, M. Igashira et al., Journal of Nuclear Science and Technology **49**, 708 (2012)

[14] K. Shibata, O. Iwamoto, T. Nakagawa, N. Iwamoto, A. Ichihara, S. Kunieda, S. Chiba, K. Furutaka, N. Otuka, T. Ohsawa et al., Journal of Nuclear Science and Technology **48**, 1 (2011)

The Covariance of PFNS Results from the Chi-Nu Experiment

Keegan J. Kelly[1,*], *John* M. O'Donnell[1], and *Matthew* Devlin[1]

[1]Los Alamos National Laboratory, Los Alamos, NM, 87544, USA

Abstract. The prompt fission neutron spectrum (PFNS) from neutron-induced fission is a fundamental quantity for the behavior of nuclear reactors, and has been measured many times on a wide variety of nuclei and covering different ranges of incident and emitted neutron energies. However, results from past measurements are frequently called into question in modern nuclear data evaluations because of a lack of thorough experimental documentation and incomplete uncertainty analyses. The Chi-Nu experiment at Los Alamos National Laboratory was designed to produce high-precision measurements of the PFNS of major actinides over a wide range of incident and emitted neutron energies, and with the documentation and covariance analysis required to ensure that the results of this experiment maintain their impact long into the future, thereby avoiding this pitfall of past measurements. In this work we describe the Chi-Nu experiment along with summaries of the treatment of and methods developed to address two important components of the analysis of Chi-Nu data: random-coincidence backgrounds and MCNP simulations. Furthermore, we describe the first results for correlations not just between all data points collected on a single target nucleus, but also between all data points from separate Chi-Nu measurements on ^{235}U and ^{239}Pu. These correlations are important for accurately calculating ratios of the PFNS from one actinide to another, which are rare and can be informative for nuclear data evaluation efforts.

1 Introduction

Measurements of the prompt fission neutron spectrum (PFNS) began near the discovery of nuclear fission [1, 2], spanning a wide range of nuclei and coverages in incident and emitted neutron energy from fission. Despite the multitude of impressive PFNS measurements carried out over the last seven decades, the lack of documentation, reliability, and detailed uncertainty analyses has caused many of these expensive and potentially useful data sets to either have uncertainties increased to reflect potential errors in the results or to be discarded entirely in nuclear data evaluations [3, 4]. The Chi-Nu experiment at Los Alamos National Laboratory (LANL) was designed to measure the PFNS for major actindes over a wide range of incident and emitted neutron energies from fission, including a detailed documentation and covariance analysis and thorough documentation. In this way, results from Chi-Nu experiments are in position to avoid the same pitfalls rendering many past measurements either less impactful or altogether irrelevant.

In this manuscript, we describe the Chi-Nu experimental environment in Sec. 2, focusing on efforts to reduce the potential for bias in the experimental results. Two important components of the analysis and sources of systematic uncertainty for results from a Chi-Nu experiment are discussed in Secs. 3 and 4: random-coincidence background measurements and MCNP®-based corrections for neutron interactions in the environment, respectively [5, 6]. Methods developed to define covariances from each of these sources of uncertainty are described in the subsections within Secs. 3 and 4. Correlation definitions not just between different incident and emitted energies of a single target nucleus, but between results from entirely different target nuclei in separate experiments are described in Sec. 5. Finally, concluding remarks are given in Sec. 6.

2 Experimental Setup

The incident neutron beam for Chi-Nu experiments is the result of 800 MeV proton beam bursts approximately 150 ps wide and typically separated by 1.78889 μs from neighboring beam pulses incident on a tungsten target. Proton-induced spallation reactions in the tungsten target yield, among other products of the reaction, neutrons with a continuous energy spectrum spanning the practical energy range of roughly 0.7–20 MeV or more. The Chi-Nu experiment at flight path 15L of the Weapons Neutron Research (WNR) facility allows neutrons to travel a 21.5 m path length before impinging on a parallel-plate avalanche counter (PPAC) target [7] containing a series of deposits of the actinide of interest, typically totaling approximately 100 mg of target material with roughly 400 μg/cm^2 target thickness on each deposit. Fission events are detected in the PPAC targets, and neutrons from fission are detected in one of two neutron detector arrays in separate experiments, both of which can be seen in Fig. 1. A 22-detector Li-glass array is used to detect neutrons with energies below approximately 1.5 MeV, and a 54-detector liquid scintillator array is used to detect neutrons with energies above 0.89 MeV. Incident neutron energies are assigned based

*e-mail: kkelly@lanl.gov

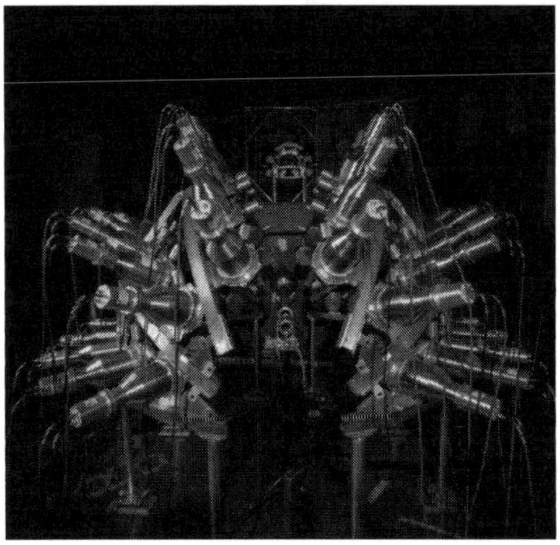

Figure 1: The Chi-Nu 54-detector liquid scintillator array is shown in the center of this image. The Li-glass array used in separate a experiment is seen above the shielding walls in the upper center of this image.

on the time of flight from the proton spallation reaction and a fission detection. Similarly, outgoing neutron energies are assigned based on the difference between neutron detection and fission detection, though it should be noted that the outgoing neutron energy from time of flight does not necessarily correspond to the neutron energy as it was emitted from the fission event (see Sec. 4). From these detector arrays and this white incident neutron beam, Chi-Nu experiments have published results for the PFNS of ^{239}Pu and ^{235}U [8–10], with ^{238}U and ^{240}Pu results forthcoming.

The Chi-Nu experimental environment was specifically designed to reduce potential bias on results from common sources of spectrum distortion, specifically neutron scattering in the environment. Perhaps the most impactful effort to reduce scattering was the creation of a "get-lost basement" beneath the neutron detection area. This consists of a thin Al floor on which the PPAC and neutron detectors sit, with 2 m of empty space below that. The walls of the experimental environment were also placed at least 2–3 m from the detection environment. Both of these features place hydrogenous (and therefore highly scattering) concrete walls far enough away from neutron detectors that a neutron scattering from the concrete walls and floors will be outside the fission-neutron time coincidence window by the time it gets back to the experimental environment [11]. In terms of analysis development, the multi-year effort to include in the MCNP simulation all materials within the experimental environment that have any impact on environmental neutron scattering prior to detection has been of fundamental importance for properly interpreting data from Chi-Nu experiments. For more details of the Chi-Nu experimental setup, see Refs. [9–11].

3 Random-Coincidence Background Measurements and Uncertainties

The majority of the data observed in terms of fission-neutron time coincidences is from random coincidences, i.e., neutrons and fissions that are measured within a coincidence time window of each other, but that are not necessarily correlated with one another. These random coincidences could come from fission coincidence with neutrons from another neutron-emitting reaction on a target, epithermal neutrons in the environment, or another source of background neutrons. Alternatively, if the target actinide of interest also emits a substantial amount of α particles, which typically can not be completely separated from fission detections, then neutrons measured to be in coincidence with these α particles would also create a substantial random-coincidence background. The processes among others contribute to the data observed within the 1.8 μs and 500 ns coincidence windows defined for measurements of incident and outgoing neutrons, subject to the $\approx$1 ns at 1σ time resolution of the full detection setup.

To account for this background, and method was developed to use the singles (pre-coincidence) neutron and fission data directly from the experiment to determine the rate of random-coincidence detection [12]. In short, the Poisson-distributed fission and neutron detection rates as a function of time relative to the creation of the incident neutron beam allows for the calculation of the probability of (1) detecting a fission even at a time t_f, (2) not detecting a neutron over the fission-neutron coincidence time difference, and (3) detecting a coincident neutron at a time t_n. This method produces background spectra with negligible statistical uncertainty compared with the statistics of the foreground coincidence data.

While the statistical uncertainty of results from this random-coincidence background measurement method is negligible compared with the statistics of the coincidence data themselves, it is possible for the background to be subject to a source of systematic error rendering the background spectrum less accurate. However, it can be difficult to determine the accuracy of the background based on comparisons with the coincidence data alone (e.g., if the background is lower than the data, are there really more counts above background than expected, or is there a deficiency in the background?). Thus, understanding how experimental conditions impact the accuracy of the random-coincidence background is fundamentally important. Most recently, the impact of detection rate changes and correlations between fission and neutron detection rates were investigated as a potential source of systematic biases in this background spectrum [13]. From this work, a few main conclusions were reached:

1. The rates of all detectors of the same type can be summed prior to calculation of the background spectrum without introducing an error

2. A single detection rate can change in any way without introducing an error in the background as long as the other relevant rates remain constant

3. If two detection rates are changing, then it is almost always true that an error in the random-coincidence background will be generated

4. Each rate can follow a Poisson distribution without introducing an error in the background provided that the entire experiment is long enough to obtain the Poisson mean value of each rate

5. A positive correlation between detection rates from, e.g., an external particle beam, causes the calculated background to be systematically lower than the correct background spectrum, and the shape is distorted as well

Conclusion (1) is potentially extremely useful for time spectra because it reduces the total time required for background spectrum calculation by a factor of the product of the number of each detector type. However, when energy spectra are needed for the final result, this is less useful if the distance from each detector of either type is different (implying the need to calculate the time-dependent background for each combination before conversion to energy). Conclusions (2) and (3) above can be calculated analytically, and are not too surprising, though it is useful that variations in a single detector rate can be handled without issue. Conclusion (4) is comforting in that it is an inevitable reality of virtually all experiments. On the other hand, conclusion (5) is concerning, and is believed to be a dominant source of systematic errors (as opposed to uncertainties) in the random-coincidence background spectra calculated for Chi-Nu experiments.

In Ref. [9], a method to correct the random-coincidence background spectrum was developed and applied. This method was later described in more detail in Ref. [13]. In short, when a time shift is introduced into the coincidence analysis for one detector type (typically the fission detection time here) such that the possibility of true coincidence detection is negligibly small but the rate correlations between detector types are approximately maintained, then the difference between the measured data with the time shift and the random-coincidence background calculated for the same analysis can be used as a correction to the random-coincidence background. This approach reduces the potential for systematic error in the background to a negligible level, but at the cost of increasing the statistical uncertainty in accordance with the statistics of the coincidence data in the time-shifted analysis. However, multiple time shifts can be averaged together to reduce the statistical uncertainty to the desired level. For more details on this method and the study of sources of systematic error from this method, see Refs. [12, 13].

4 MCNP®-based Corrections

Nearly all PFNS experiments rely on measurements of the ^{252}Cf spontaneous fission PFNS for results on any other nucleus [1]. Some measurements use ^{252}Cf in a direct ratio to the data from another actinide to obtain the desired PFNS result, and others use ^{252}Cf to measure an efficiency that is then used in the calculation of the desired PFNS, but these and all similar methods reduce down to the same method of measuring the PFNS as a ratio to the ^{252}Cf PFNS. Results from the Chi-Nu experiment are some of the only data sets to *not* measure the PFNS as a ratio to ^{252}Cf. Instead, the Chi-Nu team relies on highly-detailed and vetted MCNP simulations of the experimental environment [3, 14]. While the use of MCNP simulations allows for a detailed understanding of how neutrons interact with the experimental environment, simulations themselves invite a series of other systematic uncertainties from, e.g., uncertainties in post-processing parameters required to match the simulation to the data, and uncertainties in the nuclear data libraries used as input to the simulation. This latter source of uncertainty, generally termed MCNP input nuclear physics uncertainty here, is a particularly challenging source of uncertainty to quantify, and is the focus of the remainder of this section.

A simple approach to the problem of quantifying the impact of this uncertainty source on a result utilizing MCNP simulations would be to vary the relevant cross sections, and run a series of sequential simulations. However, given that there are potentially a large number of cross sections to vary, each cross section may require hundreds of variations to acquire sufficient statistics, and each simulation can take hours or more, this approach quickly becomes intractable. There are also a variety of methods available within MCNP to test the sensitivity of a simulation to input nuclear data, but none of these methods allow the use to vary data according to the covariance of the input nuclear data, or allow for total freedom of choice of the user to vary different numbers of various cross sections in selected cells, for example. As such, the Chi-Nu team has attempted to develop some flexible methods of quantifying this source of uncertainty with reduced calculation time. The three most useful methods are described below.

4.1 Calculation of a Bayes-like Factor

For the purposes of Chi-Nu experiments, the output of a simulation can be summarized in a matrix of neutron energy as it was emitted from the fission event (i.e., the initial emitted neutron energy) versus the neutron energy measured using time of flight. An example of this response matrix for the Li-glass detector array is shown in Fig. 2, which is reproduced from Fig. 10 of Ref. [9]. The structures seen in this matrix are representative of nuclear cross sections relevant to the neutron transport in the environment, as well as physical objects in the experimental area.

Using only two simulations, one with the original nuclear data input and another with varied nuclear data input, one can calculate a quantity, $\mathcal{B}$, that resembles a term known as the Bayes factor [15]. In short, the factor calculated here quantifies the relative likelihood between two different models describing the data, calculated separately for each bin of the response matrix. For the case of Chi-Nu PFNS measurements, the two models are (a) the change in input nuclear data did *not* alter the contents of the bin of the response matrix, and (b) the change in input nuclear data *did* alter the contents of the bin of the response matrix. It might initially seem that this information could be

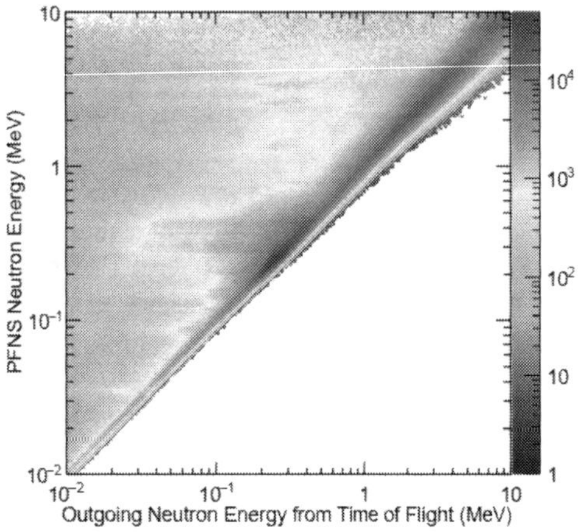

Figure 2: The neutron response matrix for the ^{239}Pu Chi-Nu PFNS experiment using the Li-glass detector array. This figure is a reproduction of Fig. 10 of Ref. [9].

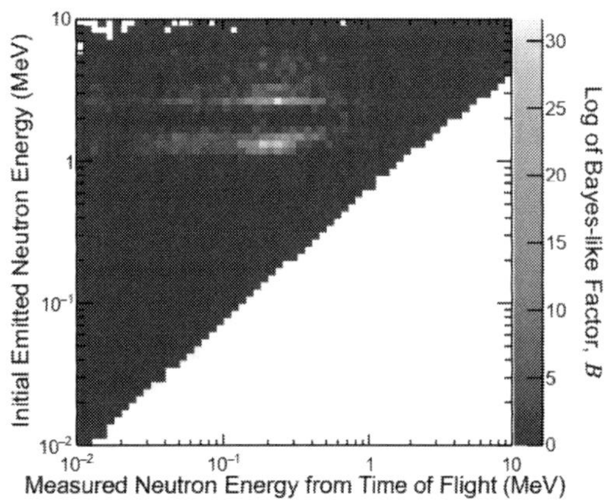

Figure 3: The matrix of the log of Bayes factors for increasing the ^{27}Al(n,n) cross section by 5%.

obtained in a simpler way by, for example, just dividing or subtracting the response matrices before and after altering the input nuclear data, but methods like this don't work well for bins with low statistics because variations are expected given the Poisson distributions from which the counts observed in each bin are drawn.

In short, the desired information is whether the counts in each bin of the response matrix were changed enough to support a model where the Poisson mean value, λ, has changed because of the change in the input nuclear data. Assuming unity for the prior probability for the model, the Bayesian posterior probability for the case where the Poisson mean value λ is assumed to have not changed can be expressed as

$$P(\lambda|c_1c_2) = \frac{\mathcal{P}_\lambda(c2)\mathcal{G}_\lambda(c_1+1,1)}{\int \mathcal{P}_\lambda(c2)\mathcal{G}_\lambda(c_1+1,1)d\lambda}, \tag{1}$$

where $\mathcal{P}_\lambda(c)$ is a Poisson distribution with mean value λ and observed counts c, $\mathcal{G}_\lambda(\alpha,\beta)$ is a gamma distribution with input parameters α and β defined as in the equation above, and c_1 and c_2 are the counts from the simulations before and after changing the input nuclear data. Note that the gamma distribution is included as the conjugate prior for the Poisson distribution. For the case where the λ is not constant, we instead define two Poisson mean values before and after the simulation, λ_1 and λ_2 and calculate the posterior probability as

$$P(\lambda_1\lambda_2|c_1c_2) = \frac{\mathcal{P}_{\lambda_2}(c2)\mathcal{G}_{\lambda_1}(c_1+1,1)}{\int \mathcal{P}_{\lambda_2}(c2)\mathcal{G}_{\lambda_1}(c_1+1,1)d\lambda_1 d\lambda_2}. \tag{2}$$

In this scenario, the ratio of the evidence (denominator) in the posterior probabilities can be calculated to yield, i.e.,

$$\mathcal{B} = \frac{\int \mathcal{P}_{\lambda_2}(c2)\mathcal{G}_{\lambda_1}(c_1+1,1)d\lambda_1 d\lambda_2}{\int \mathcal{P}_\lambda(c2)\mathcal{G}_\lambda(c_1+1,1)d\lambda}. \tag{3}$$

This factor, while similar to a Bayes factor, is not a formally correct derivation of a Bayes factor because, in part, the denominator of Bayesian posterior probabilities is not generally integrable. The reader is referred to Ref. [15] for formal discussions of the Bayes factor. Nevertheless, this factor $\mathcal{B}$ fulfills the needs of the Chi-Nu experiment in that it exposes which bins of the response matrix are impacted by variations in the input nuclear data.

An example matrix of $\mathcal{B}$ factors for increasing the ^{27}Al(n,n) elastic neutron scattering cross section by 5% is shown in Fig. 3. As can be seen, two horizontal bands in a region of counts in the matrix representing high levels of neutron downscattering (i.e., energy measured via time of flight is less than the initially emitted neutron energy) appear to be the dominant region impacted by this Al cross section. Similar studies have been done for the Chi-Nu environment on a suite of nuclei present in the experimental environment. While this method is extremely useful for determining to which nuclear data an experiment is sensitive and has been applied successfully in that respect to the Chi-Nu experiment, unfortunately no clear method of directly relating this factor to an rigorous uncertainty on input nuclear data was developed, nor is it possible to accurately reflect input nuclear data covariances with only one variation of input nuclear data. Thus, other methods are required for the final quantification of this uncertainty source.

4.2 Implicit Capture Simulation Manipulation

In MCNP, the user can choose, among many other options, between analog and implicit capture simulation techniques. Analog simulations proceed in a step-by-step process similar to how one might imagine a real experiment occurring. Alternatively, implicit capture simulations force a capture reaction to occur whenever it can, but with a weight corresponding to the probability that the capture reaction would have occurred. Thus, a portion of the neutron, in this case, is captured and the surviving portion of the neutron carries on with a reduced weight.

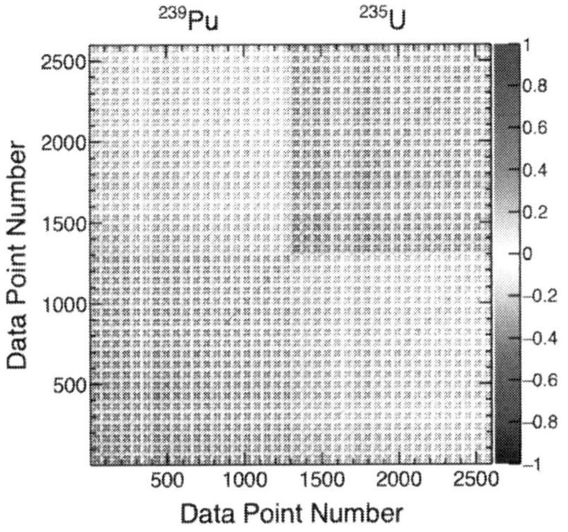

Figure 4: The cross-nuclide correlation matrix describing the correlations between Chi-Nu ^{239}Pu and ^{235}U PFNS results for all incident and emitted neutron energies. See the text for a description of this figure.

In Ref. [16] a method was developed to utilize the output of implicit capture simulations in order to vary input nuclear data cross sections in a post-processing calculation using the result from only the original simulation without no nuclear data variations. This is done by varying the weights of the surviving nuclei according to the altered probabilities for capture reactions occur if a cross section is changed. This method is powerful enough to incorporate covariances for input nuclear data and fast enough to allow for hundreds or thousands of nuclear data variations, but has a limited range of applicability to nuclei with reasonably-high capture cross sections. The ^{6}Li(n,t) cross section, which is the neutron detection mechanism for ^{6}Li-enriched Li-glass detectors, is ideally suited for this purpose. The reader is referred to Ref. [16] for more details on this method.

4.3 Direct Track Weight Variations

As opposed to relying on implicit capture cross simulations from MCNP, it is possible to simply write out all data from every interaction and surface crossing event from an MCNP simulation using the PTRAC output format [17]. With this information, one can again calculate varied neutron weights by altering three probabilities calculated during the course of an analog simulation: (1) the probability for a neutron to travel a path length l in a material, (2) the probability for interacting with a chosen nucleus in the material, and (3) the probability for choosing the nuclear physics reaction that occurred in the simulation. These probability variations can be described in terms of three factors altering the weight of the neutron at each step in the simulation, f_l, f_i, and f_σ, respectively.

For a given isotope, i, with a total macroscopic cross section, $\Sigma_{t,i}$, in a material with integrated macroscopic total cross section Σ_T, we otbain

$$f_l = \left(1 - e^{-\Sigma_T' l}\right)\left(1 - e^{-\Sigma_T l}\right)^{-1}, \quad (4)$$

and

$$f_i = \left[\Sigma_{t,i}'/\Sigma_T'\right]\left[\Sigma_{t,i}/\Sigma_T\right]^{-1}, \quad (5)$$

where l is the path length traveled by the neutron, and a prime (as in $\Sigma_{t,i}'$) indicates a quantity calculated after a cross section was changed in the isotope of interest. The value of the factor f_σ depends on the cross section being varied, but is a straightforward ratio before and after variation. These factors reflect the altered probabilities for the neutron to travel along the path taken during the simulation. This method, while it has not been applied yet to Chi-Nu PFNS results, has the potential to be used for variations of any nucleus and including covariance input. Among other potential applications, it could also be a method for carrying out optimizations for a forward analysis approach for extracting cross sections from an experiment, though this possibility has not yet been explored fully.

5 Correlations between PFNS Data Points Across Target Nuclei

Results from Chi-Nu PFNS measurements are unique among any other PFNS measurement in that careful covariance calculations were carried out to describe correlations between not just emitted neutron energy data points within a single incident energy range, but between all emitted energy points of all incident energies. This information then yields a single covariance matrix describing the entire data set. Correlations between data points of different incident energy ranges arise primarily from the similar analysis methods and corrections applied to each incident energy range, most of which contain correlations between different emitted energy points as well. In order to properly report the covariance of the PFNS shape, the PFNS results for each incident energy range are also normalized with the Li-glass and liquid scintillator data scaled to obtain the same data integral in the overlap region of approximately 0.89–1.5 MeV, which also has the effect of creating an anticorrelation between high and low emitted neutron energies. See Refs. [9] and [10] for more details on these covariances.

In Fig. 4, we report for the first time a calculation of not just a single covariance matrix for all data points of a single PFNS measurement, but a single matrix describing correlations between all data points of all published Chi-Nu PFNS measurements [9, 10]. Data point number, along both the x and y axes, is an index for outgoing neutron energy points of a PFNS result. Each PFNS result from a Chi-Nu experiment contains 65 data points, and we typically report 20 incident energies yielding 1300 total data points for the complete data set of the ^{239}Pu and ^{235}U PFNS. Data points 1–1300 on each axis correspond to the ^{239}Pu results, and data points 1301–2600 correspond to ^{235}U. Thus, while the matrix from x = 1–1300 and y = 1-1300 corresponds to the published ^{239}Pu correlation matrix [9] and the matrix from x = 1301–2600 and y = 1302–2600 corresponds to ^{235}U [10], the off-diagonal blocks (i.e., x = 1-1300 and y = 1301–2600, and its reflection over the line x = y) describes the correlations between ^{239}Pu and ^{235}U PFNS results from the Chi-Nu experiment. These correlations arise primarily from the similarities in the MCNP simulations used for analysis, since the simulations differ only in target nucleus identity and orientation of the PPAC target chamber. Thus these cross-nuclide correlations are estimated based on the magnitude of uncertainty and correlation created by MCNP-based aspects of the analysis between experiments for each nucleus. These correlations are necessary for properly calculating ratios of, e.g., the ^{239}Pu to the ^{235}U PFNS for various incident energy ranges.

6 Conclusions

The loss of reliability of results from PFNS measurements over the last seven decades is an unfortunate side effect of experiments that are not well documented and did not have a complete or reliable uncertainty and covariance analysis at the time analysis was being completed. The Chi-Nu experiment to measure the PFNS for major actinides was designed to produce results that stand the test of time and preserve reliability of PFNS results through thorough documentation and detailed covariance analysis. In this manuscript we describe details of two of the most important aspects of data analysis for Chi-Nu experiments: random-coincidence backgrounds and MCNP simulations. For random-coincidence backgrounds, a summary of recent work investigating the potential for systematic errors and uncertainties in the random-coincidence background measurement based on changes in detection rates one might encounter in an experiment, with important conclusions regarding correlations between detection rates provided. The uncertainty on the use of MCNP from uncertainties in input nuclear data to the simulation was also discussed in terms of three methods developed for during the course of the development of the Chi-Nu experiment: a Bayes-like factor, implicit-capture simulation output variations, and direct track weight variations. While the factor calculated for the Chi-Nu experiment is not a formally rigorous formulation of a true Bayes factor, the concept and application described in this work to nuclear physics experiments is not common, and the other methods for varying input nuclear data cross sections to MCNP were developed by the Chi-Nu team for this experiment.

Lastly, for the first time, a single correlation matrix describing the relation of each emitted energy point in each incident energy range for each target nucleus in different Chi-Nu experiments was shown. This represents an advancement in PFNS experiments beyond what has been done in previous measurements, and is essential for accurately calculating informative ratios of the PFNS of one nucleus to another. In addition to ^{239}Pu and ^{235}U, Chi-Nu measurements of the ^{238}U and ^{240}Pu PFNS have been completed as of writing this manuscript, and thus these measurements will follow the same detailed analysis methods employed for previous Chi-Nu PFNS measurements.

7 Acknowledgements

This work was supported by the U.S. Department of Energy through Los Alamos National Laboratory. Los Alamos National Laboratory is operated by Triad National Security, LLC, for the National Nuclear Security Administration of the U.S. Department of Energy (Contract No. 89233218CNA000001)

References

[1] R. Capote, Y.J. Chen, F.J. Hambsch, N.V. Kornilov, J.P. Lestone et al., Nucl. Data Sheets **131**, 1 (2016)
[2] M.B. Chadwick, Nucl. Technology **207**, 3 (2021)
[3] T.N. Taddeucci, R.C. Haight, H.Y. Lee, D. Neudecker, J.M. O'Donnell et al., Nucl. Data Sheets **123**, 135 (2015)
[4] D. Neudecker, T.N. Taddeucci, R.C. Haight, H.Y. Lee, M.C. White et al., Nucl. Data Sheets **131**, 289 (2016)
[5] X-5 Monte Carlo Team, Los Alamos National Laboratory Repport LA-UR-03-1987 (April, 2003)
[6] C.J. Werner, J. Armstrong, F.B. Brown, J.S. Bull, L. Casswell et al., LA-UR-17-29981 (2017)
[7] C.Y. Wu, R.A. Henderson, R.C. Haight, H.Y. Lee, T.N. Taddeucci et al., Nucl. Instrum. and Methods A **794**, 76 (2015)
[8] K.J. Kelly, T. Kawano, J.M. O'Donnell, J.A. Gomez, M. Devlin et al., Phys. Rev. Lett. **122**, 072503 (2019)
[9] K.J. Kelly, M. Devlin, J.M. O'Donnell, J.A. Gomez, D. Neudecker et al., Phys. Rev. C **102**, 034615 (2020)
[10] K.J. Kelly, J.A. Gomez, M. Devlin, J.M. O'Donnell, D. Neudecker et al., Phys. Rev. C **105**, 044615 (2022)
[11] M. Devlin, J.A. Gomez, K.J. Kelly, R.C. Haight, J.M. O'Donnell et al., Nucl. Data Sheets **148**, 322 (2018)
[12] J.M. O'Donnell, Nucl. Instrum. Methods A **805**, 87 (2016)
[13] K.J. Kelly, J.M. O'Donnell, J.A. Gomez, M. Devlin, E.A. Bennett, Nucl. Instrum. Methods A **1045**, 167531 (2023)
[14] K.J. Kelly, J.M. O'Donnell, J.A. Gomez, T.N. Taddeucci, M. Devlin et al., Nucl. Instrum. and Methods A **866**, 182 (2017)
[15] P.M. Lee, *Bayesian Statistics: An Introduction* (Wiley, 2012)
[16] K.J. Kelly, J.A. Gomez, J.M. O'Donnell, M. Devlin, R.C. Haight et al., Nucl. Instrum. and Methods A **954**, 161411 (2020)
[17] C.J. Solomon, C. Bates, J. Kulesza, LA-UR-17-21779 (March 30, 2017)

Experimental spectrum averaged cross sections (SACS) in ^{252}Cf(sf) neutron field and its impact on the evaluation of neutron standards

R. Capote[1],* *G.* Schnabel[1], *A.D.* Carlson[2], *V.G.* Pronyaev[1], *G.* Noguere[3], and *D.* Neudecker[4]

[1]NAPC–Nuclear Data Section, International Atomic Energy Agency, PO Box 100, A-1400 Vienna, Austria
[2]National Institute of Standards and Technology, Gaithersburg, MD, USA
[3]Commisariat a l'Energie Atomique, Cadarache, DEN/DER 13108 Saint-Paul-Lez-Durance, France
[4]Los Alamos National Laboratory, Los Alamos, NM, USA

Abstract. A new evaluation of spectrum averaged cross sections (SACS) of ^{235}U, ^{238}U, and ^{239}Pu measured in the ^{252}Cf(sf) reference neutron field is presented and found to be consistent with the original Mannhart SACS evaluation in IRDF-2002. The comprehensive vetted experimental database that includes SACS ratio and absolute SACS measurements of major actinides is being used to update the SACS database employed as input of the new GMApy code to derive the Neutron Standards. An update of the current neutron standards based on Time Projection Chamber (TPC) shape data, a new comprehensive uncertainty quantification, and revised SACS experimental database is proposed which result in a 0.7% increase of the evaluated ^{239}Pu(n,f)/^{235}U(n,f) cross-section ratio in the 1–5 MeV energy region. The increase is due to a 0.3% reduction of the standard ^{235}U(n,f) cross section and a 0.4% increase of the ^{239}Pu(n,f) reference cross section in the 1–5 MeV energy region. Those changes are well within estimated USU fission cross-section uncertainties of 1.2%, but are relevant for the evaluated mean values.

1 Introduction

Evaluation of Neutron Standards have been a critical component of ENDF/B library releases. The Neutron Standards have been coordinated by the IAEA since early 2000s and have been incorporated into the ENDF/B-VII.0 [1] and ENDF/B-VIII.0 releases [2]. The latest IAEA Neutron Data Standards were released in 2017 (will be called STD 2017 in this paper) [3, 4] and were adopted for the ENDF/B-VIII.0 library. Since very early on, the Neutron Standards have been derived using the GMA code originally developed by W. Poenitz with an extensive database of vetted experimental data [5, 6] and represent our best-known cross sections derived by a non-model evaluation [7]. The GMA standard input database also included the simplest integral data, namely spectrum averaged cross sections (SACS) of ^{239}Pu(n,f) and ^{235}U(n,f) reactions measured in the ^{252}Cf(sf) neutron field.

Note that the reference ^{252}Cf(sf) neutron spectrum was originally evaluated by Mannhart [8] based on time-of-flight measurements and has been validated as a reference neutron field in IRDFF-II library [9, 10]. We can use the reference ^{252}Cf(sf) neutron spectrum [9, 10] to convolute evaluated Neutron Standard cross sections to derive SACS and corresponding uncertainties. At the same time Mannhart evaluated the experimentally measured SACS in the ^{252}Cf(sf) spectrum for the IRDF-2002 library [11]. A comparison of evaluated fission SACS measured in the ^{252}Cf(sf) spectrum for dosimetry reactions ^{235}U(n,f), ^{238}U(n,f), and ^{239}Pu(n,f) versus fission SACS derived from Neutron Standards is given in Table 1.

The fission spectral indexes (SI) ^{239}Pu(n,f)/^{235}U(n,f) and ^{238}U(n,f)/^{235}U(n,f) in the ^{252}Cf(sf) neutron spectrum are listed in Table 2 with the corresponding C/E values showing an underestimation larger than 2%. Such underestimation is close to the lower bound of the estimated uncertainty of the C/E ratios assuming the calculated and measured values to be independent. This underestimation of the SI compared to measured values deserves an investigation.

Additionally, it was pointed out by Casperson [12] that a similar underestimation is observed in C/E of fission SI measured in the neutron field of well-understood fast critical assemblies Godiva and Flaptop-U [13, 14]. These assemblies show an excellent C/E for more than 20 different dosimetry reactions [9] demonstrating our very good understanding of the neutron fluxes in these critical assemblies in a broad energy range. The comparison of measured versus calculated SI in Godiva and Flaptop-U critical assemblies are shown in Tables 3 and 4, respectively. There is an agreement in the observed underestimation of fission SI in ^{252}Cf(sf) and critical-assemblies' neutron fields.

Finally, a 2% normalization difference between the measured ^{239}Pu(n,f)/^{235}U(n,f) absolute cross-section ratio using a Time Projection Chamber (TPC) detector to the Neutron Standard was recently discussed [15][1].

*e-mail: r.capotenoy@iaea.org

[1]Note that the TPC ratio data were treated as relative ratios (shape data) in a recent ^{239}Pu re-evaluation.

Table 1: Fission SACS calculated C in a ^{252}Cf(sf) reference neutron spectrum [10] versus measured data E (represented by the Mannhart evaluation [11]), and the corresponding C/E with uncertainties.

Reaction	IAEA STD 2017 $\equiv C^1$ [barn]	Mannhart eval. [11] $\equiv E^2$ [barn]	C/E^3
^{235}U(n,f)	1.2267(1.2%)	1.210 (1.2%)	1.014(1.70%)
^{238}U(n,f)	0.3215(1.3%)	0.3257(1.6%)	0.987(2.06%)
^{239}Pu(n,f)	1.7978(1.3%)	1.812 (1.4%)	0.992(1.91%)

[1] Evaluated ^{239}Pu(n,f), ^{238}U(n,f), and ^{235}U(n,f) cross sections assumed fully correlated, which is an approximation.
[2] Manhhart evaluated correlations [11]:
corr(^{239}Pu,^{235}U)=0.28, corr(^{238}U,^{235}U)=0.15.
[3] Calculated C/E uncertainties derived assuming C and E are independent, which is an approximation.

Table 2: Fission SI in ^{252}Cf(sf) neutron spectrum.

Spectral Index (SI)	IAEA STD 2017 $\equiv C^1$ [barn]	Mannhart eval. [11] $\equiv E^2$ [barn]	C/E^3
^{239}Pu(n,f)/^{235}U(n,f)	1.465(1.2%)	1.498(1.7%)	0.978(2.0%)
^{238}U(n,f)/^{235}U(n,f)	0.262(1.2%)	0.269(1.8%)	0.973(2.2%)

[1] SI C (ratio) uncertainties of IAEA STD 2017 were derived considering full correlation of numerator and denominator as stated in Table 1, which is an approximation.
[2] SI E evaluated by Mannhart include evaluated correlations as listed in Table 1.
[3] SI C/E uncertainties assumed uncorrelated C and E quantities, which is an approximation.

Table 3: Fission SI in Godiva (HEU-MET-FAST 001 [14]) critical assembly.

Reaction	IAEA STD 2017 $\equiv C^1$ [barn]	Measured SI [13] $\equiv E$ [barn]	C/E^1
^{239}Pu(n,f)/^{235}U(n,f)	1.3841(1.7%)	1.4152(1.0%)	0.978(2.0%)
^{238}U(n,f)/^{235}U(n,f)	0.1584(1.7%)	0.1643(1.0%)	0.972(2.0%)

[1] C/E uncertainties derived assuming numerator C and denominator E are independent, which is an approximation.

Table 4: Fission SI in Flaptop-U (HEU-MET-FAST 028 [14]) critical assembly.

Reaction	IAEA STD 2017 $\equiv C^1$ [barn]	Measured SI [13] $\equiv E$ [barn]	C/E
^{239}Pu(n,f)/^{235}U(n,f)	1.3615(1.7%)	1.3847(0.9%)	0.985(1.9%)
^{238}U(n,f)/^{235}U(n,f)	0.1450(1.7%)	0.1492(1.1%)	0.972(2.0%)

[1] C/E uncertainties derived assuming numerator C and denominator E are independent, which is an approximation.

Observed differences may be due to deficiencies in the Neutron Standards input data, in the Neutron Standard evaluation methodology or in both. Biases in measured SI in critical assemblies and in the measured absolute TPC cross-section ratio are also possible. The goal of this work is to study the sources of these differences, and whether we can improve Neutron Standards fission cross sections and associated reference cross sections like the ^{239}Pu(n,f).

2 New evaluation of measured SACS in the ^{252}Cf(sf) neutron spectrum

The updated experimental database of ^{235}U(n,f), ^{238}U(n,f), and ^{239}Pu(n,f) cross sections and corresponding ratios measured in the ^{252}Cf(sf) neutron spectrum are compiled in Table 5. The corresponding experimental covariance matrix was constructed from the analysis and uncertainty quantification of the SACS experimental data similarly as done by Mannhart [11]. In fact, Mannhart provided in Ref. [11] most of the listed coefficients, we only needed to add Schröder data. Note that we also had the original Mannhart input used in his SACS evaluation provided by Mannhart himself. That makes his evaluation fully traceable. Corresponding experimental correlation coefficients are zero by default[2] except those listed in Tables 6, 7, 8, 9, and 10. The most important difference from Mannhart evaluation [11] is the addition of newly measured data by Schröder at NIST in 1985 [16].

With the updated experimental SACS database listed above we have used Mannhart BAYES code (employed in his original evaluation) to produce a new evaluation with results listed in Table 11 and corresponding evaluated correlations listed in Table 12. Both evaluations show good agreement well within their one-sigma quoted uncertainties. An small absolute increase of evaluated values is observed. Ratios of evaluated cross sections agree extremely well. Evaluated ratio uncertainty are reduced due to the very strong positive correlations derived for the ratio numerator and denominator. The obtained agreement shows the robustness of Mannhart SACS evaluation. Therefore, the newly updated SACS experimental database was validated, and can be used in the new GMA fit aiming at improving the normalization of GMA fission cross section fits, as will be shown in next section.

3 Update of the Neutron Standard evaluation with new experimental SACS data

The IAEA standard cross sections 2017 [3] are taken as the reference for the updated evaluation of the fission cross sections which will be discussed in this Section. All presented calculations were done with the new GMApy code developed by G. Schnabel at the IAEA and fully validated vs the original GMA FORTRAN code.

The GMApy evaluation was carried out in the full energy range of data; discussions in the following are restricted to the impact of the SACS experimental inputs on the normalization of the evaluated fission cross sections. Note that the new GMApy code is necessary to use measured SACS ratios that the original GMA code can not use directly. We focus on the fast energy region from 1 up to 5 MeV of incident neutron energy for comparison, being the neutron energy region that includes the maximum of the ^{252}Cf(sf) reference neutron spectrum [8, 10] and covers a significant percentage of the outgoing neutrons.

In Figs.1–3 the ratios to the IAEA reference cross sections are shown in the energy region from 1 MeV up to 5 MeV for the following cases:

1. **std2017-input**: Same input as in STD 2017, but recalculated with the GMApy code correcting identified glitches.

2. **additional-input**: New GMApy fit based on updated uncertainty quantification of ^{239}Pu data [17–19] and new TPC experimental shape data [15, 20, 21]. The fitted ^{239}Pu cross sections are used in the INDEN ^{239}Pu evaluated file pu239p55 available at the INDEN webpage (https://nds.iaea.org/INDEN/).

[2] Reflecting assumed lack of correlations by Mannhart [11].

Table 5: Experimental ^{235}U(n,f), ^{238}U(n,f), and ^{239}Pu(n,f) SACS in ^{252}Cf(sf) neutron spectrum: updated database.

Reaction index	Reaction	SACS [barn]	SACS uncert. [%]	Reaction label	Reference
1	^{235}U(n,f)	1216	1.62	U5F	Grundl 1983 [22, 23, 25]
2	^{235}U(n,f)/^{238}U(n,f)	3.73	1.20	U5F/U8F	Grundl-Gilliam 1983 [22, 24, 28], inverse = 0.2681
3	^{235}U(n,f)/^{239}Pu(n,f)	0.666	0.90	U5F/PU9F	Grundl-Gilliam 1983 [22, 24], inverse = 1.502
4	^{239}Pu(n,f)/^{235}U(n,f)	1.500	1.60	PU9F/U5F	Heaton 1976 [25]
5	^{238}U(n,f)/^{235}U(n,f)	0.2644	1.32	U8F/U5F	Heaton 1976 [25]
6	^{238}U(n,f)/^{235}U(n,f)	0.269	1.20	U8F/U5F	Schröder 1985 [16]
7	^{239}Pu(n,f)/^{235}U(n,f)	1.500	0.80	PU9F/U5F	Schröder 1985 [16]
8	^{235}U(n,f)	1234	1.45	U5F	Schröder 1985 [16]
9	^{235}U(n,f)	1215	1.79	U5F	Davis/Knoll 1978 [26]
10	^{239}Pu(n,f)	1790	2.26	PU9F	Davis/Knoll 1978 [26]
11	^{238}U(n,f)/^{235}U(n,f)	0.2741	1.66	U8F/U5F	Adamov 1977 [27]
12	^{239}Pu(n,f)/^{235}U(n,f)	1.475	1.50	PU9F/U5F	Adamov 1977 [27]

Table 6: Experimental correlations estimated by Mannhart [11] for Grundl-Gilliam SACS data [22–25].

Reaction or Spectral Index	U5F [%]	SI U5F/U8F [%]	SI U5F/PU9F [%]
^{235}U(n,f)	100		
^{235}U(n,f)/^{238}U(n,f)	23	100	
^{235}U(n,f)/^{239}Pu(n,f)	-9	36	100

Table 7: Experimental correlations for Heaton et al. [25].

Spectral Index	SI PU9F/U5F [%]	SI U8F/U5F [%]
^{239}Pu(n,f)/^{235}U(n,f)	100	
^{238}U(n,f)/^{235}U(n,f)	60	100

Table 8: Experimental correlations for Schröder et al. [16].

Spectral Index	SI PU9F/U5F [%]	SI U8F/U5F [%]
^{239}Pu(n,f)/^{235}U(n,f)	100	
^{238}U(n,f)/^{235}U(n,f)	39	100

Table 9: Experimental correlations for Davis et al. [26].

Spectral Index	SI PU9F/U5F [%]	SI U8F/U5F [%]
^{239}Pu(n,f)/^{235}U(n,f)	100	
^{238}U(n,f)/^{235}U(n,f)	59	100

Table 10: Experimental correlations for Adamov et al.[27].

Spectral Index	SI PU9F/U5F [%]	SI U8F/U5F [%]
^{239}Pu(n,f)/^{235}U(n,f)	100	
^{238}U(n,f)/^{235}U(n,f)	39	100

Table 11: Evaluated experimental SACS in ^{252}Cf(sf) neutron spectrum vs Mannhart original evaluation [11]. These values together with correlations listed in Table 12 represent our best experimental knowledge and are independent of the evaluated ^{252}Cf(sf) neutron spectrum.

Reaction	This work [barn]	Mannhart eval. [11] [barn]	new/old eval.
^{235}U(n,f)	1.221 (0.91%)	1.210 (1.2%)	1.009
^{238}U(n,f)	0.327 (1.06%)	0.3257(1.6%)	1.004
^{239}Pu(n,f)	1.826 (1.03%)	1.812 (1.4%)	1.008

Table 12: Evaluated experimental SACS correlations in ^{252}Cf(sf) neutron spectrum of major-actinide fission cross sections corresponding to mean values and uncertainties listed in Table 11.

Reaction	U5F [%]	U8F [%]	PU9F [%]
^{235}U(n,f)	100		
^{238}U(n,f)	80	100	
^{239}Pu(n,f)	87	82	100

3. **additional-input + updated SACS**: In addition to new data the vetted SACS experimental database (see Sec.3) was used as input in the GMApy fit.

Case 1 in Figs.1–3 uses exactly the same input deck as the STD 2017 fit (labelled **std2017-input** and represented by the red line). The goal is to assess the impact of the new GMApy code and of few coding glitches which were found and corrected. The impact on ^{235}U(n,f) cross sections is up to +0.1%, on ^{238}U(n,f) cross sections – +0.2% and up to +0.15% on ^{239}Pu evaluated fission cross sections. The overall impact of the new code and corrected bugs was found to be very small.

Case 2 in Figs.1–3 represents the advanced uncertainty quantification of ^{239}Pu(n,f) GMA input database [17–19] and the addition of new TPC experimental data with very low uncertainties [15, 20, 21] used as shape data (labelled

additional-input and represented by the blue line). The impact of these changes on ^{235}U(n,f) cross sections is a reduction of up to 0.2% in the evaluated fission cross sections, on ^{238}U(n,f) cross sections – a reduction of 0.25% is observed, and an increase of up to 0.35% on ^{239}Pu evaluated fission cross sections.

Case 3 in Figs.1–3 represents the addition of the updated SACS experimental database listed in Table 5 and corresponding correlations listed in Tables 6, 7, 8, 9, and 10. It is labelled as **additional-input+SACS** and represented by the orange line. The new SACS data further decrease the ^{235}U(n,f) evaluated cross section by an additional 0.15%. That results in a total change of ^{235}U evaluated cross section by -0.35% relative to the IAEA STD 2017. No change in evaluated cross sections due to the use of updated SACS is practically observed for ^{238}U(n,f) and ^{239}Pu(n,f) reactions.

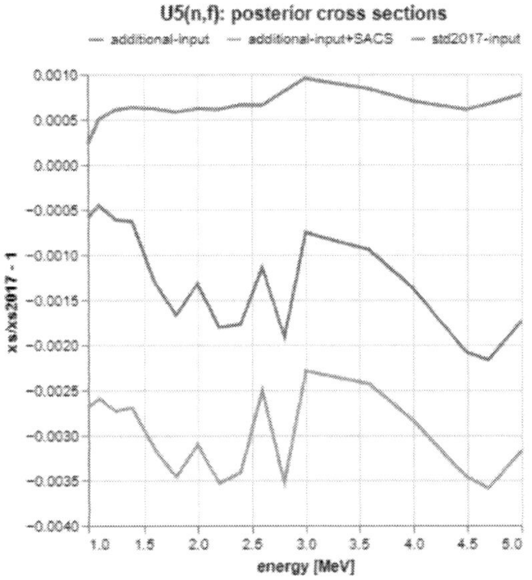

Figure 1: Evaluated ^{235}U(n,f) cross sections relative to the IAEA STD 2017 [3] from 1 MeV up to 5 MeV.

Summarizing, a reduction of 0.35% is observed of the evaluated fission cross sections of ^{235}U, whilst a 0.35% increase is obtained for the evaluated ^{239}Pu fission cross section. These changes led to a ^{239}Pu/^{235}U evaluated fission cross-section ratio increase of almost 0.7% as shown in Fig. 4. This ratio increase improves the agreement with published TPC absolute ^{239}Pu/^{235}U ratio [15]. Without the update in the experimental SACS database the absolute ^{239}Pu/^{235}U ratio increase is reduced to 0.45%. The corresponding change in the ^{238}U/^{235}U evaluated fission cross-section ratio is smaller going from 0.1% up to 0.3% in the energy region corresponding to the first chance fission plateau as shown in Fig. 5.

Newly evaluated GMApy cross sections that include the TPC ratio measurements as shape data and a revised experimental SACS database can be convoluted with the reference ^{252}Cf(sf) neutron spectrum [8, 10] to calculate

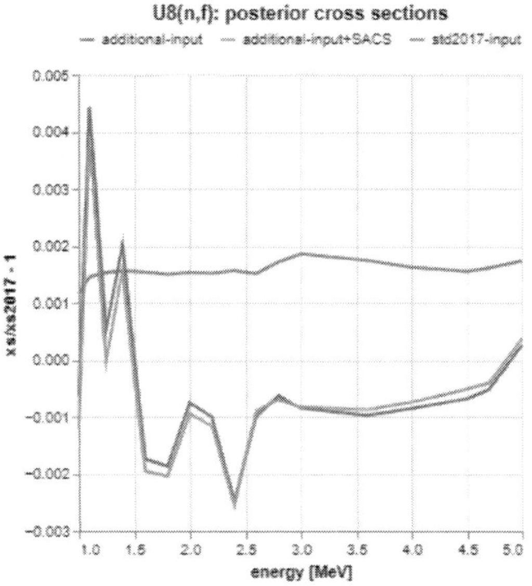

Figure 2: Evaluated ^{238}U(n,f) cross sections relative to the IAEA STD 2017 [3] from 1 MeV up to 5 MeV.

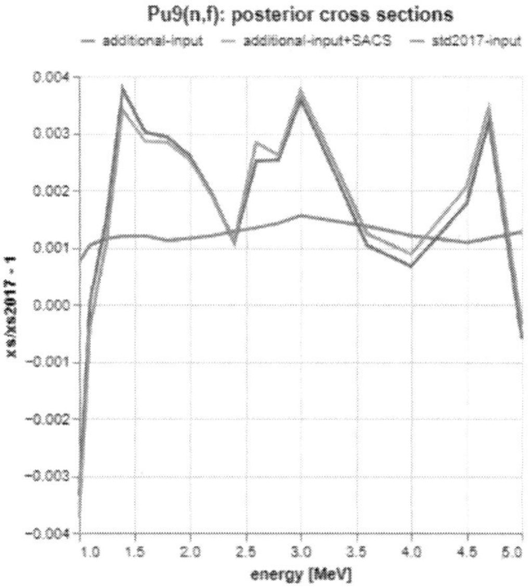

Figure 3: Evaluated ^{239}Pu(n,f) cross sections relative to the IAEA STD 2017[3] from 1 MeV up to 5 MeV.

the SACS which can be compared with the SACS evaluated directly from measured data listed in Table 11. This comparison is shown in Table 13. It can be seen that the agreement of SACS derived from standard and reference cross sections is already within the uncertainty of the updated SACS evaluation based on direct SACS measurements; small underestimation of the evaluated SACS average values remain to be investigated in future works. Additional review of measured absolute cross sections and

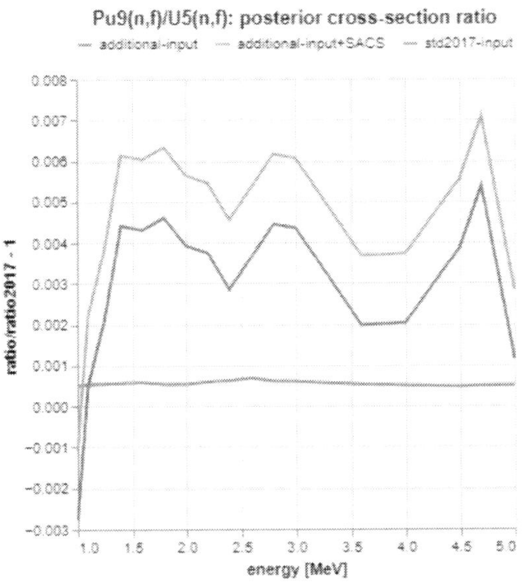

Figure 4: Evaluated ^{239}Pu(n,f)/^{235}U(n,f) cross-section ratio relative to the IAEA STD 2017 cross-section ratio [3].

absolute cross-section ratios involving ^{239}Pu targets is ongoing to further refine the experimental GMApy input database.

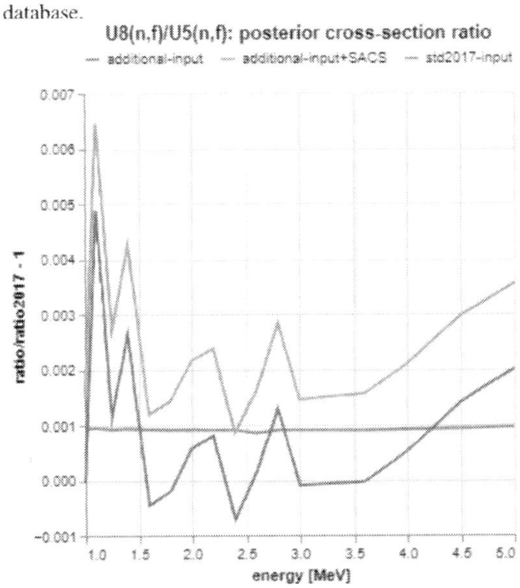

Figure 5: Evaluated ^{238}U(n,f)/^{235}U(n,f) cross-section ratio relative to the IAEA STD 2017 cross-section ratio [3].

Finally, it should be noted that the evaluated total uncertainties of the major-actinide fission cross sections remains practically unchanged from the uncertainties evaluated in the IAEA STD 2017 without considering the Unrecognized Sources of Uncertainties (USU) [3, 29]. Newly evaluated uncertainties are very weakly affected by the use of the updated experimental database and updated uncertainty quantification including the new experimental SACS input database. For the ^{235}U(n,f) cross section the total uncertainty in the first fission plateau (1–5 MeV) is about 0.5%; for the ^{239}Pu(n,f) – about 0.65–0.7%. Additional studies on the impact of energy-dependent USU on evaluated cross sections are warranted. However, energy-dependent USU are not expected to affect evaluated SACS in the ^{252}Cf(sf) spectrum.

Table 13: Calculated C fission SACS (based on GMApy fit) in a ^{252}Cf(sf) reference neutron spectrum [8, 10] vs evaluated SACS data based on experimental data E (see Table 11). The C/E values are given in the last column.

Reaction	This work GMApy $\equiv C$ [barn]	This work, SACS exp. $\equiv E$ [barn]	C/E[1]
^{235}U(n,f)	1.221 (0.3%)	1.221 (0.91%)	1.000 (0.96%)
^{238}U(n,f)	0.323 (0.4%)	0.327 (1.06%)	0.988 (1.13%)
^{239}Pu(n,f)	1.803 (0.4%)	1.826 (1.03%)	0.987 (1.10%)

[1] C/E uncertainties derived assuming C and E are independent quantities.

4 Conclusions

A vetted experimental database of ^{235}U(n,f), ^{238}U(n,f), and ^{239}Pu(n,f) spectrum averaged cross sections (SACS) measured in the ^{252}Cf(sf) reference neutron field [8, 10] is developed (see Table 5 and correlations in Tables 6–10). The experimental SACS database includes SACS ratio and absolute SACS measurements of neutron-induced fission reactions on the three major actinides using the new capabilities of the GMApy code. The new database was used to derive a non-model SACS evaluation in the ^{252}Cf(sf) neutron field that agrees within 1% with the original Mannhart SACS evaluation (see evaluated results in Tables 11 and 12). This agreement is well within the quoted experimental uncertainties. The vetted experimental SACS database is also used to update the input of the GMApy code employed to derive the Neutron Data Standards.

An update of the current neutron standards based on new TPC shape data, advanced uncertainty quantification of ^{239}Pu(n,f) reactions, and vetted SACS experimental database is carried out. The update results in a reduction of 0.3% and 0.25% of the evaluated ^{235}U(n,f) and ^{238}U(n,f) cross sections, respectively. A 0.45% increase is obtained for the evaluated ^{239}Pu fission cross section. Therefore, a significant 0.7% increase is derived for the ^{239}Pu(n,f)/^{235}U(n,f) ratio normalization. These changes are within Standard USU fission cross-section uncertainty of 1.2% [3, 29], but the obtained changes improve the C/E of fission spectral indexes in Godiva (HEU-MET-FAST 001) and Flaptop (HEU-MET-FAST 028) critical assemblies, and also improve the agreement of standard cross sections with published TPC ^{239}Pu(n,f)/^{235}U(n,f) absolute ratio data [15].

New SACS evaluations of fission cross sections in the ^{252}Cf(sf) neutron field obtained in this work are listed in Table 13. Our SACS value for the ^{235}U(n,f) of 1221 mb, newly derived from the GMApy fit, is exactly equal to the one derived by Otsuka and Iwamoto [30, 31]. For the

^{238}U(n,f) SACS value, a +2.2% difference to Otsuka and Iwamoto evaluation remains; our evaluated SACS value of 323 mb is consistent with the experimental SACS evaluated value of 327 mb. Finally, for ^{239}Pu(n,f), our new SACS value of 1803 mb is 0.7% below the evaluated value by Otsuka and Iwamoto of 1814 mb, which represents about 1.5 times our evaluated uncertainty without USU. Our newly evaluated ^{239}Pu(n,f) SACS value is 1.3% lower than the experimental SACS evaluation value of 1826 mb, which represents about 3 times our evaluated uncertainty without USU. Further investigation of the ^{239}Pu(n,f) absolute and absolute cross-section ratio experimental database used in the GMApy fit and a better uncertainty quantification are warranted to address these discrepancies.

Acknowledgements

The IAEA is grateful to all participating laboratories for their assistance in the collaborative INDEN network and for supporting the IAEA meetings and activities. Work described in this paper would not have been possible without contributions from the IAEA Member States. Authors would like to thanks D.L. Smith for his guidance and always sensible advices.

References

[1] M.B. Chadwick, P. Obložinský, M. Herman *et al.*, "ENDF/B-VII.0: next generation evaluated nuclear data library for science and technology," Nucl. Data Sheets **107**, 2931–3060 (2006).

[2] D.A. Brown, M.B. Chadwick, R. Capote *et al*, "ENDF/B-VIII.0: The 8th Major Release of the Nuclear Reaction Data Library with CIELO-project Cross Sections, New Standards and Thermal Scattering Data," Nucl. Data Sheets **148**, 1–142 (2018).

[3] A. D. Carlson, V.G. Pronyaev, R. Capote *et al.*, "Evaluation of neutron data standards," Nucl. Data Sheets **148**, 142–187 (2018).

[4] A. D. Carlson, V.G. Pronyaev, R. Capote *et al.*, Numerical data available oonline at the IAEA Standard webpage https://nds.iaea.org/standards/.

[5] W.P. Poenitz, "Data interpretation, objective evaluation procedures and mathematical techniques for the evaluation of energy-dependent ratio shape and cross section data," Report **BNL-NCS-51363**, Vol. 1, p.249 (BNL, Brookhaven 1981).

[6] W.P. Poenitz, S.E. Aumeier, "The simultaneous evaluation of the standards and other cross sections of importance for technology," Report **ANL/NDM-139** (ANL, Argonne 1997).

[7] R. Capote, D.L. Smith, and A. Trkov, "Nuclear data evaluation methodology including estimates of covariances," EPJ Web of Conf. **8**, 04001 (2010).

[8] W. Mannhart, "Status of the Cf-252 Fission Neutron Spectrum Evaluation with Regard to Recent Experiments,"IAEA Consult. Meet. on "Physics of Neutron Emission in Fission," Mito, Japan, 24–27 May 1988, Report **INDC(NDS)-0220**, pp. 305–336, IAEA, Vienna (1989). Available online at https://www-nds.iaea.org/publications/indc/indc-nds-0220/.

[9] A. Trkov, P.J. Griffin, S.P. Simakov, L.R. Greenwood, K.I. Zolotarev, R. Capote, D.L. Aldama, V. Chechev, C. Destouches, A.C. Kahler, C. Konno, et al., "IRDFF–II: A New Neutron Metrology Library," Nucl. Data Sheets 163 (2020) 1–108.

[10] W. Mannhart, Numerical data of the ^{252}Cf(sf) reference neutron spectrum are available from the IAEA standard webpage: https://nds.iaea.org/standards/ref-spectra/PFNS-Cf252sf-ENDF.txt, and also from the IRDFF-II webpage (MT9861) https://nds.iaea.org/IRDFF/IRDFF-II_sp_ENDF.zip.

[11] W. Mannhart, in International Reactor Dosimetry File 2002 (IRDF-2002), **Tech. Report Series 452**, IAEA (2006); numerical data: https://www-nds.iaea.org/irdf2002/.

[12] R.J. Casperson, "The benefit of adjusting with criticality and reaction rate data,"mini-CSWEG 2021, private communication. See also R. J. Casperson, ND2022 contribution.

[13] M.B. Chadwick, M. Herman, P. Obložinský *et al.*, "ENDF/B-VII.1 Nuclear Data for Science and Technology: Cross Sections, Covariances, Fission Product Yields and Decay Data," Nucl. Data Sheets **112**, 2887–2996 (2011).

[14] ICSBEP 2020: International Handbook of Evaluated Criticality Safety Benchmark Experiments, Nuclear Energy Agency, OECD, Paris (2020). See list online at *www.oecd-nea.org/science/wpncs/icsbep/handbook.html*

[15] L. Snyder et al., "Measurement of the ^{239}Pu(n,f)/^{235}U(n,f) Cross-Section Ratio with the NIFFTE fission Time Projection Chamber," Nucl. Data Sheets **178**, 1–40 (2021). Also in arXiv 2107.02881 [nucl-ex] (2021).

[16] I.G. Schroder, Li Linpei, D.M. Gilliam, E.D. McGarry, C.M. Eisenhauer, "Measurement of absolute fission cross sections for Cf-252(sf) neutrons," Trans. Amer. Nucl. Soc. **50** (1985) 154–155.

[17] D. Neudecker, B. Hejnal, F. Tovesson *et al.*, "Template for estimating uncertainties of measured neutron-induced fission cross-sections," EPJ Nucl. Sci. Tech. **4**, 21 (2018).

[18] D. Neudecker, "Experimental Data and Covariances for Evaluating the Neutron-induced ^{239}Pu Fission Cross-section," Technical report **LA-UR-18-20767** (2018).

[19] D. Neudecker, D.L. Smith, F. Tovesson, R. Capote, M.C. White, N.S. Bowden, L. Snyder, A.D. Carlson, R.J. Casperson, V. Pronyaev, S. Sangiorgio, K.T. Schmitt, B. Seilhan, N. Walsh, W.Younes, "Applying a Template of Expected Uncertainties to Updating ^{239}Pu(n,f) Cross-section Covariances in the Neutron Data Standards Database," Nucl. Data Sheets **163**, 228–248 (2020).

[20] D. Neudecker, V. Pronyaev, L. Snyder, "Including ^{238}U(n,f)/^{235}U(n,f) and ^{239}Pu(n,f)/^{235}U(n,f) NIFFTE

fission TPC cross-sections into the Neutron Data Standards Database," Technical report **LA-UR-21-24093** (2022), https://doi.org/10.2172/1788383.

[21] R.J. Casperson et al., "Measurement of the normalized ^{238}U(n,f)/^{235}U(n,f) cross section ratio from threshold to 30 MeV with the NIFFTE fission Time Projection Chamber," Phys. Rev. **C97**, 034618 (2018).

[22] IAEA Consultants' Meeting on "the U-235 fast-neutron fission cross-section, and the Cf-252 fission neutron spectrum," Smolenice, Czechoslovakia, 28 March - 1 April 1983. IAEA technical report **INDC(NDS)-146**, Vienna, Austria (1983).

[23] J.A. Grundl, et al: Memorandum for Allan Carlson on "Revised value for the Cf-252(sf) fission-spectrum averaged cross-section for U-235 fission," IAEA Technical report **INDC(NDS)-146** [22], p.237.

[24] J.A. Grundl, D.M. Gilliam, "Fission Cross-Section Measurements in Reactor Physics and Dosimetry Benchmarks," IAEA Technical report **INDC(NDS)-146** [22], p.241.

[25] H. T. Heaton II, D. M. Gilliam, V. Spiegel, C. Eisenhauer, and J. A. Grundl, "Fission Cross Sections of ^{235}U, ^{238}U, and ^{239}Pu Averaged Over the ^{252}Cf(sf) Neutron Spectrum," Technical report **ANL-76-90** p.333, Proc. NEANDC/NEACRP Specialist Meeting on Fast Neutron Fission Cross Sections of U-233, U-235, U-238 and Pu-239, Argonne National Laboratory (June 1976).

[26] M.C. Davis and G.F. Knoll, "Fission cross sections of ^{235}U and ^{239}Pu averaged over Cf-252(sf) neutron spectrum," Ann. Nucl. En. **5** (1978) 583–588.

[27] V.M. Adamov, B.M. Aleksandrov, I.D. Alkhazov, L.V. Drapchinskiy, S.S. Kovalenko, O.I. Kostochkin, G.Ju. Kudryavcev, L.Z. Malkin, K.A. Petrzhak, L.A. Pleskachevskiy, A.V. Fomichev, and V.I. Shpakov, "Absolute fission cross section of heavy elements by fast neutrons," Technical report **INDC(CCP)-114**, 8–15 (1977).

[28] J.A. Grundl, V. Spiegel, Jr., and C. Eisenhauer, "Measurement of U238 and U235 fission cross sections for Cf-252(sf) neutrons," Trans. Amer. Nucl. Soc. **15**, 945–946 (1972).

[29] R. Capote, S. Badikov, A.D. Carlson, I. Duran, F. Gunsing, D. Neudecker, V.G. Pronyaev, P. Schillebeeckx, G. Schnabel, D.L. Smith, and A. Wallner, "Unrecognized Sources of Uncertainties (USU) in Experimental Nuclear Data," Nucl. Data Sheets **163**, 191–227 (2020).

[30] N. Otsuka and O. Iwamoto, "EXFOR-based simultaneous evaluation of neutron-induced uranium and plutonium fission cross sections for JENDL-5," J. Nucl. Sci. Tech. **59**, 1004-1036 (2022).

[31] N. Otsuka and O. Iwamoto, "Simultaneous evaluation of uranium and plutonium fast neutron fission cross sections up to 200 MeV for JENDL-5 and its updates," presented at ND2022, Sacramento, CA, USA; submitted to Conference Proceedings (2022).

Uncertainty evaluation of peak energy of giant dipole resonance propagated from uncertainty of parameters of effective interaction

Tsunenori Inakura[1]*

[1] Tokyo Institute of Technology, Ookayama, Meguro, Tokyo 152-8550, Japan

Abstract. We evaluate uncertainty of giant dipole resonance (GDR) peak energy propagated from the uncertainty of effective interaction parameter. The Monte Carlo calculation of microscopic random phase approximation using randomized parameter sets is performed. Under the condition that correlations between the parameters are considered, the calculated GDR peak energy has an uncertainty of ~1 MeV irrespective of nuclear mass and is strongly correlated with the parameters, in the present calculations. Our result serves as a guide for a new parametrization of effective interaction.

1 Introduction

Random phase approximation (RPA) calculation with effective interaction is one of the standard theoretical tools to calculate photoabsorption cross section in MeV region from the degree of freedom of nucleons, and has been applied to systematic calculations of nuclei covering wide range of the nuclear chart [1,2]. However, it is known that the RPA calculation with Skyrme effective interaction underestimates peak energies of giant dipole resonances (GDRs) in light nuclei by a few MeV.

The parameters of effective interactions are determined to reproduce limited sets of experimental data of ground states and nuclear matter properties. This limitation raises the question of its predictive power especially for excited states. Therefore, uncertainty estimation of the calculated values and feedback to the parameters are strongly desired. In the last decade, many theoretical studies have estimated uncertainty of calculated values and correlations of calculated values and the parameters [3-10]. However, these results are not used to improve the parameters of the effective interactions. This is because the relations between the parameters and the calculated values are not known clearly.

In this study, we performed Monte Carlo calculation of RPA using randomized parameters in Skyrme interaction. We evaluated the uncertainty of the GDR peak energy, propagated from the uncertainties of Skyrme parameters. By explicit taking the correlations of the parameters in randomization, we found that there are strong correlations between the Skyrme parameters and the calculated GDR peak energy.

2 Method

We employ Skyrme SLy5-min parameter set as the effective interaction. The SLy5-min parameters and their uncertainties Δp_i [8] are listed in Table **1**. Note that parameters x_2, W_0, and α are fixed when the SLy5-min parameters were determined and thus they do not have uncertainties. Table **2** shows a part of correlation matrix of the SLy5-min parameters. Other matrix elements of correlation $\boldsymbol{C}$ are listed in Ref. [11].

Table 1. Parameter name $\boldsymbol{p}$, their optical value $\boldsymbol{p_0}$, uncertainty $\boldsymbol{\Delta p}$ for SLy5-min parameters [8].

p	$p_0 \pm \Delta p$	Unit
t_0	-2475.408±149.455	MeV fm^3
t_1	482.842±55.537	MeV fm^5
t_2	-559.374±144.534	MeV fm^5
t_3	13697.07±1672.93	MeVfm$^{3+\alpha}$
x_0	0.741185±0.189191	
x_1	-0.146374±0.468173	
x_2	-1: fixed	
x_3	1.162688±0.340537	
W_0	126: fixed	MeV fm^5
α	1/6: fixed	

* Corresponding author: inakura.t.aa@m.titech.ac.jp

Table 2. Part of correlation matrix elements of SLy5-min parameters, $\boldsymbol{C}$.

	t_0	t_1	t_2	t_3
t_0	1.0000	0.9837	0.9854	-0.9997
t_1	0.9837	1.0000	0.9575	-0.9870
t_2	0.9854	0.9575	1.0000	-0.9863
t_3	-0.9997	-0.9870	-0.9863	1.0000

For the Monte Carlo calculation using randomized SLy5-min parameter sets, we generate $n = 1000$ random samples, $\boldsymbol{X}^{(j)}$ ($j = 1,2,\dots,n$), which satisfy the correlation $\boldsymbol{C}$. Before generating the random samples, we introduce a one-dimensional normal distribution $N(\mu,\sigma)$, where μ and σ are mean value and standard deviation of the distribution, respectively. And in preparation, the correlation matrix $\boldsymbol{C}$, which is positive semidefinite matrix, is factorized as $\boldsymbol{C} = \boldsymbol{Q}^T\boldsymbol{Q}$ by the singular value decomposition.

First, we generate n sets of random independent samples $\overline{\boldsymbol{X}}^{(j)} = \left(\overline{x}_1^{(j)}, \overline{x}_2^{(j)}, \dots \overline{x}_m^{(j)}\right)^T$ from a normal distribution $N(0,1)$. Namely, each variable is $N(0,1)$ and is uncorrelated with each others. Secondly, we act the factorized matrix $\boldsymbol{Q}$ on each $\overline{\boldsymbol{X}}^{(j)}$, i.e., creating correlated samples $\left(\boldsymbol{Q}\overline{\boldsymbol{X}}\right)^{(j)}$ satisfying the correlation $\boldsymbol{C}$. Thirdly, we shift the mean value and standard deviation, $X_i^{(j)} = p_{0i} + \left(Q\overline{X}\right)_i^{(j)} \Delta p_i$, employing $\boldsymbol{p_0}$ and $\boldsymbol{\Delta p}$ in Table 1. These are the correlated random parameter sets of SLy5-min.

Using these randomized parameter sets $\boldsymbol{X}^{(j)}$, we perform the RPA calculations n times. The RPA solver is Skyrme-rpa [12]. The Skyrme-rpa solves the RPA equation for spherical nuclei in the coordinate representation. The calculation space is a sphere with radius 25 fm and mesh span 0.1 fm. The photoabsorption cross section is computed from the resulting dipole strength functions and smeared with a width $\Gamma = 2$ MeV. From this photoabsorption cross section, we extract the position of the GDR peak top of jth random parameter set, $E_{GDR}^{(j)}$. This calculated $E_{GDR}^{(j)}$ is different form one that the original SL5-min parameter set $\boldsymbol{p_0}$ produces because different parameter set $\boldsymbol{X}^{(j)}$ is used. These are repeated $n = 1000$ times. The mean value and the standard deviation of the calculated GDR peak energies are obtained by statistical process from these 1000 results. This is the uncertainty evaluation of E_{GDR} propagated from the uncertainties of the parameters.

3 Results and discussion

The photoabsorption cross sections in ^{40}Ca calculated with the random parameter sets are plotted in **Figs. 1**. In **Fig. 1(b)**, the red line shows the photoabsorption cross section calculated with the original SLy5-min parameter set, and black lines are those with the randomized parameter sets. **Figure 1(a)** shows histogram of the randomized GDR peak energies. Arrow denotes mean value of the peak energies and its standard deviation, 17.9 ± 1.5 MeV. Remarkably, the randomized GDR peak energies have two regions separated in their energy. In upper energy region, the randomized peak energies scatter around $E_{GDR} = 18.3$ MeV that the original SLy5-min results. In the lower energy region at ≤ 17 MeV, the bunch is clearly separated from the upper one, while the number of the low-energy randomized peaks is ~ 200. These peaks originate from a shoulder at 17 MeV of the original cross section denoted by led line in **Fig. 1(b)**.

Figure 1(c) shows averaged cross section and its standard deviations. If the parameter uncertainties are small, $\frac{|p_i - \Delta p_i|}{|p_i|} \ll 1$, the averaged cross section is expected to be close to the original cross section. However, the averaged cross section is different from the original one. The GDR peak height is lowered, its width is broadened

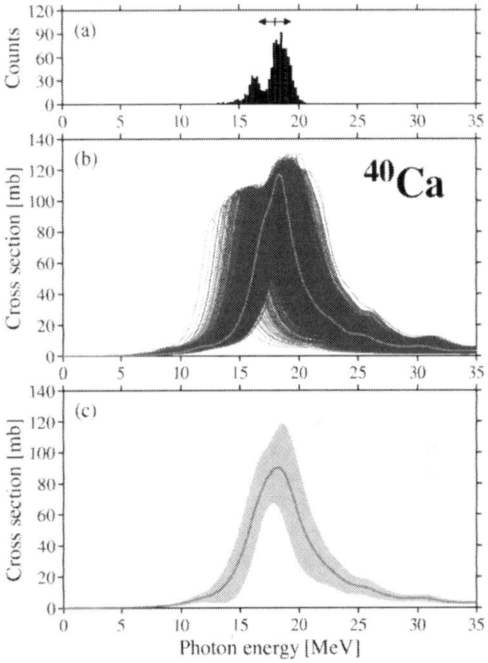

Fig. 1 ^{40}Ca photoabsorption cross sections with randomized parameter sets of SLy5-min interaction. (a) histogram of peak energies. Arrow denotes the mean value of randomized peak energies and standard deviation. (b) photoabsorption cross section calcauated with the original SLy5-min interatrction (red) and those with randomized parameters (black). (c) Averaged cross section (red) with standard deviation (shaded red area).

To see which parameter affects the GDR peak energy, we calculate the Pearson correlation coefficient of the GDR peak energies and the parameters p_i. **Figures 2** show relations of the randomized parameters

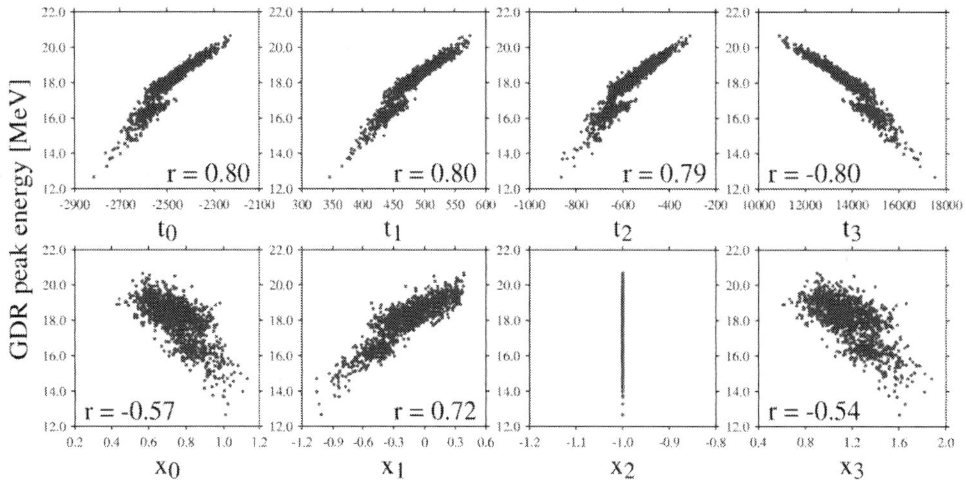

Fig. 2 Relations between the radomized GDR peak energies and paramers, $t_0, t_1, t_2, t_3, x_0, x_1, x_2,$ and t_3, and Pearson correlation coefficients.

and the corresponding randomized GDR peak energies. Note that the parameters x_2, W_0, and α are fixed in the SLy5-min parameter set, and therefore the relations of W_0 and α are not plotted and always $x_2 = -1$. The randomized GDR peak energies have strong correlations with t_0, t_1, t_2, and t_3, and their Pearson correlation coefficients calculated from all randomized GDR peak energies are 0.8 or -0.8. Since the randomized GDR peak energies has two separated bunches [See **Fig. 1(a)**], the plotted relations are also separated into upper- and lower-energy regions. If we pick up the upper-energy region only, the correlations become stronger.

Similar values of these Pearson correlation coefficients are related to the strong correlations between t_0, t_1, t_2, and t_3, shown in **Table 2**. The moduli of the correlations between t_0, t_1, t_2, and t_3 are larger than 0.95. This means that uncertainties of t_1, t_2, and t_3 are almost redundant and can be approximated by the uncertainty of t_0. Similar is seen in the relations of the randomized GDR peak energies and the parameters x_0, x_1, and x_3. The moduli of the correlations between x_0, x_1, and x_3 are larger than 0.93. The strong correlations between the randomized GDR peak energies and the parameter enable us to improve the calculated GDR peak energy by tuning the parameters.

Same Monte Carlo calculations are performed for spherical nuclei, ^{16}O, ^{90}Zr, and ^{208}Pb. **Figures 3** show the averaged cross sections and the standard deviations in ^{16}O, ^{90}Zr, and ^{208}Pb. The averaged GDR peak energies and their standard deviations are listed in **Table 3**. The standard deviations of the GDR peak energies are 1.0-1.5 MeV, irrespective of nuclear mass. The Pearson correlation coefficients between the randomized GDR peak energies and the randomized parameters $t_{0,1,2,3}$ in ^{16}O, ^{90}Zr, and ^{208}Pb are strong, similar to those in ^{40}Ca. These moduli are approximately 0.83, 0.86, 0.65, respectively. Also, the correlation coefficients of $x_{0,1,3}$ are almost same as those in ^{40}Ca. The uncertainty propagation from the parameters to the GDR peak energies are not sensitive to nuclear mass.

Table 3. Averaged GDR peak energies and standard deviations for ^{16}O, ^{40}Ca, ^{90}Zr, and ^{208}Pb, obtained by Monte Carlo calculation using randomized SLy5-min parameter sets.

	Averaged E_{GDR} and deviations
^{16}O	19.10 ± 1.21 MeV
^{40}Ca	17.89 ± 1.46 MeV
^{90}Zr	15.70 ± 1.08 MeV
^{208}Pb	12.93 ± 1.19 MeV

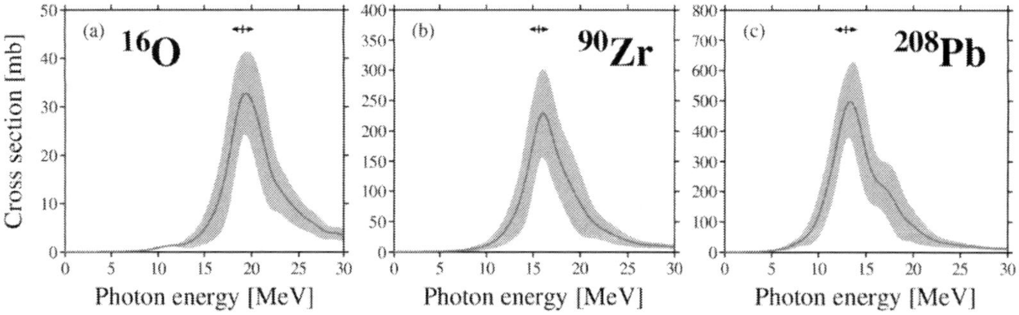

Fig. 3 Same as Fig. 1(c) but for (a) ^{16}O, (b) ^{90}Zr, and (c) ^{208}Pb.

4 Conclusion

We performed the Monte Carlo calculation to evaluate the uncertainty of the GDR peak energy, propagated from the uncertainties of effective interaction parameters. The RPA calculation with randomized parameters is applied to spherical nuclei, ^{16}O, ^{40}Ca, ^{90}Zr, and ^{208}Pb. In the case that SLy5-min parameters is employed with the correlations between the parameters, the standard deviations of the GDR peak energies E_{GDR} are ~ 1 MeV, irrespective of nuclear mass. The GDR peak energy has strong correlations with the parameters. These correlations enable us to handle the calculated GDR peak energy by tuning the parameters.

One of the promising ways to tune the parameters is by shifting some parameters in directions that calculated E_{GDR} becomes higher, such as $t_0 \rightarrow t_0 + \frac{f\Delta t_0}{A}$, with satisfying the correlation matrix (**Table 2**). Here f is a free parameter. We introduce mass dependence to push up E_{GDR} larger (smaller) for lighter (heavier) nuclei. We have the plan to propose a new parameter set that reproduces E_{GDR} properly.

References

1. J. Terasaki, J. Engel. Self-consistent description of multiple strength: Systematic calculations. Phys. Rev C **74**, 044301 (2006).
2. T. Inakura, T. Nakatsukasa, K. Yabana. Emergence of pygmy dipole resonance: Magic numbers and neutron skins. Phys. Rev. C **84**, 021302 (R) (2011).
3. P.-G. Reinhard, W. Nazarewicz. Information content of a new observable: The case of the nuclear neutron skin. Phys. Rev. C **81**, 051303(R) (2010).
4. M. Kortelainen, T. Lensinski, J. More, W. Nazarewicz, J. Sarich, N. Schunck, M.V. Stoitsov, S.M. Wild. Nuclear energy density optimization. Phys. Rev. **C** 82, 024313 (2010).
5. F.J. Fattoyev, J. Piekarewicz. Accurate calibration of relativistic mean-field models: Correlating observables and providing meaningful theoretical uncertainties. Phys. Rev. C **84,** 064302 (2011).
6. M. Kortelainen, J. McDonnell, W. Nazarewicz, P.-G. Reinhard, J. Sarich, N. Schunck, M.V. Stoitsov, S.M. Wild. Nuclear energy density optimization: Large deformations. Phys. Rev. C **85**, 024304 (2012).
7. M. Kortelainen, J. Erler, W. Nazarewicz, N. Birge, Y. Gao, E. Olsen. Neutron-skin uncertainties of Skyrme energy density functionals. Phys. Rev. C **88**, 031305(R) (2013)
8. X. Roca-Maza, N. Paar, G. Colo. Covariance analysis for energy density functionals and instabilities. J. Phys. G: Nucl. Part. Phys. 42 (2015) 034003.
9. J. Erler and P.-G. Reinhard. Error estimates for the Skyrme-Hartree-Fock model. J. Phys. G: Nucl. Part. Phys. **42** (2015) 034026.
10. V. Kejzlar, L. Neufcourt, W. Nazarewicz, P.-G. Reinhard. Statistical aspects of nuclear mass models. J. Phys. G: Nucl. Part. Phys. **47** (2020) 0-94001.
11. X. Roca-Maza. Covariance analysis and an example: SLy5-min, Materials of workshop on Information and Statistics in Nuclear Experiment and Theory. https://indico.cern.ch/event/253381/attachments/442168/613405/correlations.pdf.
12. G. Colo, L. Cao, N.V. Giai, L, Capelli. Self-consistent RPA calculations with Skyrme-type interactions: The Skyrme-rpa program. Comp. Phys. Commu. **184** (2013) 142-161.

Covariance evaluation of neutron cross sections in CENDL

Ruirui Xu[1*], *Zhigang* Ge[1], *Yuan* Tian[1], *Xi* Tao[1], *Yongli* Jin[1], *Yue* Zhang[1], *Duan* Wang[2], *Xiaodong* Sun[1], *Zhi* Zhang[1], *Jimin* Wang[1], *Dongdong* Wang[1] and *Zihao* Wei[3]

[1] China Nuclear Data Centre, China Institute of Atomic Energy, 102413 P.O.Box, Beijing, China
[2] China Institute of Atomic Energy, 102413 P.O.Box, Beijing, China
[3] Graduate School of China National Nuclear Corporation, China Industry College, 102413 P.O.Box, Beijing, China

Abstract: The covariance evaluation for neutron cross sections in CENDL is briefly introduced in this work. The methodology for evaluation contains the nuclear reaction theoretical model-dependent approach and the non-model dependent one according to the amount of experimental data. Both approaches are based on the Generalized Least-Squares (GLSQ) method. To obtain more reliable uncertainties from experimental measurement, the analysis of the sources of experimental uncertainties (ASEU) is used rigorously in the evaluation. Moreover, machine learning (ML) methods which can deal with the data mining with a more automatic way are employed to evaluate the cross sections in a large-scale nuclear mass region to compensate the uncertainties on some nuclides and reactions, lack of experimental data for, e.g., unstable nuclei and fission products. The covariance files for 70 fission product nuclei are obtained through the model-dependent method in CENDL-3.2, and the covariances for U and Pu isotopes have also been finished with high fidelity, which will be released as part of the next CENDL.

1 Introduction

Reasonable estimation for nuclear data covariance is very important in the study of sensitivity and uncertainty analysis, the nuclear data adjustment and so on for the advanced reactors [1]. Besides the early covariance databases such as COMMARA-2.0 [2], AFCI-1.2 [3] et al., the newly evaluated covariance files for MF=32, 33, 34, 35 are also released in the various latest evaluated nuclear data libraries, ENDF/B-VIII.0[4], JENDL-5[5], JEFF-3.3[6], CENDL-3.2[7] and TENDL-2021[8]. The total amount and quality of covariance data in these libraries have been greatly improved in the past decade. The nucleus and files involved in each library are show in Table 1. It is shown that covariances have been provided for average number of neutrons per fission (NU), resonance parameters (RES), cross sections (SIG), angular distributions (DA) and energy distributions of secondary particles (DE) in a large number. CENDL-3.2 also contains COV/SIG for 76 nuclei, which are mainly located in the fission product mass region and several light nuclei. Details will be introduced in Section 3.

Table 1. Status of covariance evaluations in international evaluated libraries.

ENDF MF	ENDF/B-VIII.0	JENDL-5	JEFF-3.3	CENDL-3.2	TENDL-2021
COV/NU	72	79	136	0	432
COV/RES	118	43	352	0	1
COV/SIG	221	102	442	76	2808
COV/DA	108	97	359	0	386
COV/DE	65	79	36	0	431

It is almost impossible to hand down the uncertainty information of experiments and theories among the nuclear data pipeline to covariance strictly. In order to achieve covariance data with better physics, great effort has been made in the past decades and the deterministic (Generalized Least-Squares, GLSQ) and stochastic (Monte Carlo, MC) approaches were developed, each with their advantages and disadvantages. They are both applied in the real covariance evaluation, LS approach is mainly adopted in ENDF/B, JENDL and CENDL, and MC is used in JEFF and TENDL. In addition to the traditional methods above, some innovative techniques, such as Machine Learning (ML) and Artificial Intelligence (AI), are also expected to join the nuclear data evaluation to automate evaluation work and uncover helpful information to improve data quality [6].

The total evaluation scheme and the derived covariance in CENDL will be briefly presented in this paper. The content following is laid out in three sections. Section 2 will provide the overview of methodology with LS. Section 3 will present some discussions on our derived cross sections covariance in CENDL-3.2 and the future CENDL. the summary of the conclusions and future work will be outlined in Section 4.

2 Methodology

* Corresponding author: xuruirui@ciae.ac.cn

2.1 Covariance evaluation based on the traditional GLSQ approach

The systematic technique to derive the covariance of cross sections for CENDL has been developed progressively in the past ten years [7]. The GLSQ approach is used in the whole process to perform the uncertainty propagation from the applied nuclear data experiment and theory. According to the abundance of experimental data for each physical quantity, we build the model dependent and non-model dependent approaches respectively, which have been described in [7, 8] and the details will not be talked too much in this paper.

It is worth mentioning that the central and most time-consuming process in GLSQ is to perform the reliable assessment of experimental data and their experimental covariance, particularly to deal with the multiple measurements to the same quantity from different laboratories. The analysis of the sources of experimental uncertainties (ASEU) is used rigorously in the evaluation. Noted that we try to follow the reported experimental errors in the references, but we also perform some variation to the experimental errors, such as when the data are measured for a long time, some important uncertainties are obviously missing in the report, and the pseudo-uncertainty will be supplemented. In other case, we rarely modify the experimental data error directly. In addition, some subjective effects are inevitable when determining the uncertainties from background subtraction, data normalization, detector counting statistics, detector efficiency as well as fundamental energy resolution of the machine in a cross-section measurement experiment, especially when analysing the measurements reported with scarce information in publications. In order to make compensation to the tractional GLSQ scheme, some innovative approaches such as machine learning are also involved in nuclear data study to reduce the subjectivity during covariance evaluation and make good predictions for the data without measurements.

2.2 Machine learning in the experimental cross sections analysis

Machine learning is a well-known promising data analysis technology that has been widely used in various scientific reach fields in recent years. It can convert the disordered data into useful information through some advanced computational algorithm, and is very effective to deal with complex nuclear data problems.

In the real nuclear cross section evaluation, one always meets with the cases the experimental information available is not enough to constrain all nuclear reaction theoretical parameters accurately to obtain the good description to nuclei cross sections without measurements. Therefore, we introduce the Artificial Neural Network (ANN) method to learning the experimental data available of a certain reaction (such as (n,2n), (n,3n)) systematically and to predict the nuclear cross sections (such as (n,2n)) without measurements; and then, we apply the predicted data in the theoretical calculation to further support the nuclear reaction model calculation to get better results.

In our scheme, a modified Bayesian Neural Network (BNN) with the Local Reparameterization Trick [9] is employed in the experimental data analysis and predictions. The designed BNN in our work is shown as Figure 1, which has the similar Topology structure with the so-called Variational Autoencoder (VAE) generator model [10]. Two intermediate variables ($\mu(X)$ and $\log(\sigma(X)^2)$) are first calculated and the output can be local reparametrized through the following relation,

$$output = \mu(X) + e^{\frac{1}{2}log(\sigma(X)^2)} \times \epsilon$$

where $\epsilon \sim N(0, 1)$ is the normal distribution. In our scheme, the predicted cross sections of BNN are taken as an effective supplement measurement in the traditional GLSQ covariance evaluation scheme.

In the real calculations, eight physical parameters including proton number, mass number, single neutron separation energy, single proton separation energy, shell effect, level density parameter, pairing correlation, neutron incident energy are adopted in the input layer. Two hidden layers with each 128 nodes are contained in the network. The single task output means the obtained cross sections by BNN, and the uncertainties of output cross sections can be produced self-consistently.

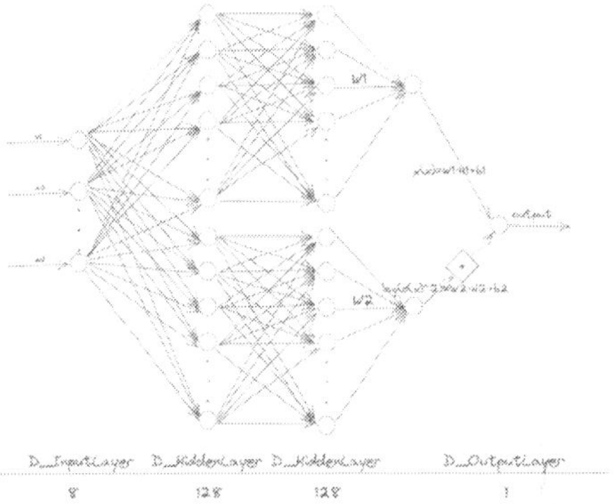

Fig. 1. Topology structure of BNN in this work.

We take as example of ^{23}Na to ^{209}Bi (n,2n) cross sections, and predict them with a database of 5000 experimental data points. During BNN learning and validation processes, the learning set is built by randomly selecting 80% of the data, while the validation set is made up by the remaining 20%. Figures 2 and 3 are the derived (n,2n) cross sections of ^{89}Y and ^{209}Bi. It can be observed that after making full use of the existing experimental data, the BNN predictions are good enough to describe the physical curves and the relevant BNN uncertainty looks also reasonable regarding to the measurements. Both of the information will provide key input in GLSQ to improve the reliability of covariance, especially to the nuclear reactions without measurements.

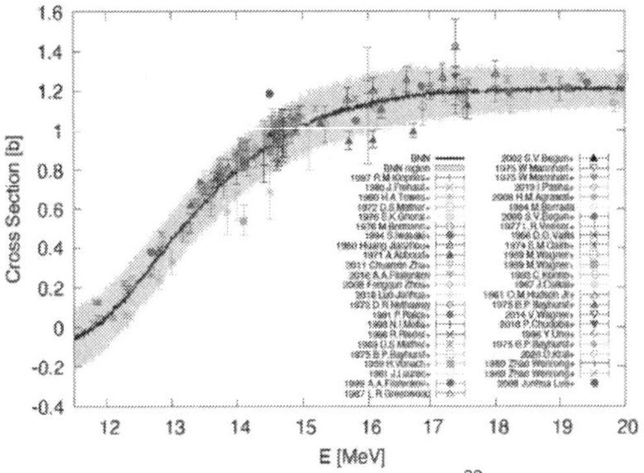

Fig. 2. The comparison of the obtained ^{89}Y(n,2n) cross sections and uncertainty and the relevant measurements.

3 Evaluated covariance in CENDL

Based on the methodology in Section 2, the covariance files of nuclear reaction cross sections are systematically generated for the light, medium heavy and fission nucleus, among which, 76 nuclei are released in CENDL-3.2 as listed in Table 2, and eight neutron reactions (n,total), (n,elastic), (n, γ), (n,inelastic), (n,p), (n,2n), (n,np) and (n,n α) are concerned for the medium heavy mass nucleus, two more reactions (n,fission) and (n,3n) are treated for the fission nucleus.

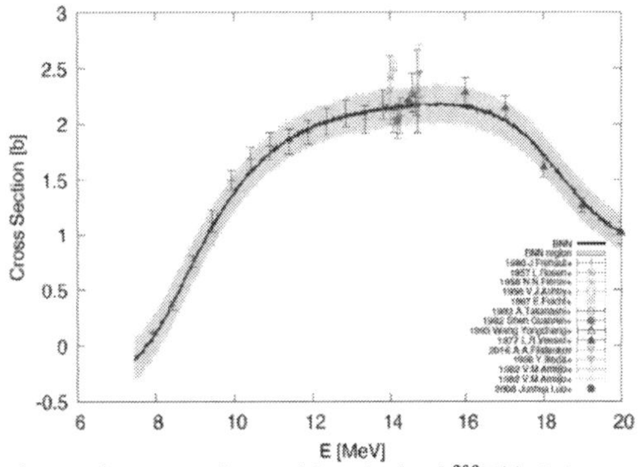

Fig. 3. The comparison of the obtained ^{209}Bi(n,2n) cross sections and uncertainty and the relevant measurements.

Table 2. The nucleus of covariance evaluation in CENDL-3.2.

Nucleus	Num.	Elements
Light	4	^{2,3}H, ^{3}He, ^{19}F
Structure	2	^{55}Mn, ^{56}Fe
Fission products	70	^{69,71}Ga, ^{71,73,74,75,76,77,78}Ge, ^{75,77,79}As, ^{89,91}Sr, ^{93,95}Zr, ^{95}Nb, ^{99}Tc, ^{99,100,101,103,104,105}Ru, ^{103,105}Rh, ^{105,108}Pd, ^{113}Cd, ^{121,125}Sb, ^{127,129,135}I, ^{124,132,134,135}Xe, ^{133,135,137}Cs, ^{130,134,135,136,137,138}Ba, ^{139}La, ^{141,144}Ce, ^{141}Pr, ^{143,145,146,148}Nd, ^{147,148,149}Pm, ^{150,151}Sm, ^{151,153,155}Eu, ^{154,155,156,157,158,160}Gd

The covariance of cross sections of 70 fission products are newly derived via the model-dependent methods in CENDL-3.2. The current model dependent covariance is mainly constructed from the uncertainty of theoretical model parameters, in the meantime, the outputs based on BNN are also partly referred in this work to compensate the scarcity of measurements in this mass region.

In total, the assessed uncertainties for the nuclear reactions for 70 fission produce nuclei are about less than 10% for (n,total); 20% for (n,inelastic); 20% for (n,2n); 30% for (n, γ); 80% for (n,n α); 50% for (n,np), which can be evidenced by Figure 4-6. In these figures, the isotopes of Ge are sampled and the uncertainties of neutron cross sections look all reliable in magnitude.

To further compare the current results with other libraries, ^{95}Zr(n,inl) and (n,2n) reactions are plotted with the evaluated uncertainties available in ENDF/B-VIII.0, JEFF-3.3 and CENDL-3.2. The uncertainties in ENDF/B-VIII.0 are the largest one comparing to the others, because data are directly adopted from COMMARA-2.0, which contains evaluated low-fidelity covariances [11]. The data in JEFF-3.3 are generated by the Monte Carlo TASMAN code in TALYS, the uncertainties mainly rely on the MC samples and build the margin of the theoretical parameter, and the absolute values look identical to CENDL-3.2. The deviation illuminates the obtained covariance through different approaches, and the covariance passed from the real nuclear data evaluation process is always more reliable to show the confidence interval of evaluated nuclear cross sections.

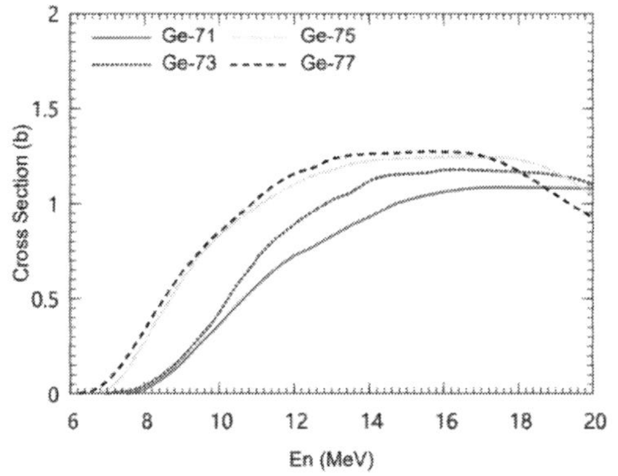

Fig. 4. The evaluated cross sections of ^{71,73,75,77}Ge(n,2n) in CENDL-3.2.

Besides the new evaluations for fission produce nuclei, the covariance for actinide nuclei is also built together with the evaluation for cross sections. Figure 7 shows the data for ^{235}U(n,2n) and (n3n), and the results are achieved through the combination of model dependent and non-model dependent methods. It can be seen that the experimental data are almost in the uncertainty bands. In addition, the correlated coefficient between the different reactions are also shown in Figure 8, which provide more complete information to the nuclear application.

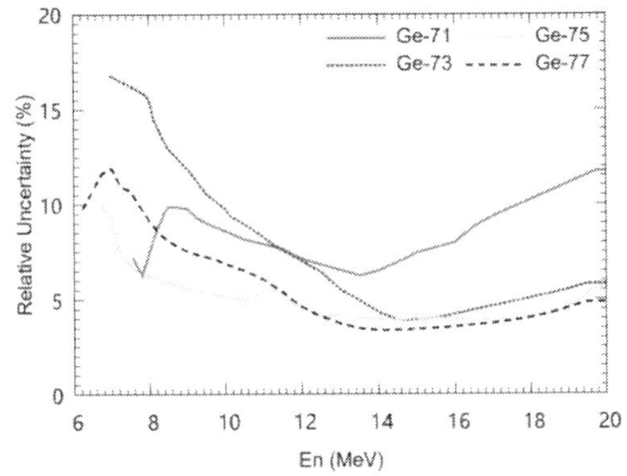

Fig. 5. The evaluated uncertainties relevant to the cross sections in Fig. 4.

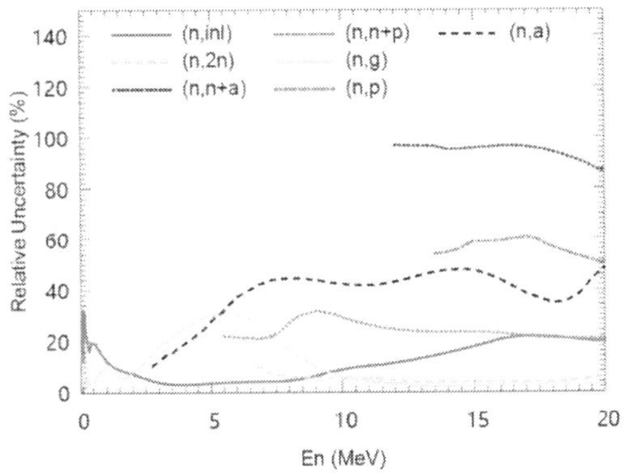

Fig. 6. The uncertainties of seven neutron cross sections of ^{75}Ge CENDL-3.2.

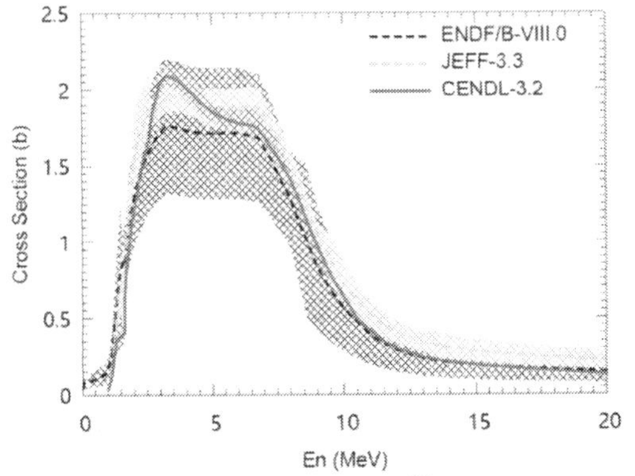

Fig.7. The comparison of evaluated ^{95}Zr(n,inl) cross sections and uncertainties in ENDF/B-VIII.0, JEFF-3.3 and CENDL-3.2.

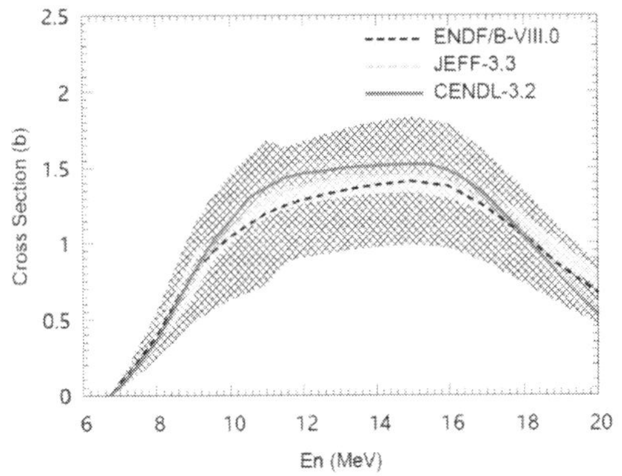

Fig.8. The comparison of evaluated ^{95}Zr(n,2n) cross sections and uncertainties in ENDF/B-VIII.0, JEFF-3.3 and CENDL-3.2.

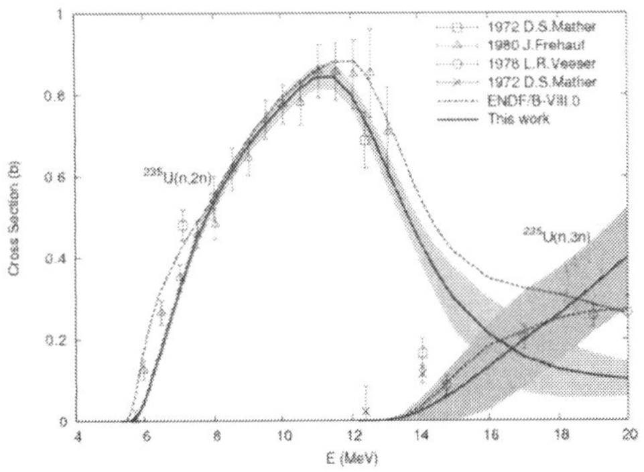

Fig.9. The comparison between the evaluated ^{235}U(n,2n) and (n,3n) cross sections and uncertainties and the measurements [8].

4 Conclusion and perspective

The methodology of nuclear cross section covariance evaluation in CENDL is introduced, meanwhile the new evaluated results for fission products and actinides are also discussed. As a result, the current covariance in CENDL seems confident to describe the uncertainties of most important neutron data, which could supply nuclear application with a good database. In addition, the new covariance for actinides will be released in the next CENDL.

To involve more information from experimental data, the machine learning approach BNN is employed in an exploratory study. Joining the BNN outputs of cross sections and uncertainties in GLSQ are promising to increase the reliability of the traditional covariance results in future.

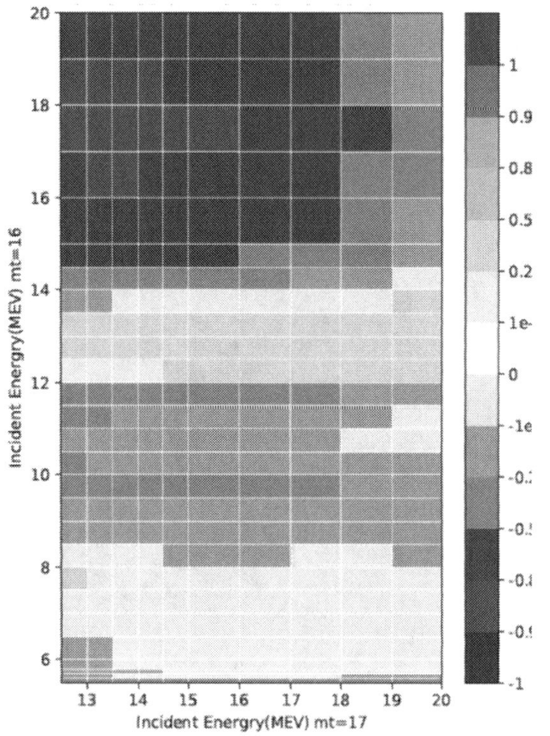

Fig. 10. The correlated coefficient of cross sections between ^{235}U(n,2n) and (n,3n) reactions [8].

The authors thanks Prof. *Yingxun* Zhang from China Institute of Atomic Energy for his kind discussion about application of machine learning in nuclear physics. This work has been supported by the National Natural Science Foundation of China (Grants No. 11875321, No. 11790320, No. 12005303).

References

1. V. Sobes, W. J. Marshall, D. Wiarda, F. Bostelmann, A. Holcomb, B. T. Rearden, ORNL/TM-2018/1037 (2019)
2. M. Herman, P. Oblozinsky, C.M. Mattoon, M. Pigni, S. Hoblit, S.F. Mughabghab, A. Sonzogni, BNL- 94830-2011 (2011)
3. P. Oblozinsky, Report BNL-90897 (2009)
4. D. A. Brown, et al., Nuclear Data Sheets 148, 1–142 (2018)
5. O. Iwamoto, N. Iwamoto, K. Shibata, A. Ichihara, S. Kunieda, F. Minato, and S. Nakayama, EPJ Web of Conferences, 239, 09002_1-6 (2020)
6. Plompen, A.J.M., Cabellos, O., De Saint Jean, C. et al., JEFF-3.3. Eur. Phys. J. A 56, 181 (2020).
7. Zhigang Ge, Ruirui Xu, Haicheng Wu et al., EPJ Web of Conferences 239, 09001 (2020).
8. A.J. Koning, D. Rochman, J.-Ch. Sublet, N. Dzysiuk, M. Fleming, and S. van der Marck, Nuclear Data Sheets 155, 1–55 (2019).
9. https://www-nds.iaea.org/exfor/endf.htm
10. R. Capote, D.L. Smith, A. Trkov, EPJ Web of Conferences 8, 04001 (2010)
11. P. Vicente-Valdez, L. Bernstein, M. Fratoni, Annals of Nuclear Energy 163, 108596 (2021)
12. R. Xu, Q. Zhang, Y. Zhang, T. Liu, Z. Ge, H. Lu, Z. Sun, B. Yu, and G. Tang, Nuclear Data Sheets 123, 16–20 (2015)
13. Z. Zhi, Y. Zhang, Y. Tian, R. Xu, J.M. Wang, X. Tao, Y.L. Jin, X.D. Sun, Z.G.Ge, Atomic Energy Science and technology, 056, 1599(2022)
14. D. P. Kingma, T. Salimans, arXiv:1506.02557v2 (2015)
15. D. P. Kingma, M. Welling, arXiv:1312.6114 (2013)
16. M.T. Pigni, M. Herman, P. Oblozinsky, BNL-79985-2008-IR (2008)

AUTHOR INDEX